Reviews and Advances in
Functional Materials

Reviews and Advances in Functional Materials

Editors

Xian Jun Loh
Institute of Materials Research and Engineering, Singapore
Nanyang Technological University, Singapore
National University of Singapore, Singapore

Ady Suwardi
The Chinese University of Hong Kong (CUHK), Hong Kong

NEW JERSEY • LONDON • SINGAPORE • GENEVA • BEIJING • SHANGHAI • TAIPEI • CHENNAI

Published by

World Scientific Publishing Co. Pte. Ltd.
5 Toh Tuck Link, Singapore 596224
USA office: 27 Warren Street, Suite 401-402, Hackensack, NJ 07601
UK office: 57 Shelton Street, Covent Garden, London WC2H 9HE

British Library Cataloguing-in-Publication Data
A catalogue record for this book is available from the British Library.

REVIEWS AND ADVANCES IN FUNCTIONAL MATERIALS

Copyright © 2025 by World Scientific Publishing Co. Pte. Ltd.

All rights reserved. This book, or parts thereof, may not be reproduced in any form or by any means, electronic or mechanical, including photocopying, recording or any information storage and retrieval system now known or to be invented, without written permission from the publisher.

For photocopying of material in this volume, please pay a copying fee through the Copyright Clearance Center, Inc., 222 Rosewood Drive, Danvers, MA 01923, USA. In this case permission to photocopy is not required from the publisher.

ISBN 978-981-98-0639-3 (hardcover)
ISBN 978-981-98-0640-9 (ebook for institutions)
ISBN 978-981-98-0641-6 (ebook for individuals)

For any available supplementary material, please visit
https://www.worldscientific.com/worldscibooks/10.1142/14135#t=suppl

Desk Editors: Yong Qi Soh/Muhammad Ihsan Putra

Typeset by Stallion Press
Email: enquiries@stallionpress.com

Contents

Recent Progress in Metasurfaces: An Introductory Note from
Fundamentals and Design Methods to Applications   1
  Wei Chen, Jiaqing Shen, Yujie Ke, Yuwei Hu, Lin Wu,
  Jinfeng Zhu and Zhaogang Dong

A Review of Self-Sensing in Carbon Fiber Structural
Composite Materials   35
  D. D. L. Chung

Silicon-Containing Additives in Encapsulation of Phase
Change Materials for Thermal Energy Storage   79
  Johnathan Joo Cheng Lee, Natalie Jia Xin Lim,
  Pei Wang, Hongfei Liu, Suxi Wang, Chi-Lik Ken Lee,
  Dan Kai, Fengxia Wei, Rong Ji, Beng Hoon Tan,
  Shaozhong Ge, Ady Suwardi, Jianwei Xu,
  Xian Jun Loh and Qiang Zhu

Comparison of the Properties of Additively Manufactured
316L Stainless Steel for Orthopedic Applications: A Review   167
  M. Nabeel, A. Farooq, S. Miraj, U. Yahya, K. Hamad
  and K. M. Deen

Cortisol Biosensors: From Sensing Principles
to Applications   243
  Yuki Tanaka, Nur Asinah binte Mohamed Salleh,
  Khin Moh Moh Aung, Xiaodi Su and Laura Sutarlie

Recent Advances in Ocular Therapy
by Hydrogel Biomaterials   287

   Lan Zheng, Yi Han, Enyi Ye, Qiang Zhu, Xian Jun Loh, Zibiao Li and Cheng Li

Green Nanotechnology and Phytosynthesis of Metallic Nanoparticles: The Green Approach, Mechanism, Biomedical Applications and Challenges   319

   Abdulrahman Alomar, Tabarak Qassim, Yusuf AlNajjar, Alaa Alqassab and G. Roshan Deen

Functional Coatings for Built Environments Based on Nanotechnology   369

   Pin Jin Ong, Suxi Wang, Ady Suwardi, Jing Cao, Fuke Wang, Xuesong Yin, Pei Wang, Fengxia Wei, Dan Kai, Enyi Ye, Jianwei Xu, Ming Hui Chua, Warintorn Thitsartarn, Qiang Zhu and Xian Jun Loh

Direct Growth of Lithium Niobate Thin Films for Acoustic Resonators   417

   Zhen Ye, Qibin Zeng, Celine Sim, Baichen Lin, Hui Kim Hui, Anna Marie Yong, Chee Kiang Ivan Tan, Seeram Ramakrishna and Huajun Liu

Recent Advances in Bio-Based Concrete Materials: A Critical In-Depth Review   449

   Abirami Manoharan, Rajapriya Raja, Himanshu Sharma and Sanjeev Kumar

Design and Gene Delivery Application of Polymeric Materials in Cancer Immunotherapy   489

   Ying Chen, Lingjie Ke, Xian Jun Loh and Yun-Long Wu

Recent Advances in Thermal Interface Materials 515
Jing Cao, Tzee Luai Meng, Xikui Zhang, Na Gong, Rahul Karyappa, Chee Kiang Ivan Tan, Ady Suwardi, Qiang Zhu and Hongfei Liu

Recent Progress in Metasurfaces: An Introductory Note from Fundamentals and Design Methods to Applications*

Wei Chen [†,‡], Jiaqing Shen [†,‡], Yujie Ke [‡], Yuwei Hu [‡], Lin Wu [§,¶,**,‡‡], Jinfeng Zhu [†,††,§§] and Zhaogang Dong [‡,‖,‡‡,§§]

[†]Institute of Electromagnetics and Acoustics and
Key Laboratory of Electromagnetic Wave Science and Detection
Technology Xiamen University, Xiamen, Fujian 361005, P. R. China
[‡]Institute of Materials Research and Engineering (IMRE)
Agency for Science, Technology and Research (A*STAR)
2 Fusionopolis Way, Innovis #08-03, Singapore 138634, Singapore
[§]Science, Mathematics, and Technology (SMT)
Singapore University of Technology and Design (SUTD)
8 Somapah Road, Singapore 487372, Singapore
[¶]Institute of High Performance Computing
A*STAR (Agency for Science, Technology, and Research)
1 Fusionopolis Way, No. 16-16 Connexis, Singapore 138632, Singapore
[‖]Department of Materials Science and Engineering,
National University of Singapore
9 Engineering Drive 1, Singapore 117575, Singapore
[**]lin_wu@sutd.edu.sg
[††]jfzhu@xmu.edu.cn
[‡‡]dongz@imre.a-star.edu.sg

[§§]Corresponding authors.
*To cite this article, please refer to its earlier version published in the *World Scientific Annual Review of Functional Materials*, Volume 2, 2430002 (2024), DOI: 10.1142/S2810922824300022.

Metasurfaces refer to the sub-wavelength nanostructures that are capable of manipulating the amplitude, phase, polarization, and other characteristics of light, to enable diverse applications across the ultraviolet, visible, infrared, and terahertz spectra. In this review paper, we aim to provide an introductory note on metasurfaces, from fundamentals and design methods to applications, such as biosensing, environmental monitoring, metalenses, optical cloaking, electromagnetic scattering, structural color, miniaturized devices, and others. Moreover, we also identify the key challenges and limitations of metasurfaces, such as fabrication, integration, optimization, tuning, signal processing, and analysis, and suggest possible directions and solutions for future research. At last, we envision several emerging and promising trends for metasurfaces, such as new materials and structures, new phenomena and mechanisms, machine learning, and artificial intelligence techniques. The review is expected to inspire future development in this exciting and rapidly evolving field of metasurface devices.

Keywords: Metamaterial; metasurface; deep learning; nanophotonics.

1. Introduction

Natural materials exhibit limited capabilities to achieve arbitrary manipulation of electromagnetic waves.[1-20] As artificially aligned arrays at microscopic and nanoscales, metamaterials offer a unique opportunity for tunable properties of electromagnetic, mechanics, acoustics, thermodynamics, and nanophotonics, as shown in Fig. 1(a).[21-45] Through deliberate designs and precise arrangements of subwavelength elements, metamaterials can exhibit various exotic electromagnetic properties, such as negative refractive index, perfect light absorption, and nanoscale electromagnetic manipulation, which are rarely found in nature.[46-70] The typical configuration of metamaterials is volumetric, with their design and fabrication being highly complex, making miniaturization and integration challenging.

In 2011, Yu *et al.* proposed a two-dimensional array of optical resonators with spatially varying phase response and subwavelength separation, which could imprint such phase discontinuities on propagating light as it traverses the interface between two media.[71] After that, metasurfaces have emerged as a revolutionary technology, promising precise control over light at the subwavelength scale.[72-81] In general, metasurfaces are regarded as the two-dimensional counterpart of metamaterials, consisting of artificial structures on a

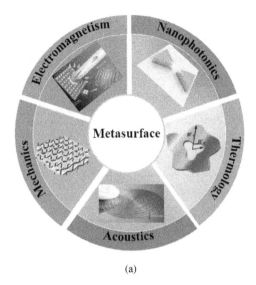

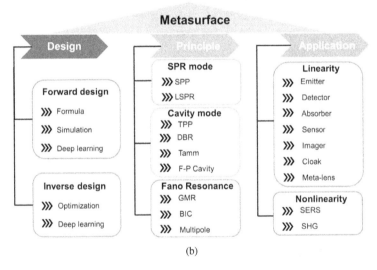

Fig. 1. Overview of metasurface research. (a) Applications of metasurfaces in various disciplines, including mechanics, acoustics, thermodynamics, nanophotonics, electromagnetism, etc.[25-28] (b) Overview of various design methods, principles, and applications for optical metasurfaces.

Source: Panel (a) Copyright 2018, Nature. Applying for a license of figures. Copyright 2023, Wiley. Applying for a license of figures. Copyright 2021, Elsevier. Applying for a license of figures. Copyright 2022, Elsevier. Applying for a license of figures.

flat surface in periodic or aperiodic arrangements.[81–92] This field has developed rapidly since 2011, giving rise to an increasing versatility and potential of metasurfaces in diverse applications, including the generation of orbital angular momentum, metalens architecture, optical cloaking, anomalous reflection, quantum information processing, and electromagnetic scattering.[93–110] Metasurfaces have experienced a fast development in design methodologies, underlying principles, and numerous applications spanning various disciplines. In addition, the continuous advancement of nanofabrication technology also contributes to discovering some novel physical properties of metasurfaces.[111] Significant progress has been made in realizing extreme light confinement and control in non-centrosymmetric phonon-polaritonic crystals.[112] Notable developments include out-of-plane hyperbolicity in uniaxial crystals, enabled by generating hyperbolic phonon polaritons in van der Waals materials. Additionally, surface anisotropy in cut crystals has generated ghost polaritons and leaky polaritons. Furthermore, investigating phonon polaritons in orthorhombic crystals with biaxial optical response has revealed valuable insights. The concept of twist-optics, exploiting the control of the propagation of phonon polaritons through twisted bilayer and trilayer van der Waals stacks, has also been thoroughly discussed.[112,113] In addition, some coupling modes (e.g. bound states in the continuum (BIC)) and multipole resonances (e.g. toroid dipole and anapole) are also explored.[114,115] These new physical properties significantly enhance the ability of light to interact with matter and expand the means of electromagnetic regulation. Therefore, it is very important to summarize recent progress in metasurface design.

Our goal is to provide readers with an introductive and comprehensive understanding of the latest advancements in metasurface research, covering fundamental theoretical frameworks, design methods, and practical implementations. Figure 1(b) outlines the review content with a systematic examination of the fundamental design principles that underpin metasurface application and engineering. We introduce the two main design methods of forward and reverse designs, summarize their technical means, and discuss their advantages and disadvantages. The intricate interplay between design parameters (materials, structures, and design methods) and optical functionalities is examined. Then, we introduce various physical principles, including surface plasmonic resonance (SPR), cavity resonance, BIC, etc. By explaining the physical principles governing metasurface behavior, we aim to equip researchers and engineers with knowledge for the informed design towards specific applications, encompassing the fields ranging from communication, sensing, and imaging to energy harvesting. Whether applied to beam shaping, polarization manipulation, or designing ultrathin optical components, metasurfaces are providing disruptive technologies towards realizing compact and efficient photonic solutions. At last, we will share our opinions regarding the future directions for metasurface research.

2. Resonance Mechanisms

Resonances are the cornerstone of photonics,[80] since the resonance mechanisms are critical for metasurfaces and related optical engineering applications, as shown in Fig. 2. One of the classic resonance mechanisms is the surface Plasmon Resonance (SPR) as supported on metal/dielectric nanostructures, including surface plasmon polariton (SPP) and localized surface plasmon resonance (LSPR).[116] SPR is an electromagnetic phenomenon occurring at the interface between a metal and a dielectric, in which electromagnetic waves interact with oscillations of free electrons on metal surface. SPP is formed through the resonant oscillations between incident light and free electrons along a metal surface, when light is incident onto the metal surface at an appropriate angle. The frequency of this resonance is determined by the refractive index of the metal and the wavelength of the incident light. In addition, LSPR is a localized form of plasmon resonance occurring at the confined surface of metal nanostructures. When the sizes of metal nanoparticles are smaller than the wavelength of the incident light, LSPR can be supported around the sharp corners or tips around the nanostructures, inducing strongly enhanced localized electromagnetic

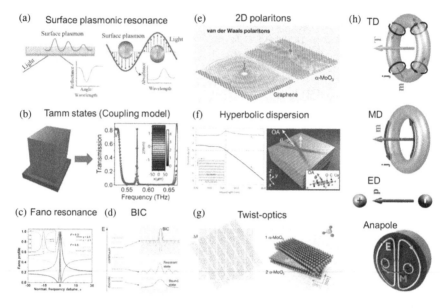

Fig. 2. Different physical mechanisms of metasurfaces. (a) Surface plasmonic resonance, including surface plasmonic polaritons (left) and local surface plasmonic resonance (right).[116] (b) Tamm states, a physical phenomenon associated with surface waves.[118] (c) Fano resonance usually occurs when a system has two different coupling modes.[119] (d) Bound states in a continuum, a kind of resonant state.[120] (e) Van der Waals polaritons.[121] (f) Hyperbolic dispersion via artificial metamaterials (left) and natural materials (right).[112,122] (g) Twist-optics: controlling the propagation of phonon polaritons with twisted bilayer and trilayer stack.[121] (h) Different modes of multipole resonance, including toroid dipole, magnetic dipole, electric dipole, and anapole.[115,127]

Source: Panel (a) Copyright 2023, Beilstein-Institut. Applying for a license of figures. Panel (b) Copyright 2021, Elsevier. Applying for a license of figures. Panel (c) Copyright 2018, Nature. Applying for a license of figures. Panel (d) Copyright 2016, Nature. Applying for a license of figures. Panel (e) Copyright 2021, Nature. Applying for a license of figures. Panel (f) Copyright 2023, Nature and 2016, Nature. Applying for a license of figures. Panel (g) Copyright 2021, Nature. Applying for a license of figures. (h) Panel Copyright 2010, AAAS and 2015, Nature. Applying for a license of figures.

fields. This phenomenon promotes the enhanced absorption and/or scattering in certain wavelengths,[117] and therefore SPR has wide applications in biosensing, optical imaging, energy, and nanophotonics fields.

At the interface of the metal film and a one-dimensional photonic crystal, an electromagnetic surface state emerges, known as the Tamm plasmon polariton (TPP), like the electronic surface state envisaged by Tamm.[118] The dispersion characteristics of TPP enable direct excitation using both TE- and TM-polarized light from free space without the need for external momentum compensation. Due to its robust localization and slow light attributes, TPP offers significant potential for diverse applications, such as perfect absorbers, solar cells, narrow-band thermal emission, nonlinear optical effects, and laser technologies.[118]

The third resonance that we introduced here is Fano resonance, which is a quantum mechanical phenomenon being characterized by an asymmetric spectral line shape, originally proposed by Italian physicist Giovanni Fano in 1961.[119] This resonance occurs when a discrete quantum state interferes with a continuum of states, resulting in a distinctive asymmetric profile in the resonance curve. Fano resonance includes several key features, such as asymmetric line shape, interference phenomenon, and unique signatures of having a narrow peak with a broad background, where a compressive review on Fano resonance in optics is by Luk'yanchuk *et al.*[120] Besides optics, Fano resonance is also encountered in other physical systems, including quantum mechanics, and condensed matter physics. Taking advantage of its peculiar spectral characteristics, it has practical applications in photonic fields, such as sensors and light filters. It's worth noting that the recently popular bound-states-in-the-continuum (BIC) resonance is also a type of Fano resonance, where the BIC resonance can be referred to more focused review and will not be repeated here.

The phonon polaritons are hybrid light–matter excitations of polar crystals and have arisen significant interest. Phonon polaritons exhibit extreme optical anisotropy and directionality, allowing new nanoscale light propagation and manipulation regimes. The role of crystal symmetry in determining the polariton response and associated phenomena is discussed, such as hyperbolic, ghost, and shear polaritons.[121,122] The paper further explores the emerging directions of polaritons in lower-symmetry materials and metamaterials, such as twist optics, topological physics, and nonlinearities.[121,123–126]

Multiple resonances describe different modes of electromagnetic oscillations in a system, being characterized by the distribution of electric and magnetic fields. Here, we present a brief introduction to some different modes of multipole resonances:

(1) Electric dipole resonance: This mode is associated with oscillating positive and negative charges in a linear or nonlinear structure. In this mode, the electric field dominates, and the charges experience a shift, leading to an oscillation that resonates with the incident electromagnetic field.
(2) Magnetic dipole resonance: This mode involves oscillating magnetic moments under an external electromagnetic field. The magnetic field is dominant in this mode, and the system behaves like a magnetic dipole to experience resonance with the incident magnetic field.
(3) Toroid dipole resonance: This mode is characterized by an oscillation of currents circulating in a toroidal (doughnut-shaped) structure. Toroidal geometries can support resonances with unique properties and are used in applications such as metamaterials and electromagnetic devices.
(4) Anapole resonance: This mode is a unique type of resonance, in which the electric and magnetic fields are arranged to cancel each other out, leading to a total absence of outgoing radiation. The term "anapole" is derived from Greek, meaning without poles. Anapole resonances are of interest in the design of low-scattering or stealthy structures. These multipole resonances have distinct characteristics and applications.[115,127] Understanding and controlling these resonances is crucial for tuning the optical and electromagnetic properties of metasurfaces.

3. Design Methods

In this section, we are providing a comprehensive overview of the design methods as facilitated by computing machinery, as shown in Fig. 3. The design methodologies are mainly classified into forward

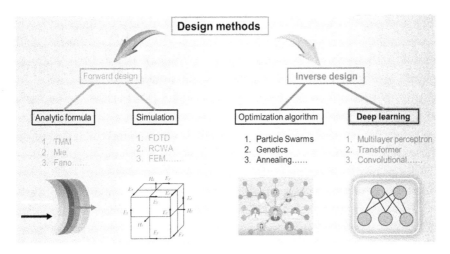

Fig. 3. Forward and inverse design methods for metasurface devices, including analytic formula, simulation tool, optimization algorithm, and deep learning.

and inverse design, e.g. from structural parameters to optical response and prediction of structural parameters based on demand. The development of nanophotonics has achieved remarkable results in both categories.

Regarding forward design, analytic formulas are based on mathematical formulas, such as the transfer matrix method (TMM) and Mie scattering theory.[128,129] TMM analyzes the light propagation of layered structures using the transfer matrix method, and Mie's theory focuses on the scattering and absorption of spherical particles. Moreover, simulation methods also rely on computer simulation, such as finite-difference time-domain (FDTD), rigorous coupled-wave analysis (RCWA), and finite element method (FEM), which use simulation tools to analyze the optical properties of nanostructures, providing researchers with flexible and intuitive means.[130,131]

On the other hand, inverse design methods aim at predicting structural parameters according to desired optical properties. In terms of optimization algorithms, particle swarm optimization, genetic algorithms, and simulated annealing are widely used Ref. 132.

These algorithms search for the optimal solution in different ways, e.g. simulating the processes of bird flocks (biological evolution) and metal annealing in nature to provide an efficient way to design nanophotonics. The advancement in metasurface inverse design has led to enduring scientific innovations. Deep learning (DL) technologies, such as artificial neural networks and generative pre-trained transformers (GPT), are flourishing in scientific discovery and intelligent design. Researchers have also utilized a variety of sophisticated neural network architectures to propel metamaterial design forward, including multilayer perceptrons, convolutional neural networks, recurrent neural networks, generative adversarial networks, and transformer networks. Deep learning, as an emerging design methodology, incorporates neural network structures like multilayer perceptrons, transformers and convolutional neural networks, facilitating smarter and faster nanostructure design processes.[133] Next, we will discuss the advances of deep learning in metasurface design.

Deep learning (DL) is crucial in metasurface design, customization, and optimization in photonics. For instance, DL is a subset of machine learning that can learn multilevel abstraction of data with hierarchically structured layers. It complements conventional physics- and rule-based methods. Zhu et al. presented a DL scheme based on a divide-and-conquer strategy for the on-demand design of circuit-analog plasmonic stack metamaterials, as shown in Fig. 4(a).[30] This method splits a spectrum into several parts and demonstrates higher precision for the spectral prediction of the plasmonic stack metamaterial (PSM) structure. The DL method uses fewer training parameters than the conventional method of the fully connected network. Using this approach, the entire bidirectional network achieves precise inverse design.

In addition, Soljačić et al. developed a method using artificial neural networks (N.N.) to approximate light scattering by multilayer nanoparticles, as shown in Fig. 4(b).[134] Only a small sampling of training data is required for the network to achieve high precision in approximating the simulation. Once trained, the neural

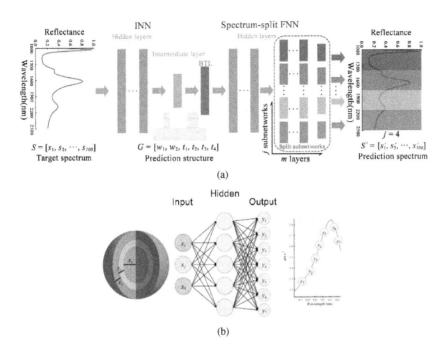

Fig. 4. Deep learning empowering and accelerating the metasurface design. (a) The architecture of divide-and-conquer DL based on a bidirectional artificial neural network.[30] (b) The N.N. architecture has as its input the thickness of each nanoparticle shell and, as its output, the scattering cross-section at different wavelengths of the scattering spectrum.[134] (c) MST architecture for the smart design of graded-refractive-index-based SMAs. (d) A deep-learning-enabled self-adaptive metasurface cloak. (e) The architecture of the transformer-based model. (f) End-to-end design pipeline for multifunctional metasurfaces.

Source: Panel (a) Copyright 2023, Wiley. Applying for a license of figures. Panel (b) Copyright 2018, AAAS. Applying for a license of figures. Panel (c) Adapted with permission Ref. 1. Copyright 2023, Wiley. 2. Applying for a license of figures. Panel (d) Adapted with permission Ref. 135. Copyright 2020, Nature. Applying for a license of figures. Panel (e) Adapted with permission Ref. 2. Copyright 2023, Wiley. Applying for a license of figures. Panel (f) Adapted with permission Ref. 136. Copyright 2022, Wiley. Applying for a license of figures.

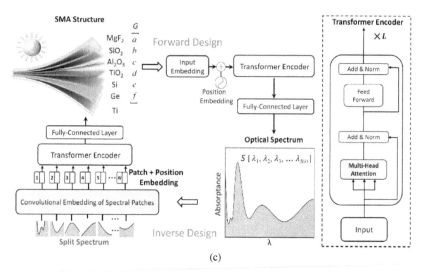

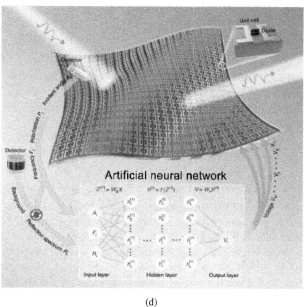

Fig. 4. (*Continued*)

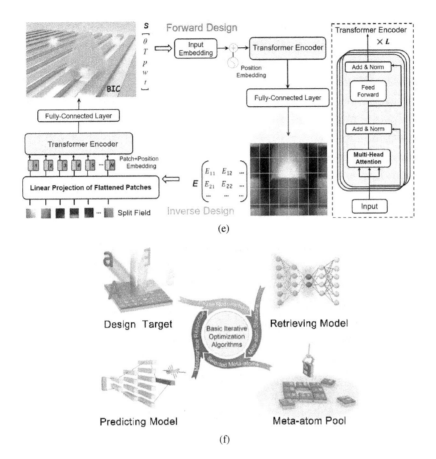

Fig. 4. (*Continued*)

network can simulate optical processes much faster than conventional methods. For instance, Chen et al. proposed a DL method based on a metamaterial spectrum transformer (MST) for the powerful design of high-performance solar metamaterial absorbers (SMAs), as shown in Fig. 4(c).[1] The MST divides the optical spectrum of metamaterial into several patches, solving the problem of overfitting in traditional DL and significantly boosting learning capability. A flexible design tool that allows user-defined specifications is developed to enable real-time and on-demand design of metamaterials with various optical functions. This scheme is applied to designing and fabricating SMAs with graded-refractive-index nanostructures, exhibiting a high average absorptance of 94% in a broad solar spectrum and significant advantages over many state-of-the-art counterparts.

Qian et al. proposed the concept of an intelligent cloak driven by DL and introduced a metasurface cloak as an example implementation, as shown in Fig. 4(d).[135] The metasurface cloak exhibits a millisecond response time to changing incident waves and the surrounding environment in experiments. This work brings to a wide range of real-time and in situ applications, such as moving stealth vehicles. Chen et al. proposed a Q-BIC based all-dielectric metasurface using a transformer-based DL network for SERS spectroscopic applications, as shown in Fig. 4(e).[2] By manipulating the incident angle, a mechanism based on strong coupling Q-BIC is achieved with a large Rabi splitting of ≈105 meV, which creates a bandgap and forms an anti-crossing behavior. Compared to conventional methods, the strong coupling Q-BIC scheme provides an extraordinary SERS enhancement factor of ~10^{17} and extends the field-enhancing scale up to 10 times. Ma et al. applied statistical machine learning in designing and optimizing complex multifunctional metasurfaces, as shown in Fig. 4(f).[136] In contrast to the traditional two-step approach, which separates phase retrieval and meta-atom structural design, the proposed end-to-end framework effectively utilizes the prescribed design space and pushes the multifunctional design capacity to its physical limit. With a single-layer structure, metasurface focusing lenses and holograms are experimentally demonstrated in the near-infrared region. It exhibits up to eight controllable responses for different combinations of working frequencies and linear polarization states, which are impossible by using conventional physics-guided approaches.[136]

4. Advanced Applications Based on Metasurfaces

In this section, diverse applications of metasurfaces are discussed, showcasing the multifaceted utility and burgeoning research in this field (Fig. 5). These applications include linear and nonlinear

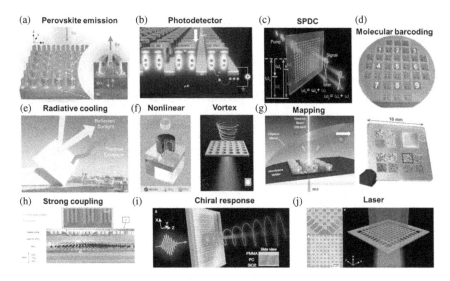

Fig. 5. Various applications enabled by metasurfaces. (a) Q-BIC antenna for engineering the perovskite QD emission characteristics. (b) Miniaturizing color-sensitive photodetectors via hybrid nanoantennas toward submicrometer dimensions. (c) Resonant metasurfaces for generating complex quantum states. (d) Imaging-based molecular barcoding with pixelated dielectric metasurfaces. (e) Radiative cooling engineering. (f) Subwavelength dielectric resonators for nonlinear nanophotonics and ultrafast control of vortex microlasers. (g) Nanoscale mapping of optically inaccessible bound-states-in-the-continuum. (h) Strongly enhanced light-matter coupling. (i) Intrinsic chiral BIC. (j) Lasing action from photonic BIC.

Source: Panel (a) Adapted with permission Ref. 139. Copyright 2024, Wiley. Applying for a license of figures. Panel (b) Adapted with permission Ref. 140. Copyright 2022, AAAS. Applying for a license of figures. Panel (c) Adapted with permission Ref. 141. Copyright 2022, AAAS. Applying for a license of figures. Panel (d) Adapted with permission Ref. 142. Copyright 2021, Wiley. Applying for a license of figures. Panel (e) Adapted with permission Ref. 139. Copyright 2018, AAAS. Applying for a license of figures. Panel (f) Adapted with permission Refs. 144 and 145. Copyright 2020, AAAS and 2020, AAAS. Applying for a license of figures. Panel (g) Adapted with permission Ref. 62. Copyright 2022, Nature. Applying for a license of figures. Panel (h) Adapted with permission Ref. 146. Copyright 2023, Nature. Applying for a license of figures. Panel (i) Adapted with permission Ref. 147. Copyright 2022, AAAS. Applying for a license of figures. Panel (j) Adapted with permission Ref. 148. Copyright 2017, Nature. Applying for a license of figures.

optical engineering, such as perovskite emission engineering and controlling, second harmonic generation, surface-enhanced Raman spectroscopy, miniaturized lasers, and so on Refs. 137–148. For instance, Csányi et al. reported on a method to engineer and shift the photoluminescence wavelength of $FAPbI_3$ perovskite quantum dots (QDs) by using optical nanoantenna arrays driven by quasi-BIC resonances via Purcell effect.[139] This method provides a non-invasive and non-destructive way to tune the emission properties of perovskite QDs.

Ho et al. presented a design of color-sensitive photodetectors based on hybrid silicon–aluminum nanostructures to achieve submicrometer pixel dimensions and minimize optical cross-talk.[140] These nanostructures have dual functionalities: (1) they exhibit a hybrid Mie-plasmon resonance to achieve color-selective light absorption and generate electron–hole pairs; and (2) create a Schottky barrier for photodetection and charge separation. The authors fabricated and characterized the color-sensitive photodetectors with varying nanodisk diameters and proved the photodetectors with a wavelength-dependent photoresponse and a high responsivity. The proposed design is promising to replace the traditional dye-based filters for camera sensors with ultrahigh pixel densities and extend the detection range to the ultraviolet (UV) regime.

Zhang et al. demonstrated chiral emission from resonant metasurfaces,[147] where the authors designed and fabricated a resonant metasurface consisting of a square lattice of tilted titanium dioxide bars on an indium tin oxide–coated substrate. By introducing two asymmetry parameters, they could manipulate the positions and properties of C-points, where the circular polarization conversion occurs. The authors experimentally demonstrated chiral emission and lasing from the metasurface under different excitation conditions, achieving high-quality chiral emission with a near-unity degree of circular polarization, narrow divergence angle, and large circular dichroism. They further achieved chiral lasing with single-mode operation, high Q factor, and improved directionality. The authors proposed that their resonant metasurface can serve as an efficient and controllable source for chiral light generation for nanophotonics and quantum optics. Numerous applications based on metasurfaces have reached an advanced level of maturity, offering a wide range of advanced functionalities. However, due to space constraints in this introductory review, providing a comprehensive overview of each application is challenging. Nevertheless, our teams have gained extensive experience and unique insights in specific domains, such as structural colors, biosensing, and molecular fingerprint sensing over the years. Therefore, we will proceed to present the detailed and comprehensive introductions to these representative applications.

4.1. *Structure color*

Structural color based on metasurface is an intriguing field that explores color generation through engineered structures rather than pigments, as shown in Fig. 6.[55,83,94,105,149-152] This phenomenon originates from the intricate interactions between light and nanostructured surfaces. The metasurface design for structural color involves the unique geometric configurations and periodicities of nanostructures. By tailoring the geometrical parameters, one can precisely control the wavelengths of reflected or transmitted light to create highly saturated and customizable color palettes. Therefore, a myriad of vibrant colors can be expertly crafted and brought to life through precise manipulation of nanostructures, such as sizes, shapes, arrangements, and material compositions. Color generation by nanostructures has many advantages, such as energy-efficient non-emissive displays, high-resolution color printing, and novel optical sensors. In addition, the nanostructural color pixels can be tuned actively. For example, Lu *et al.* presented a reversible tuning approach for Mie resonances in the visible spectrum.[83] Moreover, metasurface with structural color can be used for anti-counterfeiting, due to its complex and distinctive patterns. For instance, the authors designed a tri-functional metasurface that can manipulate the phase, amplitude, and luminescence of light by using anisotropic gap-plasmon structures with upconversion nanoparticles (UCNPs) in their dielectric gap.[94] This metasurface can generate a

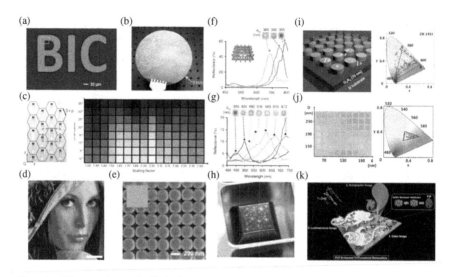

Fig. 6. Plasmonic and all-dielectric metasurfaces for structural color applications. (a) Schrödinger's red pixel by quasi-bound-states-in-the-continuum (q-BIC). (b) Spatial modulation of nanopattern dimensions by combining interference lithography and grayscale-patterned secondary exposure. (c) Versatile full-color nanopainting enabled by a pixelated plasmonic metasurface. (d) Printing color at the optical diffraction limit. Adapted with permission.[147] (e) Reversible tuning of mie resonances in the visible spectrum. Reflectance spectra and reflection-mode micrographs of woodpiles under top-down illumination conditions for (f) a_{xy} = 300–350 nm and (g) a_{xy} = 350–672 nm, respectively. Structural color three-dimensional printing by shrinking photonic crystals. (h) All colors originate solely from the surface nanotexture, which was replicated from corresponding structures on the surface of the chocolate mould. (i) Printing beyond the sRGB color gamut by mimicking silicon nanostructures in free space. (j) Reversible electrical switching of nanostructural color pixels. (k) Trifunctional metasurface enhanced with a physically unclonable function.

Source: Panel (a) Adapted with permission Ref. 61. Copyright 2022, AAAS. Applying for a license of figures. Panel (b) Adapted with permission Ref. 149. Copyright 2022, Nature. Applying for a license of figures. Panel (c) Adapted with permission Ref. 150. Copyright 2023, Nature. Applying for a license of figures. Panel (d) Copyright 2012, Nature. Applying for a license of figures. Panel (e) Adapted with permission Ref. 83. Copyright 2021, ACS. Applying for a license of figures. Panel (f) Adapted with permission Ref. 55. Copyright 2019, Nature. Applying for a license of figures. Panel (h) Adapted with permission Ref. 152. Copyright 2017, Nature. Applying for a license of figures. Panel (i) Adapted with permission Ref. 60. Copyright 2017, ACS. Applying for a license of figures. Panel (j) Adapted with permission Ref. 105. Copyright 2023, De Gruyter. Applying for a license of figures. Panel (k) Adapted with permission Ref. 94. Copyright 2023, Elsevier. Applying for a license of figures.

color image under white light, a hologram under a red laser, and a luminescence image under a near-infrared (NIR) source, depending on the polarization of the incident light. It also exhibits a physically unclonable function (PUF) feature because of the stochastic distribution of the UCNPs, which creates a unique luminescence pattern for each metasurface. Therefore, it offers a highly secure and easy-to-authenticate optical anti-counterfeiting device, which combines the overt, covert, and forensic security features in a single planar device.

4.2. *Metasurface for biosensing applications*

Metasurfaces have a strong near-field enhancement effect and are highly sensitive to changes in the refractive indices of surrounding environments. This characteristic is demonstrated by the resonance wavelength shift that occurs, when light interacts with the structures. These spectral changes can reflect microscopic information of the substances. Metasurface sensors have garnered significant attention in the academic community due to their excellent high sensitivity and fast, real-time, and label-free detection.[153-155] One can optimize the structural parameters of specific metasurface arrays and alter the incident light excitation conditions to improve the sensing performance of equipartition excitations. For instance, metasurfaces have been applied in various biosensing fields, including immunoassays,[156-159] DNA-protein interactions,[160] and real-time monitoring of antigen-antibody bindings.[161-163]

Figure 7(a) illustrates the label-free genetic screening technology proposed by Tittl *et al.*,[172] which uses high-quality factor silicon nanoantennas functionalized with nucleic acid fragments. Each antenna has a resonant quality factor of ~2,200 in a physiological buffer. Jahani *et al.*[165] developed a single-wavelength imaging biosensor, as shown in Fig. 7(b), where it can reconstruct spectral shift information using an optimal linear estimator. These chips are coupled with microfluidics and arranged as microarrays on an imaging platform to detect real-time extracellular vesicles (EVs) associated with breast cancer. They have demonstrated a figure-of-merit

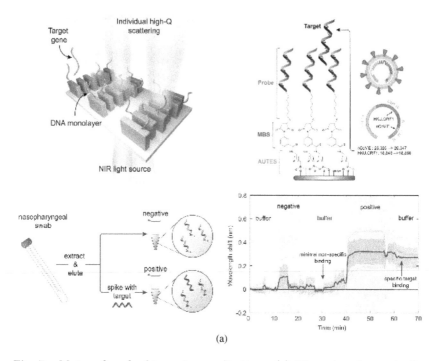

Fig. 7. Metasurface for biosensing applications. (a) Biosensing demonstration with SARS-CoV-2 gene fragment targets using metasurface arrays consisting of perturbed chains of silicon blocks interfaced with DNA probes for targeted gene detection in clinical nasopharyngeal eluates. (b) Biosensing using reconstructed spectral shift based on an aided imaging method. The new optofluidic device is applied for real-time detection of breast cancer-derived EVs by functionalizing the metasurfaces with specific antibodies. (c) Assessment of EV-drug target interactions through bio-orthogonal probe amplification and spatial patterning of molecular reactions within EV size-matched plasmonic nanoring resonators, enabling *in situ* analysis of EV-drug dynamics. (d) Detection of SARS-CoV-2 RBD variants using PMs with high-affinity monoclonal antibodies (mAbs). The PMs with multiple plasmonic modes are functionalized with diverse mAbs, and their molecular affinity characteristics to various SARS-CoV-2 RBD variants are systematically investigated.

Source: Panel (a) Adapted with permission Ref. 164. Copyright 2023, Nature. Applying for a license of figures. Panel (b) Adapted with permission Ref. 165. Copyright 2021, Nature. Applying for a license of figures. Panel (c) Adapted with permission Ref. 166. Copyright 2021, Nature. Applying for a license of figures. Panel (d) Adapted with permission Ref. 167. Copyright 2023, ACS. Applying for a license of figures.

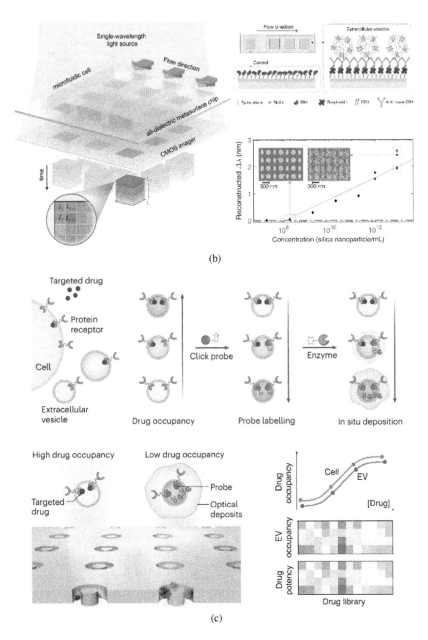

Fig. 7. (*Continued*)

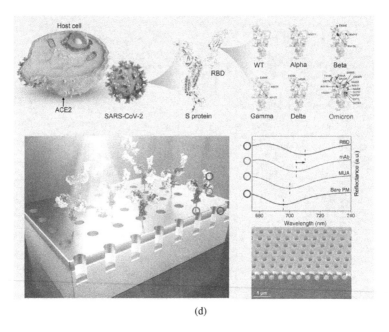

(d)

Fig. 7. (*Continued*)

of ~70 RIU^{-1}, and the system exhibits remarkable sensing capability of extracellular vesicle binding down to 0.41 nanoparticles/μm^2. Figure 7(c) shows an analytical platform that uses circulating EVs to evaluate the activity of tumor-specific drug-target interactions in patient blood samples, where they monitor EV protein expression and small-molecule chemical occupancy and track various targeted therapies. They are able to identify EV signals that correspond closely to the effectiveness of cellular treatment. Moreover, metasurfaces also have the potential applications for clinical cancer prognosis and diagnosis. Li et al.[167] presented a highly effective affinity testing approach, as shown in Fig. 7(d). They use label-free plasmonic metasurface sensors and monoclonal antibodies to test their binding affinity to 12 spike receptor binding domain (RBD) variants of SARS-CoV-2. In addition, they thoroughly analyze the molecular kinetics of the Langmuir binding equilibrium between the 12 SARS-CoV-2 RBD variants and mAb-functionalized metasurfaces.

4.3. *Metasurface for molecular fingerprint sensing applications*

Because of molecular rotations and vibrations, many complex molecules exhibit characteristic absorption in the mid-infrared to terahertz spectrum. Surface-enhanced infrared absorption has gained significant attention in recent years, due to its advantages of low-loss, and high-Q factor resonances, which can enhance samples' absorption fingerprint spectral signatures for analyte identification.[168–171] Figure 8(a) shows an imaging-based nanophotonic technique for identifying mid-infrared molecular fingerprints and its application for the chemical identification of surface-bound analytes. Tittl et al.[172] presented a metasurface device for measuring the molecular absorption signatures at various spectral points, using a two-dimensional pixelated dielectric metasurface with various ultrasharp resonances, which can be tuned to a discrete frequency. The metasurface device can convert the spectral data into an imaging-ready spatial absorption map resembling a barcode. Figure 8(b) shows a design for plasmonic nanofin metasurfaces that support symmetry-protected BIC up to the fourth order, developed by Aigner et al.[173] They obtained high-Q factor modes breaking the out-of-plane symmetry of the nanofins in parameter space. This symmetry breaking can be fine-tuned by adjusting the nanofins' triangle angle, providing a way to control the ratio of radiative to intrinsic losses precisely. Liu et al.[13] presented an inverted dielectric metagrating for enhancing the broadband THz fingerprint detection of trace analytes on a planar sensing surface, shown in Fig. 8(c). They tuned the asymmetry parameter of quasi-BIC modes, so as to boost the near-field intensity within the analytes. The multiplexing mechanism for broadband detection is demonstrated by manipulating incident angles of excitation waves and the waveguide thickness. As a result, the THz fingerprint peak values are significantly enhanced compared to conventional approaches with a factor of up to 330-fold. Lyu et al.[174] designed a frequency-selective fingerprint sensor (FSFS), as shown in Fig. 8(d). This platform achieves enhanced trace fingerprint

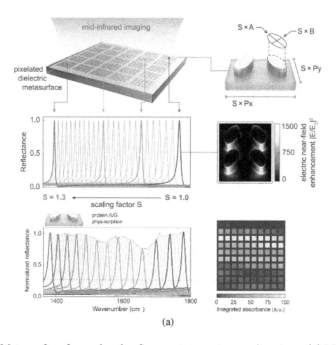

Fig. 8. Metasurface for molecular fingerprint sensing applications. (a) Molecular fingerprint detection with pixelated dielectric metasurfaces. The pixelated metasurface is composed of a two-dimensional array of high-Q resonant megapixels with resonance frequencies tuned over a target molecular fingerprint range. (b) Plasmonic nanofin metasurfaces with coupling-tailored sensing. The absorbance spectra of uncoated (grey) and 5-nm PMMA-coated (colored) metasurfaces. (c) Enhancing THz fingerprint detection. The sensing enhancement is attributed to the use of evanescent waves at the planar air-metastructure interface for a series of quasi-BIC resonance modes. The broadband fingerprint signals of trace analytes can be retrieved by tuning the reflectance peaks. (d) Enhanced absorption-induced transparency by frequency selective fingerprint sensor. Optical images of the fabricated FSFS, with 36-pixel locations (P1-P36 corresponds to L from 104 μm to 47 μm), and measured transmission spectra of 10 μm thick L-carnitine coating on the FSFS, as well as the enveloped absorbance signals of L-carnitine.

Source: Panel (a) Adapted with permission Ref. 172. Copyright 2018, AAAS. Applying for a license of figures. Panel (b) Adapted with permission Ref. 173. Copyright 2022, AAAS. Applying for a license of figures. Panel (c) Adapted with permission Ref. 13. Copyright 2022, OSA. Applying for a license of figures. Panel (d). Adapted with permission Ref. 174. Copyright 2023, Springer. Applying for a license of figures.

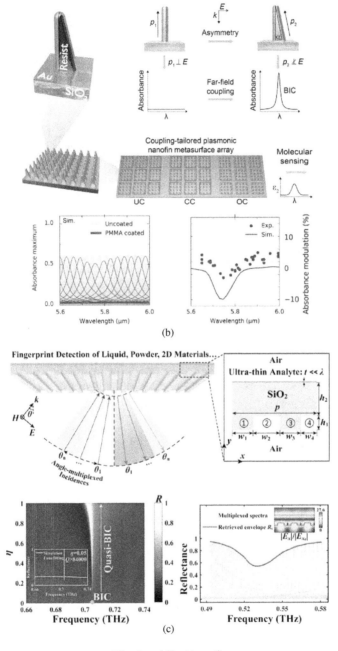

Fig. 8. (*Continued*)

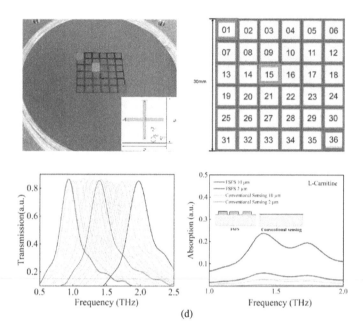

Fig. 8. (*Continued*)

detection through broadband multiplexing. Its central wavelength is linearly related to the cross-slot length, and the broadband absorption lines of trace-amount chiral carnitine were boosted by 7.3 times.

5. Insight and Outlook

Metasurfaces are emerging as a powerful and versatile tool for various applications in the UV, visible, infrared, to terahertz spectral range.[64,175–177] By exploiting the strong near-field enhancements and high-Q factor resonances, these devices can achieve multiplexing and powerful capabilities, enabling a wide range of applications in biosensing, environmental monitoring, metalenses, optical cloaking, electromagnetic scattering, structural colors, miniaturized functional devices and others. However, at the same time, metasurface devices also face several challenges and limitations that must be addressed and overcome. These challenges include: (i) The fabrication and integration of metasurfaces with microfluidics, electronics, and optics, which require high precision, scalability, and compatibility. (ii) The optimization and tuning of metasurface parameters and excitation conditions require advanced designs and simulation methods and tools. (iii) The signal processing and analysis of metasurface spectra require efficient and robust algorithms and models.

To tackle these challenges and advance the field of metasurface sensors, we envision several possible directions and solutions for future research. These directions include: (i) Developing new materials and structures for metasurfaces, such as 2D materials, phase-change materials, and reconfigurable metasurfaces, which can offer new functionalities and properties for metasurfaces. (ii) Exploring new phenomena and mechanisms for metasurfaces, such as nonlinear, quantum, and topological effects, can enhance performance. (iii) Applying machine learning and artificial intelligence techniques for metasurface design, such as deep learning, generative adversarial networks, and reinforcement learning, can improve the accuracy and efficiency of modeling. Furthermore, the frontier applications of intelligent algorithms and metasurfaces can be extended by introducing the concept of intelligent metamaterials. This involves utilizing metasurfaces to enable all-optical computing and integrating algorithms like artificial neural networks.[178-180] Metasurface devices are at the forefront of nanotechnology and optics, offering innovative solutions for advanced applications. We believe that metasurface devices have great potential and promise for advancing science and technology and benefiting society and humanity. We hope this review can provide a useful and introductory reference for researchers and practitioners interested in and working on metasurfaces and inspire them to explore new possibilities and opportunities in this exciting and rapidly evolving field.

Acknowledgments

Z. D. would like to acknowledge the funding support from the Agency for Science, Technology and Research (A*STAR) under its AME IRG (Project No. A20E5c0093), Career Development Award grant (Project No. C210112019), MTC IRG (Project Nos. M21K2c0116 and M22K2c0088), the Quantum Engineering Programme 2.0 (Award No. NRF2021-QEP2-03-P09) and DELTA-Q 2.0 (Project No. C230917005). J. Z. would like to acknowledge the funding support from NSFC (62175205), NSAF (U2130112) and the Youth Talent Support Program of Fujian Province (Eyas Plan of Fujian Province) [2022]. W. C. would like to acknowledge the funding supporting from the China Scholarship Council Scholarship (CSC No. 202306310153). J. S. would like to acknowledge the funding supporting from the China Scholarship Council Scholarship (CSC No. 202306310168). Wei Chen and Jiaqing Shen authors contributed equally.

ORCID

Wei Chen https://orcid.org/0000-0003-0103-3149
Jiaqing Shen https://orcid.org/0009-0007-2203-4454
Yujie Ke https://orcid.org/0009-0001-0720-8053
Yuwei Hu https://orcid.org/0000-0002-7282-4143
Lin Wu https://orcid.org/0000-0002-3188-0640
Jinfeng Zhu https://orcid.org/0000-0003-3666-6763
Zhaogang Dong https://orcid.org/0000-0002-0929-7723

References

1. W. Chen *et al.*, *Adv. Sci.*, 2023, **10**, 2206718.
2. W. Chen *et al.*, *Adv. Opt. Mater.*, 2023, **12**, 2301697.
3. S. Ding *et al.*, *Carbon*, 2021, **179**, 666–676.
4. M. Gao *et al.*, *Photonics Res.*, 2020, **8**, 1226–1235.
5. S. C. Huang *et al.*, *Nat. Commun.*, 2020, **11**, 4211.
6. T. X. Huang *et al.*, *Nanoscale*, 2018, **10**, 4398–4405.
7. F. F. Jiao *et al.*, *Sens. Actuators B: Chem.*, 2021, **344**, 130170.
8. C. W. Li *et al.*, *ACS Appl. Nano Mater.*, 2019, **2**, 3231–3237.
9. F. J. Li *et al.*, *Biosens. Bioelectron.*, 2022, **203**, 114038.
10. B. W. Liu *et al.*, *Adv. Mater.*, 2018, **30**, 1706031.
11. H. Liu *et al.*, *Remote Sens.*, 2016, **8**, 935.

12. J. L. Liu *et al.*, *Sensors*, 2021, **21**, 4987.
13. X. Y. Liu *et al.*, *Photonics Res.*, 2022, **10**, 2836–2845.
14. X. R. Mao *et al.*, *IEEE Photonics J.*, 2020, **12**, 5900408.
15. J. L. Qiu *et al.*, *Opt. Lett.*, 2021, **46**, 849–852.
16. L. Quan *et al.*, *Chem. Eng. J.*, 2022, **428**, 131337.
17. J. Q. Shen *et al.*, *IEEE J. Sel. Top. Quantum Electron.*, 2023, **29**, 6900408.
18. J. Liu *et al.*, *IEEE Access*, 2021, **9**, 92941–92951.
19. Z. B. Wang *et al.*, *Talanta*, 2023, **264**, 124731.
20. Y. N. Xie *et al.*, *IEEE Trans. Microw. Theory Tech.*, 2023, 72, 2368.
21. J. F. Zhu *et al.*, *Appl. Phys. Lett.*, 2018, **112**, 153106.
22. J. Liu *et al.*, *Sol. Energy Mater. Sol. Cells*, 2022, **244**, 111822.
23. X. Yao *et al.*, *ACS Appl. Mater. Interfaces*, 2020, **12**, 36505–36512.
24. J. Liu *et al.*, *IEEE Access*, 2020, **8**, 43407–43412.
25. Q. Lu *et al.*, *Engineering*, 2022, **17**, 22–30.
26. L. Zhang *et al.*, *Nat. Commun.*, 2018, **9**, 4334.
27. Y. Wang *et al.*, *Adv. Mater.*, 2023, **35**, 2302387.
28. Y. Chen *et al.*, *Compos. Sci. Technol.*, 2021, **214**, 108970.
29. J. F. Zhu *et al.*, *Biosens. Bioelectron.*, 2020, **150**, 111905.
30. J. K. Xiong *et al.*, *Laser Photonics Rev.*, 2023, **17**, 2100738.
31. W. Z. Ma *et al.*, *Results Phys.*, 2021, **31**, 105055.
32. J. F. Zhu *et al.*, *Nanoscale*, 2020, **12**, 2057–2062.
33. W. Chen *et al.*, *Appl. Sci. (Basel)*, 2020, **10**, 3276.
34. Y. S. Chen *et al.*, *Opt. Express*, 2021, **29**, 37234–37244.
35. C. Dou *et al.*, *Carbon*, 2024, **216**, 118556.
36. J. Liu *et al.*, *J. Mater. Res. Technol.*, 2020, **9**, 6723–6732.
37. O. A. M. Abdelraouf *et al.*, *Adv. Funct. Mater.*, 2021, **31**, 2104627.
38. Z. Dong *et al.*, *Nano Letters*, 2021, **21**, 4853–4860.
39. L. F. Ye *et al.*, *Nanomaterials*, 2018, **8**, 562.
40. M. Asbahi *et al.*, *ACS Appl. Mater. Interfaces*, 2019, **11**, 45207–45213.
41. K. Yang *et al.*, *Adv. Opt. Mater.*, 2021, **9**, 2001375.
42. J. Liu *et al.*, *Microchem. J.*, 2022, **183**, 108043.
43. Z. B. Wang *et al.*, *Sensors*, 2022, **22**, 133.
44. J. Liu *et al.*, *Opt. Express*, 2020, **28**, 23748–23760.
45. Z. Dong *et al.*, *Light Sci. Appl.*, 2022, **11**, 20.

46. J. F. Zhu *et al.*, *IEEE Photonics Technol. Lett.*, 2017, **29**, 504–506.
47. W. Z. Ma *et al.*, *Curr. Appl. Phys.*, 2021, **27**, 51–57.
48. D. Meng *et al.*, *Results Phys.*, 2022, **39**, 105766.
49. J. Liu *et al.*, *Infrared Phys. Technol.*, 2022, **127**, 104447.
50. Z. Dong *et al.*, *ACS Photonics*, 2015, **2**, 385–391.
51. O. A. M. Abdelraouf *et al.*, *ACS Nano*, 2022, **16**, 13339–13369.
52. M. Asbahi *et al.*, *Nanoscale*, 2017, **9**, 9886–9892.
53. J. Liu *et al.*, *Nanomaterials*, 2020, **10**, 27.
54. Z. Dong *et al.*, *Nano Letters*, 2015, **15**, 5976–5981.
55. Y. Liu *et al.*, *Nat. Commun.*, 2019, **10**, 4340.
56. M. Asbahi *et al.*, *Nanotechnology*, 2016, **27**, 424001.
57. Z. Dong *et al.*, *Nanotechnology*, 2014, **25**, 135303.
58. O. A. M. Abdelraouf *et al.*, Conf Lasers and Electro-Optics (CLEO), San Jose, CA, USA, 2021, pp. 1–2.
59. J. Liu *et al.*, *Results Phys.*, 2021, **22**, 103818.
60. Z. Dong *et al.*, *Nano Letters*, 2017, **17**, 7620–7628.
61. Z. Dong *et al.*, *Sci. Adv.*, 2022, **8**, eabm4512.
62. W. Z. Ma *et al.*, *Nanophotonics*, 2023, **12**, 3589–3601.
63. K. Huang *et al.*, *Laser Photonics Rev.*, 2016, **10**, 500–509.
64. Z. Dong *et al.*, *Nano Letters*, 2019, **19**, 8040–8048.
65. J. Gu *et al.*, *Adv. Opt. Mater.*, 2023, **11**, 2202826.
66. R. He *et al.*, *Nanoscale*, 2023, **15**, 1652–1660.
67. J. Ho *et al.*, *ACS Nano*, 2018, **12**, 8616–8624.
68. Z. Ding, W. Su and F. Hakimi, *Sol. Energy Mater. Sol. Cells*, 2020, **262**, 112563.
69. K. Huang *et al.*, *ACS Nano*, 2020, **14**, 7529–7537.
70. M. Jalali *et al.*, *Nanoscale*, 2016, **8**, 18228–18234.
71. N. Yu *et al.*, *Science*, 2011, **334**, 333–337.
72. L. Jiang *et al.*, *ACS Nano*, 2015, **9**, 10039–10046.
73. L. Jin *et al.*, *Nano Letters*, 2018, **18**, 8016–8024.
74. L. Jiang *et al.*, *Light Sci. Appl.*, 2019, **8**, 21.
75. L. Jin *et al.*, *Nat. Commun.*, 2019, **10**, 4789.
76. C. Jung *et al.*, *Chem. Rev.*, 2021, **121**, 13013–13050.
77. S. Kruk *et al.*, Conf Lasers and Electro-Optics (CLEO), San Jose, CA, USA, 2021, pp. 1–2.
78. S. S. Kruk *et al.*, *Nat. Photonics*, 2022, **16**, 561.
79. H. Liu *et al.*, *ACS Nano*, 2022, **16**, 8244–8252.
80. M. Limonov *et al.*, *Nat. Photonics*, 2017, **11**, 543–554.

81. Z. Ding, W. Su and L. Ye, *Sol. Energy*, 2024, **271**, 112449.
82. B. Wang *et al.*, Appl. Phys. Lett. 2024, 124, 151105.
83. L. Lu *et al.*, *ACS Nano*, 2021, **15**, 19722–19732.
84. H. K. Ng *et al.*, *Nat. Electron.*, 2022, **5**, 489–496.
85. R. J. H. Ng *et al.*, *Opt. Mater. Express*, 2019, **9**, 788–801.
86. S. D. Rezaei *et al.*, *ACS Photonics*, 2021, **8**, 18–33.
87. Z. Wang *et al.*, *Nat. Commun.*, 2016, **7**, 11283.
88. S. D. Rezaei *et al.*, *Adv. Opt. Mater.*, 2019, **7**, 1900735.
89. X. Yu *et al.*, *Optica*, 2016, **3**, 979–984.
90. X. Su *et al.*, *ACS Nano*, 2022, **16**, 11492–11497.
91. J. Tao *et al.*, *Opt. Express*, 2015, **23**, 7809–7819.
92. J. Trisno *et al.*, *ACS Photonics*, 2018, **5**, 511–519.
93. T. Wang *et al.*, *ACS Photonics*, 2019, **6**, 1272–1278.
94. S. D. Rezaei *et al.*, *Mater. Today*, 2023, **62**, 51–61.
95. Z. Wang *et al.*, *ACS Nano*, 2018, **12**, 1859–1867.
96. X. Xiong *et al.*, *Adv. Opt. Mater.*, 2022, **10**, 2200557.
97. J. Xu *et al.*, *Nano Letters*, 2021, **21**, 3044–3051.
98. P. Y. M. Yew *et al.*, *Mater. Today Chem.*, 2021, **22**, 100574.
99. T. Yin *et al.*, *ACS Photonics*, 2016, **3**, 979–984.
100. T. Yin *et al.*, *Nanoscale*, 2017, **9**, 2082–2087.
101. X. Yu *et al.*, *Nanoscale*, 2016, **8**, 327–332.
102. S. D. Rezaei *et al.*, *ACS Nano*, 2019, **13**, 3580–3588.
103. X. Yu *et al.*, Conf Lasers and Electro-Optics (CLEO), San Jose, CA, USA, 2016, pp. 1–2.
104. L. Zhang *et al.*, *Nanoscale*, 2015, **7**, 12018–12022.
105. S. Zhang *et al.*, *Nanophotonics*, 2023, **12**, 1387–1395.
106. D. Zhao, Z. Dong and K. Huang, *Opt. Lett.*, 2021, **46**, 1462–1465.
107. D. Zhu *et al.*, *ACS Photonics*, 2014, **1**, 1307–1312.
108. Q. Zhu *et al.*, *Mater. Today Adv.*, 2022, **15**, 100270.
109. S. Molesky *et al.*, *Nat. Photonics*, 2018, **12**, 659–670.
110. W. Ma, Z. Liu and Z. A. Kudyshev, *Nat. Photonics*, 2021, **15**, 77–90.
111. D. Lee *et al.*, *eLight*, 2022, **2**, 1–23.
112. E. Galiffi *et al.*, *Nat. Rev. Mater.*, 2024, **9**, 9–28.
113. X. R. Mao *et al.*, *Nat. Nanotechnol.*, 2021, **16**, 1099–1105.
114. L. Huang *et al.*, *Nat. Commun.*, 2023, **14**, 3433.
115. T. Kaelberer *et al.*, *Science*, 2010, **330**, 1510–1512.
116. A. R. Indhu, L. Keerthana and G. Dharmalingam, *Beilstein J. Nanotechnol.*, 2023, **14**, 380–419.

117. Y. Ke *et al.*, *Mater. Horiz.*, 2021, **8**, 1700–1710.
118. K. Zhang *et al.*, *Opt. Lett.*, 2020, **45**, 3669–3672.
119. T. T. Hoang *et al.*, *Sci. Rep.*, 2018, **8**, 16404.
120. B. Luk'Yanchuk *et al.*, *Nat. Mater.*, 2010, **9**, 707–715.
121. G. Hu *et al.*, *Nat. Photonics*, 2019, **13**, 467–472.
122. K. V. Sreekanth *et al.*, *Nat. Mater.*, 2016, **15**, 621–627.
123. G. Hu *et al.*, *Nature*, 2020, **582**, 209–213.
124. G. Hu *et al.*, *Nano Letters*, 2020, **20**, 3217–3224.
125. Q. Zhang *et al.*, *Nature*, 2021, **597**, 187–195.
126. G. Hu *et al.*, *Nat. Nanotechnol.*, 2023, **18**, 64–70.
127. A. E. Miroshnichenko *et al.*, *Nat. Commun.*, 2015, **6**, 8069.
128. T. Zhan *et al.*, *J. Phys.: Condens. Matter*, 2013, **25**, 215301.
129. M. Nieto-Vesperinas, *Dielectric Metamaterials*, Woodhead Publishing, 2020, pp. 39–72.
130. K. S. Kunz and R. J. Luebbers, The finite difference time domain method for electromagnetics, CRC Press, 1993.
131. F. L. Teixeira *et al.*, *Nat. Rev. Methods Primers*, 2023, **3**, 75.
132. M. N. Ab Wahab, S. Nefti-Meziani and A. Atyabi, *PLoS One*, 2015, **10**, e0122827.
133. L. Huang, L. Xu and A. E. Miroshnichenko, *Advances and Applications in Deep Learning*, InTechOpen, Vol. 65, 2020, pp. 1–14.
134. J. Peurifoy *et al.*, *Sci. Adv.*, 2018, **4**, eaar4206.
135. C. Qian *et al.*, *Nat. Photonics*, 2020, **14**, 383–390.
136. W. Ma *et al.*, *Adv. Mater.*, 2022, **34**, 2110022.
137. H. H. Yoon *et al.*, *Science*, 2022, **378**, 296–299.
138. J. Hu *et al.*, *Front. Phys.*, 2021, **8**, 586087.
139. E. Csányi *et al.*, *Adv. Funct. Mater.*, 2024, **34**, 2309539.
140. J. Ho *et al.*, *Sci. Adv.*, 2022, **8**, eadd3868.
141. T. Santiago-Cruz *et al.*, *Science*, 2022, **377**, 991–995.
142. A. Leitis *et al.*, *Adv. Mater.*, 2021, **33**, 2102232.
143. J. Mandal *et al.*, *Science*, 2018, **362**, 315–319.
144. K. Koshelev, S. Kruk, E. Melik-Gaykazyan, J.-H. Choi, A. Bogdanov, H.-G. Park and Y. Kivshar, *Science*, 2020, **367**, 288–292.
145. C. Huang *et al.*, *Science*, 2020, **367**, 1018–1021.
146. P. Corbae *et al.*, *Nature Mater.*, 2023, **22**, 200–206.
147. X. Zhang *et al.*, *Science*, 2022, **377**, 1215–1218.
148. A. Kodigala *et al.*, *Nature*, 2017, **541**, 196–199.

149. Z. Gan *et al.*, *Light Sci. Appl.*, 2022, **11**, 89.
150. M. Song *et al.*, *Nat. Nanotechnol.*, 2023, **18**, 71–78.
151. K. Kumar *et al.*, *Nat. Nanotechnol.*, 2012, **7**, 557–561.
152. A. Kristensen *et al.*, *Nat. Rev. Mater.*, 2017, **2**, 16088.
153. A. Azzouz *et al.*, *Biosens. Bioelectron.*, 2022, **197**, 113767.
154. S. Ferguson *et al.*, *Sci. Adv.*, 2022, **8**, eabm3453.
155. Y. Fan *et al.*, *Biosens. Bioelectron.*, 2020, **154**, 112066.
156. M. H. Jeong *et al.*, *Adv. Sci.*, 2023, **10**, 2205148.
157. C. Jiang *et al.*, *Nano-Micro Lett.*, 2021, **14**, 3.
158. C. Z. J. Lim *et al.*, *Nat. Commun.*, 2019, **10**, 1144.
159. Y. Ren *et al.*, *Biosens. Bioelectron.*, 2022, **199**, 113870.
160. A. Natalia *et al.*, *Nat. Rev. Bioeng.*, 2023, **1**, 481.
161. A. Thakur *et al.*, *Biosens. Bioelectron.*, 2021, **191**, 113476.
162. J. Wang *et al.*, *Sci. Adv.*, 2020, **6**, eaax3223.
163. Q. Wang *et al.*, *Biosens. Bioelectron.*, 2019, **135**, 129–136.
164. J. Hu *et al.*, *Nat. Commun.*, 2023, **14**, 4486.
165. Y. Jahani *et al.*, *Nat. Commun.*, 2021, **12**, 3246.
166. S. Pan *et al.*, *Nat. Nanotechnol.*, 2021, **16**, 734–742.
167. F. Li *et al.*, *ACS Nano*, 2023, **4**, 3383.
168. X. Liu *et al.*, *Nanophotonics*, 2023, **12**, 4319–4328.
169. Z. Liu *et al.*, *Nano Letters*, 2023, **23**, 10441–10448.
170. L. Sun *et al.*, *Nanoscale*, 2022, **14**, 9681–9685.
171. M. Zheng *et al.*, *Adv. Funct. Mater.*, 2022, **33**, 2209368.
172. A. Tittl *et al.*, *Science*, 2018, **360**, 1105.
173. A. Aigner *et al.*, *Sci. Adv.*, 2022, **8**, eadd4816.
174. J. Lyu *et al.*, *PhotoniX*, 2023, **4**, 28.
175. Y. Ke *et al.*, *ACS Nano*, 2017, **11**, 7542–7551.
176. Y. Ke *et al.*, *Joule*, 2019, **3**, 858–871.
177. Y. Ke *et al.*, *Opt. Express*, 2021, **29**, 9324–9331.
178. X. Ding *et al.*, *Adv. Mater.*, 2024, **36**, 2308993.
179. X. Wang *et al.*, *Nat. Commun.*, 2023, **14**, 2063.
180. Z. Wang *et al.*, *Nat. Commun.*, 2022, **13**, 2188.

© 2025 World Scientific Publishing Company
https://doi.org/10.1142/9789819806409_0002

A Review of Self-Sensing in Carbon Fiber Structural Composite Materials*

D. D. L. Chung

Composite Materials Research Laboratory, Department of Mechanical and Aerospace Engineering, University at Buffalo, The State University of New York, Buffalo, NY 14260-4400, USA
ddlchung@buffalo.edu

Sensing is a basic ability of smart structures. Self-sensing involves the structural material sensing itself. No device incorporation is needed, thus resulting in cost reduction, durability enhancement, sensing volume increase and absence of mechanical property diminution. Carbon fiber renders electrical conductivity to a composite material. The effect of strain/damage on the electrical conductivity enables self-sensing. This review addresses self-sensing in structural composite materials that contain carbon fiber reinforcement. The composites include polymer–matrix composites with continuous carbon fiber reinforcement (relevant to aircraft and other lightweight structures) and cement–matrix composites with short carbon fiber reinforcement (relevant to the civil infrastructure). The sensing mechanisms differ for these two types of composite materials, due to the difference in structures, which affects the electrical and electromechanical behaviors. For the polymer–matrix composites with continuous carbon fiber reinforcement, the longitudinal resistivity in the fiber direction decreases upon uniaxial tension, due to the fiber residual compressive stress reduction, while the through-thickness resistivity increases, due to the fiber waviness reduction; upon flexure, the tension surface resistance increases, because of the reduction in the current penetration from the surface,

---

*To cite this article, please refer to its earlier version published in the *World Scientific Annual Review of Functional Materials*, Volume 1, 2230004 (2023), DOI: 10.1142/S2810922822300045.

while the compression surface resistance decreases. These strain effects are reversible. The through-thickness resistance, oblique resistance and interlaminar interfacial resistivity increase irreversibly upon fiber fracture, delamination or subtle irreversible change in the microstructure. For the cement–matrix composites with short carbon fiber reinforcement, the resistivity increases upon tension, due to the fiber–matrix interface weakening, and decreases upon compression; upon flexure, the tension surface resistance increases, while the compression surface resistance decreases. Strain and damage cause reversible and irreversible resistance changes, respectively. The incorporation of carbon nanofiber or nanotube to these composites adds to the costs, while the sensing performance is improved marginally, if any. The self-sensing involves resistance or capacitance measurement. Strain and damage cause reversible and irreversible capacitance changes, respectively. The fringing electric field that bows out of the coplanar electrodes serves as a probe, with the capacitance decreased when the fringing field encounters an imperfection. For the cement-based materials, a conductive admixture is not required for capacitance-based self-sensing.

Keywords: Carbon fiber; polymer; cement; self-sensing; electrical resistance.

1. Introduction

Smart or multi-functional structures are the ones that can provide both structural function (for bearing loads) and one or more non-structural functions, such as sensing and response to the information sensed. The response may be movement (actuation), elastic modulus change (either stiffening or softening), energy dissipation (with the energy being mechanical or thermal), healing, etc. The deformation (strain), loading (stress), damage, temperature, humidity, etc. are among the attributes that are sensed. The strain relates to the vibration, which is of concern to the performance and reliability of structures in general. The sensing ability stems from the output of the sensor being sensitive to one or more of the attributes. Sensing (akin to feeling) is the most basic nonstructural function, since the inability of sensing a stimulus makes it impossible for the structure to respond to the stimulus.

The field of smart or multi-functional structures has grown tremendously during the past 20 years. This is partly because of the aging infrastructure, aging aircraft and structural damage caused by extreme events (such as earthquakes) causing the need for structural health monitoring (i.e. damage sensing) for the purpose of safety enhancement. In addition, the field is propelled by the desire to have the structure be able to respond automatically to the information sensed. An example of a response to damage is the healing of the damage.

The smartness in a smart structure can be attained by either the incorporation of devices in the structure or the exploitation of the inherent ability of the structural material to provide the desired nonstructural function. The former approach gives extrinsic smartness, which is derived from the devices (e.g. optical fiber), whereas the latter approach gives intrinsic smartness, which is derived from the structural material.

The devices are more costly and less rugged than the structural materials. Intrinsic smartness is advantageous over extrinsic smartness in terms of the lower cost, higher durability, larger functional volume and the mechanical properties not being compromised. The embedment of a device in a structure is intrusive and tends to degrade the mechanical properties. Due to the intrusiveness, the embedment is more suitable for new structures than existing structures. Since there are many more existing structures than new structures, it is not desirable to limit the application to new structures. The attachment of a device on a structure would not degrade the structure, but the attached devices tend to be detached during use of the structure. The repeated attachment of the devices on a structure is costly.

In spite of the advantages of intrinsic smartness, research on smart structures is dominated by extrinsic smartness, because intrinsic smartness is challenging from the scientific viewpoint. The unravelling of the nonstructural behavior of structural materials is needed for the development of intrinsic smartness. For example, the nonstructural behavior pertains to the electrical conductivity, which is negatively affected by damage. Thus, the damage can be sensed by resistivity measurement. The nonstructural behavior of structural materials has traditionally been given little attention.

Longstanding research on structures or structural materials is dominated by the study of the structural properties rather than the nonstructural properties. Moreover, the researchers in the field of structures tend to be inadequately trained in the principles and methods of measuring the nonstructural properties, such as the electrical properties. As a result, numerous published papers are flawed. Pitfalls[1] are particularly common in research on the piezoresistive behavior.[2] Piezoresistivity pertains to the effect of strain (or stress) on the electrical resistivity. Electrical engineers that are trained in measuring the electrical properties do not tend to engage in research related to structures. On the other hand, the incorporation of devices in a structure is less challenging scientifically than the unravelling of the nonstructural behavior of structural materials. As a result, intrinsic smartness has received much less attention than extrinsic smartness.

This review is focused on intrinsic smartness by addressing multi-functional structural materials. The science behind the multi-functionality is addressed with consideration of the mechanisms behind the multi-functionality. The mechanisms depend on the structure and properties of the materials involved. In relation to sensing, attention is given to the choice of attribute to be measured for the purpose of achieving the functionality and the methodology for measuring the chosen attribute.

A structural material that senses itself without any device incorporation is said to be self-sensing. Self-sensing is attained by exploiting the influence of strain, damage, temperature or other stimuli on the electrical behavior, which provides the attributes that can be measured in order to sense the stimuli. The electrical resistance (which pertains to the conduction behavior) and capacitance (which pertains to the dielectric behavior) are two electrical attributes that can be conveniently measured. Resistance-based self-sensing has received much more attention than capacitance-based self-sensing. Both avenues of self-sensing are addressed in this review.

2. Carbon Fibers

It is well known that carbon fibers exhibit low density, high strength, high elastic modulus and low coefficient of thermal expansion. The high modulus and strength enable the fiber to be effective for reinforcing a composite material. The low density is advantageous for lightweight structures. The low coefficient of thermal expansion is advantageous for high thermal fatigue resistance. However, carbon fibers are brittle. The brittleness problem can be circumvented by the embedment of the fibers in a matrix, thereby forming a composite, particularly if the matrix is less brittle than the carbon fiber. Polymers are relatively ductile, in addition to their processability, so they are most widely used for the matrix.

Carbon fibers that are continuous, as commonly supplied commercially in the form of spools, are valuable for providing reinforcement. In general, continuous fibers are much more effective as a reinforcement than the corresponding short fibers at the same volume fraction, due to the imperfect bonding at the fiber–matrix interface. Short fibers or nanotubes that cling together by van der Waals forces to form a yarn do not provide a truly continuous fiber.

Graphite, diamond and fullerene are the three allotropes of carbon. Carbon fibers are in the graphite allotrope family, though they are not ideal graphite and a carbon fiber is not a single crystal. Graphite exhibits a layered crystal structure, with the in-plane bonding involving a combination of covalent bonding and metallic bonding (with sp^2 hybridization of the carbon atoms), and the out-of-plane bonding involving van der Waals forces. The covalent bonding renders high modulus and strength in the in-plane direction. The van der Waals forces render low strength in the out-of-plane direction. The metallic bonding renders high electrical/thermal conductivity in the in-plane direction. In contrast, diamond exhibits sp^3 hybridization of the carbon atoms and is therefore nonconductive. Graphite (i.e. ideal graphite) has the carbon layers in it being perfectly flat and parallel, in addition to them being stacked in the AB sequence, so that the layers are in registry rather than being randomly positioned relative to one another in the in-plane direction.

The carbon layers in a carbon fiber are usually not perfectly flat or parallel. Furthermore, the carbon layers are stacked, such that the AB sequence is usually absent and the dimensions of the stack are limited in both the direction perpendicular to the layers and the direction parallel to the layers. In other words, the number of layers in the stack is limited and the layers extend for limited distances only. The form of carbon without the AB stacking sequence, and with carbon layers that are not perfectly flat, not perfectly parallel and limited in dimensions, is known as turbostratic carbon, which is not graphite but belongs to the graphite allotrope family. In spite of the presence of carbon layers, turbostratic carbon has no long-range periodicity in the atomic arrangement, so it is noncrystalline. Carbon fibers are typically completely turbostratic carbon.

It is possible for a carbon fiber to exhibit a limited degree of crystallinity, so that it has graphitic (crystalline) regions and is not entirely turbostratic carbon. However, partly graphitic carbon fibers are not typical and require heating at very high temperatures (e.g. 2,800°C) during fabrication in order to develop the partial crystallinity. The tendency to crystallize as the temperature increases is due to the fact that the crystallized state has lower free energy (i.e. being more thermodynamically stable) than a partially crystallized or noncrystalline state. A high temperature of the heat treatment results in more thermal energy for the carbon atoms in the fiber to move. The movement is automatically such that the degree of crystallinity is increased through the movement. The required high temperatures for the heating cause the partly graphitic fibers to be expensive compared to the conventional carbon fibers, which are not graphitic. Furthermore, the higher degree of crystallinity (with the carbon layers being flatter and more parallel) increases the electrical conductivity, thermal conductivity and modulus, but it increases the propensity for the sliding of the carbon layers relative to one another, thereby degrading the strength of the fiber. In other words, the graphite fibers exhibit low strength but high modulus compared to the conventional carbon fibers, while they exhibit high electrical/thermal conductivity compared to the conventional carbon fibers. For structural applications, carbon fibers that are not graphitic dominate, due to both high strength and low cost.

Whether the carbon fiber has graphite regions or not, the carbon layers in the various stacks in a fiber are preferentially aligned (but not perfectly aligned) along the axis of the fiber. As a result of the preferred orientation, carbon fibers are much higher in the electrical conductivity, thermal conductivity and elastic modulus in the fiber axis than in the transverse direction.

As-fabricated, continuous carbon fibers are in the form of bundles. Each bundle is known as a tow. For example, there are 12,000 fibers in a tow, which is thus known as a 12k tow. There is a large range of the number of fibers per tow. This number is dictated by the process of fiber fabrication. The diameter of a fiber is around 10 μm, typically 7 μm for the most common continuous carbon fibers, which are made from polyacrylonitrile (PAN), a polymer.

Carbon fiber by itself (not in composite form) is not piezoresistive, but just resistive.[3] Therefore, carbon fiber by itself is ineffective for the self-sensing of strain or stress.

3. Carbon Fiber Composites

Due to the electrical conductivity, carbon fiber structural materials are amenable to providing intrinsic smartness through the effect of various stimuli on the electrical resistivity of the carbon fiber composite. In contrast, glass fibers and polymer fibers are in general not conductive. Polymer fibers that are conductive are not commonly used, due to their typical high cost, low processibility and inadequate mechanical behavior. Because of their electrical nonconductivity, glass fiber composites are typically rendered conductive by the incorporation of a conductive filler (e.g. carbon black and carbon nanotube), in order to be able to exhibit piezoresistivity for the purpose of resistance-based sensing.[4,5] Although glass fibers are less expensive than carbon fibers, the conductive filler adds to the cost. Piezoresistivity should be distinguished from piezoelectricity, which pertains to a change in the amount of stored charge (rather than a change of the resistivity) with the strain.

This review is focused on multi-functional carbon fiber structural materials, including composites with polymer and cement matrices. The polymer–matrix composites are important for lightweight structures, e.g. aircraft, satellites, missiles, wheelchairs, wind turbine blades and sporting goods. Cement–matrix composites are important for the civil infrastructure, e.g. bridges, highway, railroad, buildings and underground wells. Carbon–matrix composites are resistant to high temperatures, so they are important for high-temperature structures, e.g. reentry space vehicles (e.g. the Space Shuttle) and aircraft brakes. The carbon–matrix composites containing carbon fiber reinforcement (typically continuous) are known as carbon–carbon composites. Because the fiber and matrix are both carbon and are thus similar in the resistivity, carbon–carbon composites are weak in the resistance-based self-sensing effectiveness.[6] They are not addressed in this review.

This review covers composites with continuous and short carbon fibers. Short fibers are attractive in that they can be included in a slurry, which can be subjected to molding or other processing in order to form an article with a specific shape. In addition, short fibers are less costly than the corresponding continuous fibers. Continuous fibers are used in polymer–matrix and carbon–matrix composite structures (e.g. aircraft structures) that demand excellent mechanical properties. However, short fibers are used in cement–matrix composite structures, due to their low cost and suitability for incorporation in a cement mix (a slurry). In contrast, continuous fibers cannot be incorporated to a cement mix.

A high-performance continuous fiber structural composite is typically in the form of a laminate, which is made from a stack of plies (with each ply known as a lamina; Fig. 1).[7] Each lamina is made from a sheet of parallel carbon fibers that are obtained by spreading out a number of parallel contiguous tows. In case that the

A Review of Self-Sensing in Carbon Fiber Structural Composite Materials   43

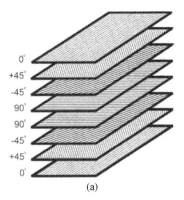

(a)

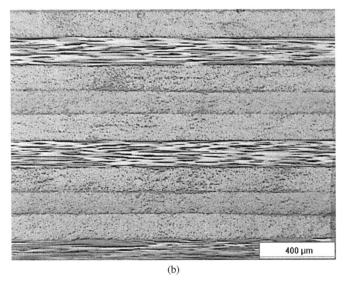

(b)

Fig. 1. (a) A common fiber lay-up configuration that involves fibers of different laminae oriented in different directions. The multi-directional character of the fibers in different laminae results in mechanical properties that are approximately isotropic two-dimensionally (see http://www.composites.ugent.be/homemade-composites/what-are-composites.html, public domain). (b) The mechanically polished cross-section of a continuous carbon fiber epoxy–matrix composite with 24 laminae and a quasi-isotropic fiber lay-up configuration $[0/45/90/-45]_{3s}$. The 0° laminae are relatively bright, as shown by optical microscopy.[7]

sheet is impregnated with the matrix or matrix precursor for facilitating composite fabrication, the sheet is known as a prepreg. The number of fibers in a lamina along the thickness of the ply is typically a few dozens, such as 40; there is a large range in this number (Fig. 2).[8]

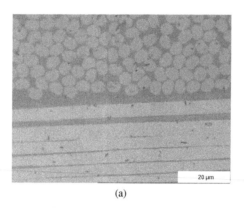

(a)

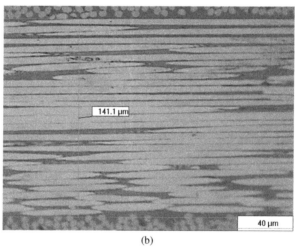

(b)

Fig. 2. The cross-section of continuous carbon fiber epoxy–matrix composites, as shown by optical microscopy. (a) Parts of two adjacent laminae and the interlaminar interface between them, as shown for a composite fabricated at a curing pressure of 2 MPa. In the top lamina, the fibers are perpendicular to the image plane; in the bottom lamina, the fibers are in the image plane. (b) A lamina and parts of two laminae adjacent to it, as shown for a composite fabricated at a curing pressure of 0.5 MPa.[8]

Within a lamina, the fibers are commonly oriented in the same direction. However, the fibers in different laminae are commonly oriented in different directions, in order for the laminate to exhibit reasonably good mechanical behavior in essentially all in-plane directions (Fig. 1). The fiber orientations in adjacent laminae are commonly either 45° or 90° apart. The configuration of fiber orientations for the various laminae is known as the fiber lay-up configuration, which should be symmetrical around the middle lamella in order to avoid asymmetrical shrinkage during fabrication and the consequent warpage of the laminate after fabrication. The shrinkage is commonly because of the process in which the matrix precursor is transformed to the matrix, in addition to the thermal contraction of the matrix during the cooling from the laminate fabrication temperature. The difference between the fiber and matrix in the thermal expansion coefficient results in more shrinkage in the matrix than the fiber, thus causing compressive residual stress in the fiber.

An alternate composite design involves using woven continuous fiber in every lamina of the laminate. The woven fiber is in fabric form. Although the handling of fabric is more convenient than that of parallel fiber sheets, the fabric suffers from the fiber bending that necessarily accompanies the weaving. The bending negatively affects the mechanical performance of the resulting laminate, since each fiber is very anisotropic, with the mechanical properties being different along the fiber direction and transverse direction of each fiber. The fiber bending would cause parts of a fiber to be exposed to stress in the weak direction of the fiber.

In a given laminate, there are fibers in various directions relative to the stress direction. Since a carbon fiber is much more conductive along the fiber axis than the transverse direction, the fiber in the stress direction mainly contributes to the resistance-based sensing.

The continuous fibers in a composite are not perfectly straight. This applies to the fibers within a lamina, even though the fibers are intended to be parallel. There is fiber waviness (also referred to as marcelling). The epoxy resin in a prepreg shrinks during its curing. This shrinkage increases the fiber waviness. The tensile strength of unidirectional laminates in the longitudinal (fiber) direction is only 51% of the value calculated based on the Rule of Mixtures and the fiber volume fraction.[9] This shows the substantial presence of fiber waviness in the laminates.

Due to the fiber waviness, the adjacent fibers touch one another at certain points along the length of the fibers. This touching of the fibers occurs along the through-thickness and transverse directions of the composite, thus enhancing the conductivity in both directions. The touching along these directions causes the electrical conductivity along these directions to be nonzero, even though the matrix has very low (close to zero) conductivity.

The fibers are in the plane of the composite. Three-dimensionally woven fiber composites are not common. With the absence of fibers in the through-thickness direction of the laminate (i.e. in the direction perpendicular to the plane of the laminate), the laminate is mechanically weak in the through-thickness direction. In other words, the interlaminar interface (i.e. the interface between the adjacent laminae) is the weak link in the composite. The weakness pertains to the mechanical strength, electrical conductivity and thermal conductivity. This interface is actually a region, the thickness of which is at least the diameter of a fiber (Fig. 2).[8] In case that a prepreg is used in the laminate fabrication, the interlaminar interface thickness depends on the thickness of the excessive matrix or matrix precursor on the prepreg surface during the prepreg fabrication. In spite of the nonzero thickness of the interlaminar interface region, some of the fibers of one lamina contact some of the fibers of an adjacent lamina, as enabled by the fact that the fibers are not completely straight. Thus, the interlaminar interface does not insulate electrically one lamina from another when the matrix is insulating. Because the interlaminar interface is the weak link, delamination (i.e. the local separation of the adjacent laminae) is the most common type of damage in a laminate. Fiber damage or fracture is less common than delamination, as it is a more severe form of damage.

4. Strain Self-Sensing Theory

This section presents the basic theories of resistance-based strain self-sensing (Sec. 4.1) and capacitance-based strain self-sensing (Sec. 4.2).

4.1. *Resistance-based strain self-sensing*

Resistance-based strain self-sensing is based on piezoresistivity. The volume resistance R relates to the volume resistivity ρ by the well-known equation

$$R = \rho l/A, \tag{1}$$

where l is the length in the direction of the resistance R, and A is the cross-sectional area. Hence

$$(\Delta R)/R = (\Delta \rho)/\rho + [(\Delta l)/l](1 + 2\nu), \tag{2}$$

where $(\Delta l)/l$ is the strain, and ν is the Poisson's ratio of the material. Equation (2) assumes that the material is isotropic in the two directions that are transverse to the stress axis. Otherwise, the term 2ν in Eq. (2) should be replaced by the sum $\nu_1 + \nu_2$, where ν_1 and ν_2 are the values of the Poisson's ratio in the two transverse directions.

Based on Eq. (2), both the change in ρ ($\Delta \rho$) and the change in l (Δl) contribute to the change in R (ΔR). In case that $(\Delta \rho)/\rho \gg (\Delta l)/l$, Eq. (2) becomes

$$\frac{\Delta R}{R} = \frac{\Delta \rho}{\rho}, \tag{3}$$

and the gage factor (abbreviated as GF), defined as the fractional change in resistance per unit strain, is given by

$$GF = \frac{\frac{\Delta R}{R}}{\frac{\Delta l}{l}} = \frac{\frac{\Delta \rho}{\rho}}{\frac{\Delta l}{l}}. \tag{4}$$

The gage factor is a common description of the sensing effectiveness.

4.2. Capacitance-based strain self-sensing

Capacitance-based strain self-sensing involves piezopermittivity, which refers to the effect of the strain on the permittivity. The capacitance C is governed by the relative permittivity κ (i.e. the permittivity relative to that of vacuum) of the material according to the well-known equation

$$C = \frac{\varepsilon_0 \kappa A}{l}, \qquad (5)$$

where l is the length between the two electrodes in the capacitance measurement direction, and A is the area in the plane perpendicular to the capacitance measurement direction. Based on Eq. (5), the fractional change in C upon a change in l is given by

$$\frac{\Delta C}{C} = \frac{\Delta \kappa}{\kappa} - \frac{\Delta l}{l}(1 + 2\nu). \qquad (6)$$

Equation (6) assumes that the material is isotropic in the two transverse directions. In accordance with Eq. (6), both the change in κ ($\Delta\kappa$) and the change in l (Δl) contribute to the change in C (ΔC).

In case that $(\Delta\kappa)/\kappa \gg (\Delta l)/l$, Eq. (6) becomes

$$\frac{\Delta C}{C} = \frac{\Delta \kappa}{\kappa}, \qquad (7)$$

and the fractional change in capacitance per unit strain (a description of the sensing effectiveness) is given by

$$\frac{\frac{\Delta C}{C}}{\frac{\Delta l}{l}} = \frac{\frac{\Delta \kappa}{\kappa}}{\frac{\Delta l}{l}}. \qquad (8)$$

5. Self-Sensing Carbon Fiber Structural Materials

Self-sensing carbon fiber structural materials include polymer–matrix composites with continuous carbon fiber reinforcement and cement–matrix composites with short carbon fiber reinforcement.

The former exhibits low density, high strength and high modulus, and is thus used for aircraft structures. The latter is not low in density, due to the relatively high density of the cement matrix compared to the polymer matrix. However, the latter is relatively inexpensive, due to the low cost of cement compared to polymer and the low cost of discontinuous fibers compared to continuous fibers. Low density is not required for typical civil infrastructures, which constitute the main area of application of the latter. The self-sensing of the strain, stress and damage is attractive for the cement-based materials in bridges, buildings, highways, rail infrastructures,[10] wells (underground wells for oil, gas and carbon dioxide that use cement for sealing[11]) and monument restoration.[12]

The resistance-based self-sensing of polymer–matrix composites with continuous carbon fiber reinforcement was discovered by Schulte and Baron in 1988–1989.[13,14] The resistance-based self-sensing of cement–matrix composites with short carbon fiber reinforcement was discovered by Chen and Chung in 1993.[15] Research by a large number of people has been reported since these discoveries, with attention to both strain sensing and damage sensing.

The resistance-based strain sensitivity is described by the gage factor, such that the strain is positive for tension and negative for compression. The gage factor is usually positive, such that the resistance increases upon tension and decreases upon compression. In case that the resistivity does not change with the strain, the gage factor approaches 2, such that the value depends on the Poisson's ratio [Eq. (2)].[2]

The capacitance-based self-sensing of polymer–matrix composites with continuous carbon fiber reinforcement was discovered by Eddib and Chung in 2018.[16] This discovery hinges on the dielectric behavior of carbon fibers,[17] the permittivity of which originates from a small fraction of the carriers (electrons and holes) interacting with the atoms in the carbon fiber.

The capacitance-based self-sensing in carbon fiber cement-based material was first reported by Fu et al. in 1997 through measurement of the reactance (imaginary part of the impedance), which is inversely related to the negative of the capacitance.[18] The reactance also includes the inductance, which is negligible.

This section covers resistance-based self-sensing in polymer–matrix composites with continuous carbon fiber reinforcement (Sec. 5.1), capacitance-based self-sensing in polymer–matrix composites with continuous carbon fiber reinforcement (Sec. 5.2), resistance-based self-sensing in cement–matrix composites with short carbon fiber reinforcement (Sec. 5.3) and capacitance-based self-sensing in cement–matrix composites with short carbon fiber reinforcement (Sec. 5.4). The self-sensing pertains to the sensing of strain/stress and damage. For the theory of strain self-sensing, readers can refer to Sec. 4.

5.1. *Resistance-based self-sensing in continuous carbon fiber polymer–matrix composites*

Because of the electrical conductivity of the carbon fiber and the nonconductivity of the polymer matrix, the resistivity of a polymer–matrix composite with continuous carbon fiber reinforcement is sensitive to the fiber arrangement. This arrangement is affected slightly by the strain experienced by the composite. For example, the degree of touching of the fibers is affected by the strain in the composite. A higher degree of touching in the through-thickness direction, as enabled by an increased degree of fiber waviness, decreases the through-thickness resistivity (Fig. 3).[7]

An increase in the degree of through-thickness touching also results in a lower longitudinal (in-plane, fiber direction) resistivity. This is because imperfections in the fibers are bound to exist, and the increased touching enables more paths for the current to detour when an imperfection is encountered.[7] This notion is in line with the report that broken fibers contribute to the conductivity of a composite with continuous carbon fiber reinforcement.[19] This touching can occur across the interlaminar interface in the through-thickness direction or within a lamina in the in-plane direction transverse to the fibers. Measurement of the in-plane resistance

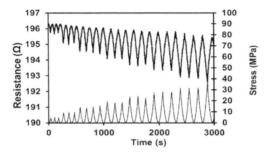

Fig. 3. The effect of through-thickness compressive stress (thin lower curve, up to a stress amplitude of 35 MPa) on the through-thickness resistance (thick upper curve) of epoxy–matrix composite with continuous carbon fiber reinforcement.[7]

change is easier and more practical than that of the through-thickness direction, due to the relatively small dimension of the laminate in the through-thickness direction. Thus, for the sake of structural monitoring, the in-plane resistance is more pertinent than the through-thickness resistance.

Flexure is practically the most common manner of loading. Elastic flexure results in the surface resistance increasing reversibly on the tension surface and decreasing reversibly on the compression surface[20] (Fig. 4). This is due to the longitudinal tension decreasing reversibly the degree of touching in the through-thickness direction, thereby decreasing reversibly the degree of current penetration from the surface electrodes. In contrast, elastic compression increases reversibly the degree of through-thickness touching and hence increases reversibly the extent of current penetration. This mechanism involving the notion of current penetration is supported by an analytical model.[21]

Under uniaxial tension, the electrical effect involves the change of the resistivity along the stress axis and that in a direction transverse to the stress axis. It is discussed below.

Consideration of the effect of uniaxial tension of a single carbon fiber embedded in epoxy is revealing scientifically. The resistivity of the fiber is above that of the corresponding single fiber

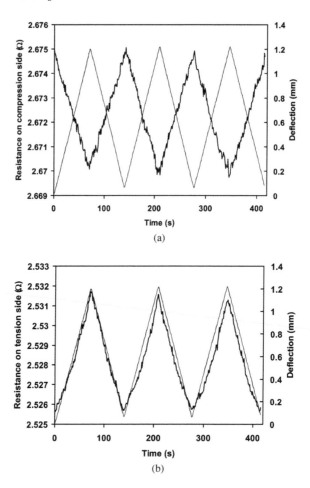

Fig. 4. The surface resistance (thick curve) during flexural cycling. The deflection (thin curve) has a maximum value of 2.098 mm (corresponding to the stress amplitude of 392.3 MPa). This is for a quasi-isotropic epoxy–matrix composite with continuous carbon fiber reinforcement and 24 laminae. (a) Compression surface. (b) Tension surface.[20]

without epoxy.[22] Moreover, the resistance of a single fiber embedded in epoxy increases during the curing of the epoxy.[23] These results are due to the residual compressive stress in the fiber after the curing of the surrounding epoxy. The residual stress results from the shrinkage of the epoxy resin during curing in the presence of the fiber and the fact that the coefficient of thermal expansion is higher for the epoxy than the fiber. In addition, the shrinkage causes fiber waviness, which is significant when the residual stress is substantial. Subsequent to the curing, tension of the epoxy-embedded carbon fiber decreases the resistivity reversibly toward the value of the fiber in the absence of epoxy.[22] The presence of residual compressive strain in a single fiber embedded in epoxy is also shown by Raman scattering,[24] which further shows that the residual compressive strain is decreased upon tension application. Therefore, the resistivity decrease upon tension stems from the decrease in the fabrication-induced residual compressive stress in the fiber. The tension also decreases the degree of fiber waviness. However, the change in the degree of fiber waviness does not affect the resistivity of a single fiber, since the radius of curvature in the waviness and the change in the degree of waviness are both small.

For a multi-lamina laminate, the polymer-rich interlaminar interface region provides a buffer zone for the fibers of each lamina to spread out to during the waviness development. With the relatively little limitation to the waviness development, the degree of fiber waviness is relatively high, thereby enabling the development of a relatively high residual compressive stress. In contrast, an interlaminar interface (a buffer zone) is absent in a single-lamina laminate. As a result, for a multi-lamina laminate, the electrical effect of the change in the residual compressive stress overshadows that of the change in the degree of fiber waviness. Thus, the longitudinal resistivity decreases upon tension for a multi-lamina laminate, as shown in Fig. 5(a) for an eight-lamina unidirectional laminate with the resistance measured by using the four-probe method (i.e. using four electrodes, with the outer electrodes for passing current and the inner electrodes for voltage measurement).[25] However, if the two-probe method (i.e. using two electrodes, with each electrode used for passing current and voltage measurement) is used for the resistance measurement, the resistance increases upon tension for the same laminate [Fig. 5(b)].[25] The resistance measured using the two-probe method includes the contact resistance at the electrodes (more exactly, at the electrode–specimen interface), so it should not be assumed to be approximately equal

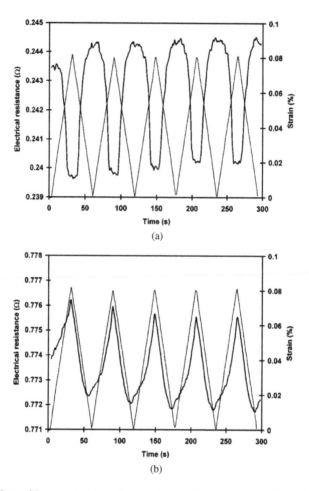

Fig. 5. Effect of longitudinal cyclic tension on the resistance of an eight-lamina unidirectional laminate. Resistance: thick line. Strain: thin line. (a) Plot of the resistance versus time and of the strain versus time, with the resistance obtained using the four-probe method, such that all four electrodes are perimetric. (b) Corresponding plot with the resistance obtained by using the two-probe method, again with perimetric electrodes.[25]

to the specimen resistance. The increase of the two-probe resistance upon tension is due to the increase in the contact resistance. The trend of the four-probe resistance decreasing upon tension is consistent with the observation for a single carbon fiber embedded in epoxy (discussed above).

For a single-lamina laminate, there is no interlaminar interface and hence no buffer zone for accommodating a high degree of fiber waviness. Hence, the residual compressive stress is low, with the electrical effect of the change in the degree of fiber waviness overshadowing that of the change in the residual compressive stress. Thus, the decrease in the degree of fiber waviness upon tension reduces the degree of fiber–fiber contact, thereby increasing the through-thickness/transverse resistivity.[26]

The above-mentioned increase in the through-thickness/transverse resistivity in turn results in an increase in the longitudinal resistivity. This is because of the above-mentioned increased deterrence of the current detour when fiber imperfections are encountered. As a result, for a single-lamina laminate, the longitudinal resistivity increases upon tension, whether without pre-tensioning[9,27] or with pre-tensioning.[28] A similar situation applies to a single-tow composite, the longitudinal resistivity of which increases upon tension, as reported without and with pre-tensioning of the tow.[29] Consistent with the above notion is that the pre-tensioning of the fibers in a single-lamina laminate (before and during resin impregnation, as well as during the subsequent resin curing) causes the effect to change from the resistance decreasing upon tension to the resistance increasing upon tension.[28]

For a single-lamina laminate, the four-probe method gives the trend of the longitudinal resistance increasing upon tension.[27] However, with the four electrodes applied after the surface polymer-rich region has been removed by mechanical polishing, the resistance is substantially lower and the resistance increases (rather than decreasing) upon tension.[27] The resistance decrease observed for the absence of mechanical polishing is due to the Poisson-related decrease in the thickness of the polymer-rich surface region upon tension. This means that the decreasing trend of the resistance is not due to a decrease in the specimen resistivity. In fact, in the absence of mechanical polishing, the resistivity increases upon tension for a single-lamina laminate.[9] The polymer-rich surface layer is presumably relatively thin in the single-lamina laminate (without mechanical polishing) that exhibits the trend of the resistance increasing upon tension.[9] This illustrates the criticality of the method of electrode application in studying the influence of stress/strain on the specimen resistivity.

Because the fibers are in the plane of the laminate rather than the through-thickness direction, the through-thickness conduction paths are few compared to the longitudinal conduction paths and are highly dependent on the degree of fiber waviness. As a consequence, the through-thickness resistivity is more sensitive to the degree of fiber waviness than the longitudinal resistivity. Upon longitudinal tension, the degree of fiber waviness is reduced, thereby resulting in less fiber–fiber touching and an increase in the through-thickness resistivity.

For a 16-lamina unidirectional laminate, it was reported that the longitudinal resistivity increases upon tension.[27] This increasing trend is opposite to the above-mentioned decreasing trend and is attributed to flaws in the electrical measurement.[30]

Fiber fracture obviously causes irreversible increase in the resistivity.[19,31–43] However, minor damage in the form of an irreversible change in the microstructure can also cause an irreversible increase in the resistivity, as described below.

Consider a single-lamina laminate serving as a sensor that is bonded to the tension surface of an epoxy–matrix laminate with continuous glass fiber (in the absence of carbon fiber) and 22 laminae. Upon flexure at progressively increasing stress amplitude within the elastic regime, the surface resistance is highly sensitive to minor damage. The baseline surface resistance (i.e. the value in the unloaded state after loading at a particular stress amplitude) on the tension surface increases irreversibly as the strain amplitude increases cycle by cycle within the elastic regime (Fig. 6).[44] This irreversible resistance increase is attributed to an irreversible microstructural change, which is probably due to an irreversible reduction in the degree of fiber–fiber contact in the through-thickness direction within the carbon fiber lamina.[44]

The electrical effect of fiber fracture is more significant than that of a microstructural change. Under uniaxial tension, the resistance in the longitudinal (fiber) direction increases irreversibly upon

A Review of Self-Sensing in Carbon Fiber Structural Composite Materials 57

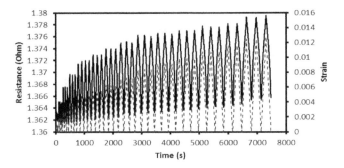

Fig. 6. Flexural sensing characteristics of a carbon fiber epoxy–matrix composite sensor with continuous carbon fiber in the form of a single lamina. The sensor is attached to the tension surface of a glass fiber epoxy–matrix composite beam with continuous glass fiber and 22 laminae. Repeated flexural loading is performed at progressively increasing strain amplitudes, such that there are three cycles for each strain amplitude. The flexural strain amplitude is up to about 0.013. The sensor resistance and flexural strain are shown by the solid curve and dashed curve, respectively. There is no change in the modulus. The strain and minor damage are indicated by the reversible and irreversible resistance changes, respectively.[44]

damage [Fig. 7(a)], due to fiber fracture.[22] During tension–tension fatigue with the applied stress in the longitudinal (fiber) direction, this damage is observed to start at 50% of the fatigue life [Fig. 7(b)].[45–47] Furthermore, the secant modulus (defined as the stress divided by strain at any point during the fatigue testing) decreases as the resistance increases [Fig. 7(b)], indicating that the resistance increase is indeed due to damage. Additional confirmation of the occurrence of damage is provided by simultaneous acoustic emission observation.[48]

Instead of measuring the longitudinal resistance, as mentioned above, the through-thickness resistance can be measured. The latter is sensitive to delamination, whereas the former is sensitive to fiber fracture. The through-thickness resistance increases irreversibly upon tension–tension fatigue, such that the stress in the longitudinal (fiber) direction, due to delamination,[45,46] which reduces the cross-sectional area of the through-thickness current path. This damage is observed to start at 33% of the fatigue life [Fig. 7(c)].[45,46]

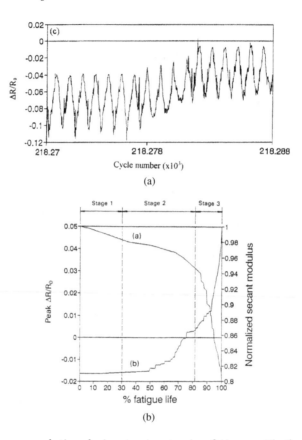

Fig. 7. Damage evolution during tension–tension fatigue, with the tension in the longitudinal (fiber) direction, as indicated by the volume resistance for a carbon fiber epoxy–matrix composite with continuous carbon fiber. The quantity $\Delta R/R_0$ is the fractional change in resistance relative to the initial resistance. (a) The resistance is in the longitudinal direction. It decreases reversibly in each loading cycle, due to strain, but the baseline resistance increases abruptly at a discrete point in the fatigue life. The first abrupt increase of the resistance corresponds to the first occurrence of fiber fracture.[46] (b) The peak value of the longitudinal $\Delta R/R_0$ in a stress cycle and the normalized secant modulus during the entire fatigue life. The variation of the resistance within a cycle is not shown. The effect shown is due to damage (fiber breakage) rather than strain.[47] (c) Variation of the through-thickness volume resistance, as shown by $\Delta R/R_0$, during fatigue. The minimum and maximum values of $\Delta R/R_0$ in a stress cycle are shown by the lower curve and upper curve, respectively. Variation of the resistance within a cycle is not shown. The effect shown is due to damage (delamination) rather than strain.[45]

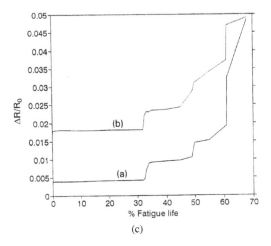

Fig. 7. (*Continued*)

Delamination increases the through-thickness resistivity, because it diminishes the degree of fiber touching across the interlaminar interface.[22,49] In addition, it increases the surface resistance, because it decreases the current penetration from the surface electrodes.[50,51] In general, delamination causes the volume resistance to increase.[46,31,52,55] In spite of the nonconductivity of the polymer matrix, matrix cracking can be sensed using suitable methods. For example, matrix cracking in the 90° lamina that is transverse to the 0° surface current direction also increases the surface resistance.[51]

Compared to delamination, interlaminar interface degradation in the absence of delamination is a more minor form of damage. Nevertheless, it also diminishes the degree of fiber touching across the interlaminar interface, thus increasing the interfacial resistivity of the interlaminar interface.[56,33] The interfacial resistivity (a form of contact resistivity) is an areal resistivity that is defined as the product of the interfacial resistance and interface (contact) area, with unit $\Omega \cdot m^2$.

Impact damage tends to be in the vicinity of the point of the impact. The sensing of impact damage can be achieved by measuring the interfacial resistivity of the interlaminar interface[56,33] or the volume resistance or surface resistance.[57] In case of measuring the

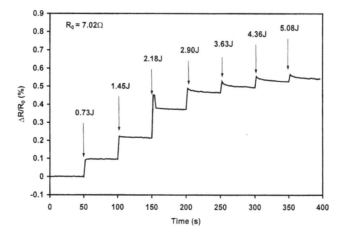

Fig. 8. Sensing the impact at progressively increasing energy for carbon fiber epoxy–matrix composite laminate with continuous carbon fiber, 16 laminae and the quasi-isotropic lay-up configuration. The quantity shown versus time is the fractional change in oblique resistance relative to the initial resistance. The times of the impacts are indicated by the arrows.[57]

surface resistance, the measurement should be directed at the part of the surface that contains the impact point, though the current spreading from the impact point may relax this requirement to a degree.

The oblique resistance is the volume resistance in an off-axis direction that is at an angle to both the in-plane and through-thickness directions. Because the current path goes through the entire thickness of the laminate and extends for a distance in the plane of the laminate, the oblique resistance is advantageous for sensing interior damage.[57,58] Figure 8 shows the increase in the oblique resistance with increasing impact energy.[57]

5.2. Capacitance-based self-sensing in continuous carbon fiber polymer–matrix composites

The permittivity is the main material property that describes the dielectric behavior. Piezopermittivity refers to the effect of elastic strain on the reversible change of the permittivity. The magnitude of the fractional change in permittivity much exceeds the strain magnitude.[59] The dielectric behavior (as described by the permittivity) and conduction behavior (as described by the conductivity) are related by the Kramers–Kronig relationship. Piezopermittivity provides the basis for the capacitance-based sensing of strain or stress. It differs from piezoresistivity, which provides the basis for the resistance-based sensing of strain or stress. Piezopermittivity also differs from the direct piezoelectric effect, which does not require a permittivity change.

The classical form of a capacitor is a parallel-plate capacitor. In this form, two electrodes, which are electrically conductive, sandwich a dielectric material. A parallel-plate capacitor can involve a fringing electric field that emanates from the capacitor and spreads to the region beyond the electrodes of the capacitor. In case that the dielectric material is conductive, such that it extends from the region sandwiched by the electrodes to the region away from the electrodes, the fringing field is particularly extensive and the measured capacitance becomes higher than the true value. The measured capacitance is affected by the presence of defects in the part of the dielectric material that extends beyond the electrode. In other words, the fringing field serves as a probe for nondestructive evaluation. The presence of defects deters the fringing of the field, thereby decreasing the effective cross-sectional area of the capacitor and decreasing the measured capacitance.

The above-mentioned parallel-plate capacitor configuration is not that suitable for structural implementation than a coplanar electrode configuration, in which two electrodes are applied to the same surface. Obviously the two electrodes do not cover the entire surface. Using this configuration, the apparent in-plane capacitance is measured. The apparent in-plane capacitance is to be distinguished from the apparent through-thickness capacitance given by the parallel-plate capacitor configuration. In the coplanar electrode configuration, the electric field lines spread between the two electrodes and bow out from the line connecting the two electrodes, thus causing the effective cross-sectional area of the in-plane capacitor to be increased and the apparent capacitance to be higher than the true capacitance.

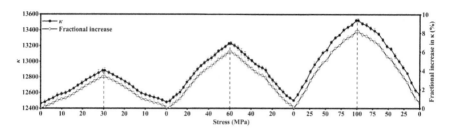

Fig. 9. The effect of tension on the apparent relative permittivity κ. The fractional increase in κ due to the stress is also shown. The tensile stress is in the elastic regime and is shown during loading and subsequent unloading at three progressively increasing stress amplitudes (30, 60 and 90 MPa). The material is a carbon fiber thermoplastic–matrix composite with biaxially woven continuous carbon fiber and a single lamina, as tested along a fiber direction.[60]

Figure 9 shows the reversible increase of the apparent relative permittivity upon elastic tension for a biaxially woven continuous single-lamina carbon fiber thermoplastic–matrix composite.[60] At the highest stress of 90 MPa, the fractional increase in the apparent relative permittivity reaches 8%.

The fringing field away from the coplanar electrodes also serves as a probe for the sensing of defects. The presence of defects deters the spreading of the field, thus decreasing the effective cross-sectional area of the in-plane capacitor and decreasing the measured capacitance.[16] The extent of the field spreading depends on the direction of the carbon fiber on the surface of the composite. The spreading is much more extensive in the fiber direction than the transverse direction. For extensive spreading, the line connecting the two coplanar electrodes is preferably normal to the fiber direction.

The capacitance is typically measured by using an LCR meter. Since the meter is not designed for the measurement of the capacitance of a conductor, an electrically insulating film must be placed between the carbon fiber composite specimen and one or both of the two electrodes, so as to increase the resistance of the material system to a level for the LCR to operate correctly.

5.3. Resistance-based self-sensing in short carbon fiber cement–matrix composites

In relation to cement-based materials, resistance-based self-sensing started with the report of this behavior in cement–matrix composites with short carbon fiber reinforcement (by Chen and Chung in 1993).[15] The short carbon fibers are in the microscale, with diameter around 10 μm. Due to the growth of the field of nanotechnology, the use of nanocarbons in place of or in combination with short carbon fiber has recently received much attention. The first report (by Fu and Chung in 1998) of resistance-based self-sensing in nanocarbon cement-based materials relates to the use of carbon nanofiber (originally known as carbon filament).[64]

The resistance-based self-sensing is not effective in cement–matrix composites without a conductive admixture, due to the inadequate sensitivity, reversibility and repeatability. Furthermore, in the absence of a conductive admixture, the resistivity is high and decreases significantly with increasing moisture content. The effect of moisture is less when the cement-based material contains a conductive admixture.[61] Carbon fiber is an electronic conductor, whereas cement is an ionic conductor.

For a significant decrease in the resistivity, the units of the short carbon fiber in the composite need to contact one another so as to form a continuous conduction path. This continuity is referred to as percolation. The percolation threshold is the fiber volume fraction above which percolation occurs. Around the percolation threshold, the resistance-based strain sensing is most effective. The use of silica fume as an additional admixture significantly enhances the fiber dispersion, because the small particles in the silica fume accentuate the mechanical agitation during the cement mixing. Sonication is commonly involved in the dispersion of nanocarbons in the cement mix, but it is not necessary for the dispersion of short carbon fiber. In spite of the fact that a longer fiber (i.e. a fiber with a higher aspect ratio) reduces the percolation threshold, the length of the short carbon fiber should be limited (e.g. 5 mm), in order to minimize the clumping of the fiber units. It is preferred that the short carbon fiber has no sizing (coating), which tends to be hydrophobic, thereby hindering the fiber dispersion in the water-based cement mix. Because of the hydrophobic character of carbon fiber, fiber surface treatment [e.g. surface oxidation by using ozone O_3 (see Ref. 62)] is used to promote the hydrophilic character. Thus, the ozone treatment promotes the fiber dispersion, thereby improving the resistance-based sensing behavior (Fig. 10).[62]

Upon strain in the elastic regime, the volume resistance of a cement–matrix composite with short carbon fiber reinforcement changes reversibly.[62-65] Upon uniaxial tension, it increases reversibly, because of the slight weakening of the fiber–matrix interface and the resulting increase in the interfacial resistivity (Fig. 10).[62] For cement paste with short ozone-treated carbon fiber, the gage factor

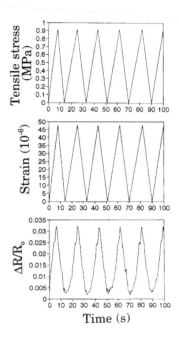

Fig. 10. The fractional change in resistance during cycle elastic tension. The resistance is in the longitudinal (stress) direction. The material is cement paste (cured) containing ozone-treated short carbon fiber (0.5% by mass of cement, 0.51 vol.%) and silica fume (15% by mass of cement).[62]

is 670 (Fig. 10). At the end of the first loading cycle, the resistance increases irreversibly and slightly. In subsequent cycles, the resistance increase is completely reversible and the extent of change in the resistance is essentially the same for the various cycles. Upon uniaxial compression, the resistance decreases reversibly,[62,65] because of the slight tightening of the fiber–matrix interface. This tightening causes the interfacial resistivity to decrease.

In the case of flexure, the surface resistance rather than the volume resistance should be measured, due to the difference in stresses between the opposite surfaces of the beam. The surface resistance, as measured using coplanar electrodes on the same surface, is separately measured on the tension and compression surfaces of the beam under flexure. During flexure, the surface resistance on the tension surface increases, whereas that on the compression surface decreases (Fig. 11).[66] If the volume resistance rather than the surface resistance is measured, an effect that represents the average over the specimen volume is observed, with the consequence that the sensing is almost ineffective.[67–69] The strain sensing mechanism mentioned above is supported by modeling.[70–73]

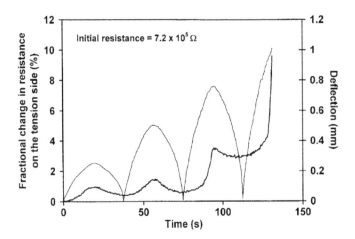

Fig. 11. Fractional change in the surface resistance (thick curve) of cement–matrix composite with short carbon fiber reinforcement at the tension side during flexural loading at progressively increasing deflection amplitudes up to failure. The mid-span deflection during flexure is shown by the thin curve.[66]

Damage increases irreversibly the resistivity of a cement–matrix composite with short carbon fiber reinforcement, whether the damage is minor (such as that related to the degradation of the fiber–matrix interface) or major (such as that related to fiber fracture or matrix cracking).[74-77] The effect of damage on the resistivity correlates with that of damage on the irreversible strain.[78] Damage diminishes the strain sensitivity, but the sensing effectiveness remains.[79]

For a cement–matrix composite with a nanocarbon reinforcement, the resistivity decreases reversibly upon compression and increases reversibly upon tension, as in the case of a cement–matrix composite with short carbon fiber reinforcement. However, most of the published works on nanocarbon cement–matrix composites address compression, but not tension. For the nanocarbon cement–matrix composites, damage increases the resistivity irreversibly, because of the irreversible decrease in the number of contacts among the nanocarbon units.

The intertwined morphology of the carbon nanofiber makes nanofiber dispersion challenging. Without silica fume and with the use of sonication, carbon nanofiber (2% by mass of cement) is more effective than similarly dispersed carbon fiber (1% by mass of cement).[80] This comparison is not totally fair, because carbon fiber needs silica fume and not sonication for its dispersion. Sonication is not practical for field implementation. Carbon nanofiber is much more costly than short carbon fiber. The combined use of short carbon fiber and a minor proportion of carbon nanofiber has been reported.[81]

The carbon nanotube is typically smaller in diameter than the carbon nanofiber and is more effective for self-sensing than the carbon nanofiber.[82] As for the carbon nanofiber, sonication helps the dispersion of the carbon nanotube.[83] The sensing mechanism is similar for the carbon nanotube cement and carbon nanofiber cement.

The addition of a minor proportion of carbon nanotube or carbon nanofiber to short carbon fiber cement may degrade the sensing performance. For example, the addition of a minor proportion of either carbon nanotube or carbon nanofiber (either 0.1% or 0.15% by mass of cement, with sonication) to short carbon fiber (ranging from 0.25% to 0.75% by mass of cement, without silica fume) decreases the strain sensitivity significantly, as shown under tension.[84] In spite of the absence of silica fume, which is important for the dispersion of the short carbon fiber, the addition of either carbon nanotube or carbon nanofiber degrades the strain sensing performance. This degradation is because the electrical connectivity of the short carbon fiber by the nanofiber or nanotube hinders the resistivity change upon stress application.

The addition of carbon nanotube to carbon fiber may help the sensing marginally, if at all.[82,85–95] For example, with the presence of silica fume, the addition of carbon nanotube (0.2% by mass of cement) to carbon fiber (0.4% by mass of cement) decreases the strain sensitivity, but increases the linearity of the resistivity decrease upon compression.[85] In another example, in the presence of silica fume, the addition of 1 vol.% carbon nanotube to 15 vol.% carbon fiber enhances the repeatability of the strain sensing.[94] However, the carbon fiber content of 15 vol.% is greatly above the percolation threshold.[94] The optimum fiber volume fraction for sensing is around the percolation threshold. For more meaningful evaluation of the effect of carbon nanotube addition, a carbon fiber content that is optimum for the sensing should have been used. The ozone treatment of the nanotube improves the self-sensing characteristics,[93] as shown for short carbon fiber earlier.[62]

5.4. *Capacitance-based self-sensing in short carbon fiber cement–matrix composites*

The capacitance-based self-sensing in carbon fiber cement-based material was discovered by Fu *et al.* in 1997.[18] Upon compressive stress application, the reactance magnitude (measured using two embedded parallel electrodes) in the stress direction increases. Because the reactance is inversely related to the capacitance, this means that the capacitance in the stress direction decreases upon compression. However, this trend of the capacitance change is inconsistent with subsequent reports, as discussed in detail below.

In 2002, an increase in the longitudinal relative permittivity (measured using parallel electrodes) upon compressive stress application was reported for cement paste without or with a conductive admixture (carbon fiber, steel fiber or carbon nanofiber).[96,97] Without a conductive admixture, the behavior is enabled by the dielectric behavior of the cement-based material, which is an ionic conductor. Similarly, in 2011, an increase in the longitudinal relative permittivity upon compression was reported for sulphoaluminate cement containing carbon fiber and fly ash.[96] In view of Poisson's effect, this is consistent with the observation that the longitudinal capacitance or permittivity decreases upon tension,[98] and that the transverse capacitance decreases upon compression (2018),[99,100] both measured by using coplanar electrodes for cement paste without admixture.[99,100]

The capacitance is AC and is associated with AC polarization. However, DC polarization is indicated by the apparent resistivity increase during constant DC current application, as enabled by the fact that the polarization results in a voltage that opposes the voltage applied for conduction.[101] In 2001, a decrease of the DC polarization in the direction transverse to the stress direction upon compressive stress application was reported for cement paste without or with a conductive admixture (namely carbon fiber).[102] In view of the Poisson's effect, this trend is also opposite to the first trend mentioned above in relation to the reactance-based sensing.

The above-mentioned inconsistency between the reactance-based sensing (based on a single report in 1997) and capacitance-based sensing (based on a number of reports since 2002) suggests that the former is less reliable. This is supported by the fact that the real part of the impedance (i.e. the AC resistance) increases upon compression, as does the imaginary part (the reactance). The increase of this AC two-probe resistance upon compression is inconsistent with the established decrease of the DC four-probe resistance upon compression (Sec. 5.3). This suggests that the two-probe method used in the reactance-based sensing is inadequate and is largely responsible for the anomalous result.

Therefore, this review contends that, whether carbon fiber is present or not, tension causes the capacitance (permittivity) in the stress direction to decrease, whereas compression causes the capacitance (permittivity) in the stress direction to increase. These changes in the permittivity are presumably because of the reversible effect of stress/strain on the microstructure. Since the capacitance (permittivity) changes are similar, whether a conductive admixture is present or not, the microstructural change occurs in the cement matrix, possibly involving a slight change in the pore structure, which affects the movement of the ions in the pore solution. The ion movement dominates the polarization mechanism, as supported by the fact that the permittivity of cement paste increases upon mild heating,[103] which increases the mobility of the ions. Carbon fiber incorporation is not required for the capacitance-based self-sensing in a cement-based material, but it is expected to reduce the undesirable sensitivity to moisture.

6. Conclusion

Sensing is the most basic ability of a smart structure. Self-sensing involves the structural material sensing itself without any sensor incorporation. Carbon fiber renders electrical conductivity to a composite material, thereby enabling sensing that is based on the effect of strain/damage on the electrical properties of the composite. Strain (stress) and damage are basic and important attributes for structural operation and structural health monitoring.

The field of self-sensing in carbon fiber structural composites is reviewed, with consideration of the formulation, microscopic structure, sensing characteristics and sensing mechanisms. These composites include polymer–matrix composites with continuous carbon fiber reinforcement (for lightweight structures such as the airframe) and cement–matrix composites with short carbon fiber reinforcement (for the civil infrastructure). The short carbon fiber serves as a conductive admixture in the cement–matrix composites. A fiber volume fraction around the percolation threshold gives most effective resistance-based self-sensing.

For the polymer–matrix composites with continuous carbon fiber reinforcement, the longitudinal resistivity decreases upon uniaxial tension, due to the decrease in the residual compressive stress in the fiber, while the through-thickness resistivity increases, because of the decrease in the fiber waviness. Upon flexure, the resistance on the tension surface increases, because of the decrease in the degree of current penetration from the surface, while the resistance on the compression surface decreases, because of the increase in the degree of current penetration. These strain effects give reversible resistance changes. Upon damage, the resistance increases irreversibly. The through-thickness resistance, oblique resistance, surface resistance and interlaminar interfacial resistivity increase irreversibly upon fiber fracture, delamination or subtle irreversible change in the microstructure.

For cement–matrix composites with short carbon fiber reinforcement, the resistivity increases upon uniaxial tension, due to the weakening (loosening) of the fiber–matrix interface, and decreases upon uniaxial compression, due to the strengthening (tightening) of the fiber–matrix interface. Upon flexure, the resistance on the tension surface increases, while that on the compression surface decreases. These strain effects result in reversible resistance changes. Upon damage, the resistance and resistivity increase irreversibly.

The incorporation of carbon nanofiber or carbon nanotube to a polymer–matrix composite with continuous carbon fiber reinforcement or a cement–matrix composite with short carbon fiber reinforcement adds much to the material and processing costs, while the improvement in the sensing performance is marginal, if any. In fact, the sensing performance can be diminished by the nanocarbon addition. The increased processing cost relates to the procedure for the dispersion of the nanofiber or nanotube in the polymer or cement matrix. In case of the polymer–matrix composite, the nanocarbon improves the effectiveness for sensing cracking in the matrix.

In case of the cement–matrix composite, the nanocarbon tends to improve the repeatability of the strain sensing.

The self-sensing can be achieved by resistance measurement or capacitance measurement, with the former dominating over the latter in research. As for resistance-based self-sensing, capacitance-based self-sensing is effective for both polymer–matrix composites with continuous carbon fiber reinforcement and cement–matrix composites with short carbon fiber reinforcement. Due to the conductivity imparted by the continuous carbon fiber to the polymer–matrix composites, the capacitance measurement should involve a thin insulating film between the specimen and an electrode in order for the LCR meter to measure the capacitance correctly. Strain and damage are indicated by reversible and irreversible capacitance changes, respectively. The fringing electric field that bows out of the coplanar electrodes serves as a probe, with the capacitance decreased when the fringing field encounters imperfections. For the cement-based materials, a conductive admixture is not required for capacitance-based self-sensing.

References

1. D. D. L. Chung, *J. Electron. Mater.*, 2022, **51**, 5473–5481, doi:10.1007/s11664-022-09857-4.
2. D. D. L. Chung, *J. Mater. Sci.*, 2020, **55**, 15367–15396, doi:10.1007/s10853-020-05099-z.
3. X. Wang and D. D. L. Chung, *Carbon*, 1997, **35**, 706–709, doi:10.1016/s0008-6223(97)86644-4.
4. F. Nanni, G. Ruscito, D. Puglia, A. Terenzi, J. M. Kenny and G. Gusmano, *Compos. Sci. Technol.*, 2010, **71**, 1–8, doi:10.1016/j.compscitech.2010.08.015.
5. Q. An, A. N. Rider and E. T. Thostenson, *ACS Appl. Mater. Interfaces*, 2013, **5**, 2022–2032, doi:10.1021/am3028734.
6. S. Wang and D. D. L. Chung, *Carbon*, 1997, **35**, 621–630, doi:10.1016/S0008-6223(97)00011-0.
7. D. Wang and D. D. L. Chung, *Carbon*, 2013, **60**, 129–138, doi:10.1016/j.carbon.2013.04.005.
8. S. Han and D. D. L. Chung, *J. Mater. Sci.*, 2012, **47**, 2434–2453.

9. D. A. Gordon, S. Wang and D. D. L. Chung, *Compos. Interfaces*, 2004, **11**, 95–103, doi:10.1163/156855404322681073.
10. S. Ding, Y. Xiang, Y. Ni, V. K. Thakur, X. Wang, B. Han and J. Ou, *Nano Today*, 2022, **43**, 101438, doi:10.1016/j.nantod.2022.101438.
11. K. Gawel, D. Szewczyk and P. R. Cerasi, *Materials*, 2021, **14**, 1235, doi:10.3390/ma14051235.
12. A. Dimou, Z. S. Metaxa, N. D. Alexopoulos and S. K. Kourkoulis, *Mater. Today, Proc.*, 2022, **62**, 2482–2487, doi:10.1016/j.matpr.2022.02.623.
13. C. Baron and K. Schulte, *Materialpruefung*, 1988, **30**, 361–366.
14. Z. Schulte and C. Baron, *Compos. Sci. Technol.*, 1989, **36**, 63–76, doi:10.1016/0266-3538(89)90016-X.
15. P. Chen and D. D. L. Chung, *Smart Mater. Struct.*, 1993, **2**, 22–30, doi:10.1088/0964-1726/2/1/004.
16. A. A. Eddib and D. D. L. Chung, *Carbon*, 2018, **140**, 413–427, doi:10.1016/j.carbon.2018.08.070.
17. X. Xi and D. D. L. Chung, *Carbon*, 2019, **145**, 734–739, doi:10.1016/j.carbon.2019.01.069.
18. X. Fu, E. Ma, D. D. L. Chung and W. A. Anderson, *Cem. Concr. Res.*, 1997, **27**, 845–852, doi:10.1016/S0008-8846(97)83277-2.
19. N. Kalashnyk, E. Faulques, J. Schjoedt-Thomsen, L. R. Jensen, J. C. M. Rauhe and R. Pyrz, *Synth. Met.*, 2017, **224**, 56–62, doi:10.1016/j.synthmet.2016.12.021.
20. S. Wang and D. D. L. Chung, *Carbon*, 2006, **44**, 2739–2751, doi:10.1016/j.carbon.2006.03.034.
21. S. Zhu and D. D. L. Chung, *Carbon*, 2007, **45**, 1606–1613, doi:10.1016/j.carbon.2007.04.012.
22. X. Wang, X. Fu and D. D. L. Chung, *J. Mater. Res.*, 1998, **13**, 3081–3092, doi:10.1557/JMR.1998.0420.
23. J. Park, D. Kim, J. Kong, S. Kim, J. Jang, M. Kim, W. Kim and K. L. DeVries, *Compos. B*, 2007, **38**, 833–846, doi:10.1016/j.compositesb.2006.12.003.
24. N. Kalashnyk, E. Faulques, J. Schjoedt-Thomsen, L. R. Jensen, J. C. M. Rauhe and R. Pyrz, *Carbon*, 2016, **109**, 124–130, doi:10.1016/j.carbon.2016.07.064.
25. S. Wang and D. D. L. Chung, *Polym. Compos.*, 2000, **21**, 13–19, doi:10.1002/pc.10160.

26. A. Todoroki, Y. Samejima, Y. Hirano and R. Matsuzaki, *Compos. Sci. Technol.*, 2009, **69**, 1841–1846, doi:10.1016/j.compscitech.2009.03.023.
27. A. Todoroki and J. Yoshida, *JSME Int. J. A, Mech. Mater. Eng.*, 2004, **47**, 357–364, doi:10.1299/jsmea.47.357.
28. C. Q. Yang, X. L. Wang, Y. J. Jiao, Y. L. Ding, Y. F. Zhang and Z. S. Wu, *Compos. B*, 2016, **102**, 86–93, doi:10.1016/j.compositesb.2016.07.013.
29. M. A. Saifeldeen, N. Fouad, H. Huang and Z. S. Wu, *Smart Mater. Struct.*, 2017, **26**, 015012:1–015012:11, doi:10.1088/1361-665X/26/1/015012.
30. N. Angelidis, C. Y. Wei and P. E. Irving, *Compos. A*, 2004, **35**, 1135–1147, doi:10.1016/j.compositesa.2004.03.020.
31. D. D. L. Chung, A critical review to elucidate the multi-faceted science of the electrical-resistance-based strain/damage/temperature self-sensing in continuous carbon fiber polymer-matrix structural composites, *J. Mater. Sci.*, 2023, **58**, 483–526, doi:10.1007/s10853-022-08106-7.
32. J. C. Abry, S. Bochard, A. Chateauminois, M. Salvia and G. Girand, *Compos. Sci. Technol.*, 1999, **59**, 925–935, doi:10.1016/S0266-3538(98)00132-8.
33. P. E. Irving and C. Thiagarajan, *Smart Mater. Struct.*, 1998, **7**, 456–466, doi:10.1088/0964-1726/7/4/004.
34. A. S. Kaddour, F. A. R. Al-Salehi and S. T. S. Al-Hassani, *Compos. Sci. Technol.*, 1994, **51**, 377–385, doi:10.1016/0266-3538(94)90107-4.
35. R. Prabhakaran, *Exp. Tech.*, 1990, **14**, 16–20, doi:10.1111/j.1747-1567.1990.tb01059.x.
36. K. Moriya and T. Endo, *J. Jpn. Soc. Aeronaut. Space Sci.*, 1990, **32**, 184–196.
37. A. Todoroki, D. Haruyama, Y. Mizutani, Y. Suzuki and T. Yasuoka, *Open J. Compos. Mater.*, 2014, **4**, 22–31, doi:10.4236/ojcm.2014.41003.
38. N. Angelidis and P. E. Irving, *Compos. Sci. Technol.*, 2007, **67**, 594–604, doi:10.1016/j.compscitech.2006.07.033.
39. O. Ceysson, M. Salvia and L. Vincent, *Scr. Mater.*, 1996, **34**, 1273–1280, doi:10.1016/1359-6462(95)00638-9.
40. D. Kwon, P. Shin, J. Kim, K. L. DeVries and J. Park, *Compos. A*, 2016, **90**, 417–423, doi:10.1016/j.compositesa.2016.08.009.

41. N. Angelidis, N. Khemiri and P. E. Irving, *Smart Mater. Struct.*, 2005, **14**, 147–154, doi:10.1088/0964-1726/14/1/014.
42. N. Muto, Y. Arai, S. G. Shin, H. Matsubara, H. Yanagida, M. Sugita and T. Nakatsuji, *Compos. Sci. Technol.*, 2001, **61**, 875–883, doi:10.1016/S0266-3538(00)00165-2.
43. R. Salvado, C. Lopes, L. Szojda, P. Araujo, M. Gorski, F. J. Velez, J. Castro-Gomes and R. Krzywon, *Sensors*, 2015, **15**, 10753–10770, doi:10.3390/s150510753.
44. A. Vavouliotis, A. Paipetis and V. Kostopoulos, *Compos. Sci. Technol.*, 2011, **71**, 630–642, doi:10.1016/j.compscitech.2011.01.003.
45. E. Jeon, T. Fujimura, K. Takahashi and H. Kim, *Compos. A*, 2014, **66**, 193–200, doi:10.1016/j.compositesa.2014.08.002.
46. H. Park, C. Kong and K. Lee, *Key Eng. Mater.*, 2012, **488–489**, 460–463, doi:10.4028/www.scientific.net/KEM.488-489.460.
47. A. Baker, N. Rajic and C. Davis, *Compos. A*, 2009, **40**, 1340–1352, doi:10.1016/j.compositesa.2008.09.015.
48. D. Wang and D. D. L. Chung, unpublished results.
49. X. Wang and D. D. L. Chung, *Compos. B*, 1998, **29**, 63–73, doi:10.1016/S1359-8368(97)00014-0.
50. X. Wang and D. D. L. Chung, *Smart Mater. Struct.*, 1997, **6**, 504–508, doi:10.1088/0964-1726/6/4/017.
51. X. Wang and D. D. L. Chung, *J. Mater. Res.*, 1999, **14**, 4224–4229, doi:10.1557/PROC-503-81.
52. T. Prasse, F. Michel, G. Mook, K. Schulte and W. Bauhofer, *Compos. Sci. Technol.*, 2001, **61**, 831–835, doi:10.1016/S0266-3538(00)00179-2.
53. X. Wang and D. D. L. Chung, *Polym. Compos.*, 1997, **18**, 692–700, doi:10.1002/pc.10322.
54. Y. Hirano, T. Yamane and A. Todoroki, *Compos. Sci. Technol.*, 2016, **122**, 67–72, doi:10.1016/j.compscitech.2015.11.018.
55. A. Todoroki, M. Tanaka and Y. Shimamura, *Compos. Sci. Technol.*, 2002, **62**, 619–628, doi:10.1016/S0266-3538(02)00019-2.
56. A. Todoroki, *Compos. Sci. Technol.*, 2001, **61**, 1871–1880, doi:10.1016/S0266-3538(01)00088-4.
57. A. Todoroki, M. Tanaka and Y. Shimanura, *Compos. Sci. Technol.*, 2003, **63**, 1911–1920, doi:10.1016/S0266-3538(03)00157-X.
58. A. Todoroki, M. Tanaka and Y. Shimamura, *Compos. Sci. Technol.*, 2005, **65**, 37–46, doi:10.1016/j.compscitech.2004.05.018.

59. A. Todoroki, Y. Tanaka and Y. Shimamura, *Compos. Sci. Technol.*, 2004, **64**, 749–758, doi:10.1016/j.compscitech.2003.08.004.
60. A. Todoroki, *Adv. Compos. Mater.*, 2014, **23**, 179–193, doi:10.1080/09243046.2013.844900.
61. S. Wang, D. P. Kowalik and D. D. L. Chung, *Smart Mater. Struct.*, 2004, **13**, 570–592, doi:10.1088/0964-1726/13/3/017.
62. S. Wang and D. D. L. Chung, *J. Mater. Sci.*, 2000, **35**, 91–100, doi:10.1023/A:1004744600284.
63. S. Wang, D. D. L. Chung and J. H. Chung, *J. Mater. Sci.*, 2005, **40**, 561–568, doi:10.1007/s10853-005-6289-6.
64. S. Wang, D. D. L. Chung and J. H. Chung, *Compos. A*, 2005, **36**, 1707–1715.
65. D. D. L. Chung and X. Xi, *Sens. Actuators A*, 2021, **332**, 113028, doi:10.1016/j.sna.2021.113028.
66. X. Xi and D. D. L. Chung, *Carbon*, 2020, **160**, 361–389, doi:10.1016/j.carbon.2020.01.035.
67. X. Fu and D. D. L. Chung, *Carbon*, 1998, **36**, 459–462.
68. S. Wen and D. D. L. Chung, *ACI Mater. J.*, 2008, **105**, 274–280.
69. X. Fu, W. Lu and D. D. L. Chung, *Carbon*, 1998, **36**, 1337–1345, doi:10.1016/S0008-6223(98)00115-8.
70. S. Wen and D. D. L. Chung, *ACI Mater. J.*, 2005, **102**, 244–248.
71. S. Wen and D. D. L. Chung, *Cem. Concr. Res.*, 2000, **30**, 1289–1294, doi:10.1016/S0008-8846(00)00304-5.
72. S. Wen and D. D. L. Chung, *Cem. Concr. Res.*, 2001, **31**, 297–301, doi:10.1016/S0008-8846(00)00438-5.
73. S. Wen and D. D. L Chung, *Carbon*, 2006, **44**, 1496–1502, doi:10.1016/j.carbon.2005.12.009.
74. M. Abedi, R. Fangueiro and A. G. Correia, *Nanomaterials*, 2022, **12**, 1734, doi:10.3390/nano12101734.
75. A. K. Roopa, A. M. Hunashyal, P. Venkaraddiyavar and S. V. Ganachari, *Mater. Today, Proc.*, 2020, **27**, 603–609, doi:10.1016/j.matpr.2019.12.071.
76. W. Dong, W. Li, K. Wang, Y. Guo, D. Sheng and S. P. Shah, *Powder Technol.*, 2020, **373**, 184–194, doi:10.1016/j.powtec.2020.06.029.
77. S. Wen and D. D. L. Chung, *Cem. Concr. Res.*, 2006, **36**, 1879–1885, doi:10.1016/j.cemconres.2006.03.029.

78. S. Zhu and D. D. L. Chung, *J. Mater. Sci.*, 2007, **42**, 6222–6233, doi:10.1007/s10853-006-1131-3.
79. A. K. Roopa and A. M. Hunashyal, *IOP Conf. Ser., Mater. Sci. Eng.*, 2021, **1070**, 012041.
80. J. Xu, W. Zhong and W. Yao, *J. Mater. Sci.*, 2010, **45**, 3538–3546, doi:10.1007/s10853-010-4396-5.
81. D. D. L. Chung, *Mater. Sci. Eng. R*, 2003, **42**, 1–40, doi:10.1016/S0927-796X(03)00037-8.
82. D. G. Meehan, S. Wang and D. D. L. Chung, *J. Intell. Mater. Syst. Struct.*, 2010, **21**, 83–105.
83. F. Reza, G. B. Batson, J. A. Yamamuro and J. S. Lee, *Concrete: Material Science to Application — A Tribute to Surendra O. Shah*, American Concrete Institute, Farmington Hills, 2002, pp. 429–437.
84. S. H. Ghasemzadeh Mosavinejad, J. Barandoust, A. Ghanizadeh and M. Sigari, *Constr. Build. Mater.*, 2018, **193**, 255–267, doi:10.1016/j.conbuildmat.2018.10.190.
85. S. Wen and D. D. L. Chung, *Carbon*, 2007, **45**, 710–716, doi:10.1016/j.carbon.2006.11.029.
86. S. Wen and D. D. L. Chung, *J. Mater. Civ. Eng.*, 2006, **18**, 355–360, doi:10.1061/(ASCE)0899-1561(2006)18:3(355).
87. F. J. Baeza, O. Galao, E. Zornoza and P. Garcés, *Materials (Basel)*, 2013, **6**, 841–855, doi:10.3390/ma6030841.
88. S. Taheri, J. Georgaklis, M. Ams, S. Patabendigedara, A. Belford and S. Wu, *J. Mater. Sci.*, 2022, **57**, 2667–2682, doi:10.1007/s10853-021-06732-1.
89. D. Yoo *et al.*, *J. Compos. Mater.*, 2018, **52**, 3325–3340.
90. B. del Moral, F. J. Baeza, R. Navarro, O. Galao, E. Zornoza, J. Vera, C. Farcas and P. Garces, *Constr. Build. Mater.*, 2021, **284**, 122786, doi:10.1016/j.conbuildmat.2021.122786.
91. S. Parveen, B. Vilela, O. Lagido, S. Rana and R. Fangueiro, *Nanomaterials*, 2022, **12**, 74, doi:10.3390/nano12010074.
92. J. Zuo, *J. Test. Eval.*, 2012, **40**, 838–843, doi:10.1520/JTE20120092.
93. F. Al-Mufadi and H. A. Sherif, *Arab. J. Sci. Eng.*, 2019, **44**, 1403–1413.
94. A. L. Materazzi, F. Ubertini and A. D'Alessandro, *Cem. Concr. Compos.*, 2013, **37**, 2–11, doi:10.1016/j.cemconcomp.2012.12.013.
95. Q. Yang, P. Liu, Z. Ge and D. Wang, *J. Test. Eval.*, 2020, **48**, 1990–2002, doi:10.1520/JTE20190170.

96. J. Luo, Z. Duan, T. Zhao and Q. Li, *Key Eng. Mater.*, 2011, **483**, 579–583, doi:10.4028/www.scientific.net/KEM.483.579.
97. A. D'Alessandro, M. Rallini, F. Ubertini, A. L. Materazzi and J. M. Kenny, *Cem. Concr. Compos.*, 2016, **65**, 200–213, doi:10.1016/j.cemconcomp.2015.11.001.
98. A. Dinesh, S. T. Sudharsan and S. Haribala, *Mater. Today, Proc.*, 2021, **46**, 5801–5807, doi:10.1016/j.matpr.2021.02.722.
99. A. D'Alessandro, M. Tiecco, A. Meoni and F. Ubertini, *Cem. Concr. Compos.*, 2021, **115**, 103842, doi:10.1016/j.cemconcomp.2020.103842.
100. B. Del Moral *et al.*, *Nanomaterials*, 2020, **10**, 807, doi:10.3390/nano10040807.
101. F. Azhari and N. Banthia, *Cem. Concr. Compos.*, 2012, **34**, 866–873, doi:10.1016/j.cemconcomp.2012.04.007.
102. S. Lee, I. You, G. Zi and D. Yoo, *Sensors*, 2017, **17**, 2516:1–2516:16, doi:10.3390/s17112516.
103. S. Wen and D. D. L. Chung, *Cem. Concr. Res.*, 2002, **32**, 1429–1433, doi:10.1016/S0008-8846(02)00789-5.
104. S. Wen and D. D. L. Chung, *Cem. Concr. Res.*, 2002, **32**, 335–339, doi:10.1016/S0008-8846(01)00682-2.
105. S. D. Wang, L. C. Lu and X. Cheng, *Adv. Mater. Lett.*, 2011, **2**, 12–16, doi:10.5185/amlett.2010.9163.
106. D. D. L. Chung and X. Xi, An assimilative review of the dielectric behavior of cured cement-based materials without poling, unpublished results.
107. D. D. L. Chung and Y. Wang, *Cem. Concr. Compos.*, 2018, **94**, 255–263, doi:10.1016/j.cemconcomp.2018.09.017.
108. K. Shi and D. D. L. Chung, *Smart. Mater. Struct.*, 2018, **27**, 105011, doi:10.1088/1361-665X/aad87f.
109. X. Xi, M. Ozturk and D. D. L. Chung, *J. Am. Ceram. Soc.*, 2022, **105**, 1074–1082, doi:10.1111/jace.18121.
110. S. Wen and D. D. L. Chung, *Cem. Concr. Res.*, 2001, **31**, 291–295, doi:10.1016/S0008-8846(00)00412-9.
111. S. Wen and D. D. L. Chung, *Cem. Concr. Res.*, 2003, **33**, 1675–1679, doi:10.1016/S0008-8846(03)00147-9

© 2025 World Scientific Publishing Company
https://doi.org/10.1142/9789819806409_0003

Silicon-Containing Additives in Encapsulation of Phase Change Materials for Thermal Energy Storage*

Johnathan Joo Cheng Lee[†], Natalie Jia Xin Lim[‡], Pei Wang[†], Hongfei Liu[†], Suxi Wang[†], Chi-Lik Ken Lee[†,‡], Dan Kai[†,‡,§], Fengxia Wei[†], Rong Ji[†], Beng Hoon Tan[†], Shaozhong Ge[¶], Ady Suwardi[†,∥], Jianwei Xu[†,§,¶,**,§§], Xian Jun Loh[†,§,∥,††,§§] and Qiang Zhu[†,‡,‡‡,§§]

[†]*Institute of Materials Research and Engineering, A*STAR (Agency for Science, Technology and Research), 2 Fusionopolis Way, Innovis, #08-03, Singapore 138634, Singapore*
[‡]*School of Chemistry, Chemical Engineering and Biotechnology, Nanyang Technological University, 21 Nanyang Link, Singapore 637371, Singapore*
[§]*Institute of Sustainability for Chemicals, Energy and Environment, A*STAR (Agency for Science, Technology and Research), 1 Pesek Road, Jurong Island, Singapore 627833, Singapore*
[¶]*Department of Chemistry, National University of Singapore, 3 Science Drive 3, Singapore 117543, Singapore*
[∥]*Department of Material Science and Engineering, National University of Singapore, 9 Engineering Drive 1, #03-09 EA, Singapore 117575, Singapore*
[**]*jw-xu@imre.a-star.edu.sg*
[††]*lohxj@imre.a-star.edu.sg*
[‡‡]*zhuq@imre.a-star.edu.sg*

[§§]Corresponding authors.
*To cite this article, please refer to its earlier version published in the *World Scientific Annual Review of Functional Materials*, Volume 1, 2230007 (2023), DOI: 10.1142/S2810922822300070.

Microencapsulated phase change materials (MEPCMs) are effective solutions for addressing the issue of leakage that phase change materials (PCMs) face in thermal energy storage devices. Their applications are ubiquitous as PCMs are utilized in industries such as logistics, construction, electronics, etc., thus, an efficient method to prevent problems such as leakage and poor thermal conductivity is to encapsulate the PCM which not only renders it leakage-proof but also impart mechanical strength and enhanced thermal properties. The application of silicon-based additives is one of the most studied methods to impart such desired properties. We discuss the silicon-containing compounds which are commonly employed in core-shell matrix of encapsulated PCMs, namely, siloxanes and silicone, silicon nitride, silicon carbide, silica/SiO_2, and other silicon-containing additives as they are able to provide synergistic improvements and exhibit enhanced physical properties. In this review, the different silicon compounds used as additives or main shell matrix are discussed, the general fabrication of the MEPCM and its thermophysical properties will be briefly highlighted. Lastly, we also examine its application and performance in thermal storage and thermal management. We hope to provide a broader perspective of silicon-containing MEPCM for those who are working in the similar field of research.

Keywords: Core-shell; microencapsulation; phase change materials; silicon; thermal energy storage.

1. Introduction

Sustainable and renewable sources for energy are imperative in worldwide industries due to the rising demand for energy in view of the rapid technological and economic advancement. The pressing need to find an alternative source of sustainable energy is also driven by environmental issues, such as pollution and climate change, due to the finite fossil fuel consumption.[1,2] Some renewable energy sources such as solar energy are the most viable option as an alternative; however, the energy conversion efficiency remains an issue yet to be improved.[3] Thermal energy storage is another potential and promising research area as thermal energy can be used as a direct source of energy. In addition, it can also be converted to other forms of energy such as electrical potential energy through photovoltaics in photoswitches,[4] or thermoelectrics.[5-7] In this review, thermal energy storage, specifically in the form of latent heat, is discussed as it has advantages of superior energy density, as well as a less significant temperature gap between the heat stored and that released.[8]

Latent heat storage materials, known as phase change materials (PCMs), can store or release energy well during the phase transition processes which are in general nontoxic, noncorrosive and nondegradable under numerous thermal cycles.[9] Solid–liquid PCMs are the most pragmatic choice for thermal energy storage since the heat involved in phase change is significantly higher than that for solid–solid PCMs, apart from other advantages such as high density, insignificant volume change, low vapor pressure and notable rate of crystallization.[10] Subcooling is mainly minimal in solid–liquid PCMs when freezing, melting or freezing at constant temperatures and at phase separation.[11] Solid–gas and liquid–gas phase changes pose too much volume and pressure difference to the system and thus, hinder its application in practical usage.[12] The practicality of the solid–liquid PCMs is evaluated based on several thermal parameters including latent heat, heat capacity, thermal conductivity and phase change temperature.[11] Generally, PCMs can be classified into three main groups namely, organic, inorganic and eutectic PCMs. Organic PCMs containing organic compounds like paraffins (Pa), fatty alcohols and sugars are an economical choice; however their numerous drawbacks like leakage and low thermal conductivity hinder their practicality in applications, as well as issues arising from supercooling[13,14] and flammability.[15,16] Inorganic PCMs consist of inorganic materials like hydrated salts, compounds and metals. Applications of inorganic PCMs are rare due to the small latent heat capacity, supercooling that leads to a longer thermal discharge time, high volume changes, corrosion and other problems, making them impractical for thermal energy storage.[17] Eutectic PCMs are mixtures that exist in the same phase of two or more constituents with congruent freezing and melting contributing to the formation of the component crystal mixture.[9] They have greatly lower latent heat and specific heat capacity compared to those of Pa and salt hydrates despite better thermal conductivity without supercooling and segregation.[18,19]

PCMs are capable of storing a large amount of latent heat and enhancing storage efficiencies during phase transitions.[20,21] They are applicable in solar thermal energy systems, textiles, buildings, waste heat recovery and buildings.[22-24] Microencapsulated PCMs (MEPCMs) and nanoencapsulated PCMs (NEPCMs) exhibit better heat transfer because of their larger surface areas.[20] Besides availability for a wide range of transition temperatures, organic PCMs are often chosen due to their facile encapsulation, high latent heat values as well as good stability.[25,26] The encapsulated PCM core and shell material affect its properties, and different combinations of the materials and additives are utilized in order to optimize the PCM properties like leakage, mechanical strength, thermal performance, and particularly thermal conductivity since it has been noted as a common issue observed in existing organic PCMs.[20]

Leakage is an existing limitation present in PCMs, especially in the case of organic PCMs.[27] Therefore, researchers have developed strategies such as encapsulation to facilitate the formation of microencapsulated and NEPCMs as alternatives to conventional PCMs. The microencapsulation and nanoencapsulation process enhances thermal stability, and furthermore, it also maintains the volume and thermal storage heat transfer area.[28] The shell material utilized for encapsulation includes organic, inorganic and organic–inorganic hybrid materials. Typically, natural and synthetic polymers constitute organic shell materials, such as poly(melamine-formaldehyde) (PMF) and urea-formaldehyde (UF), giving sealing, structural flexibility and volume change resistance from repetitive phase transformations.[29-31] Inorganic shells present an alternative to organic shells since they often possess better rigidity and subsequently superior mechanical properties, in addition to improved thermal conductivity. Silica or silicon dioxide (SiO_2) with high thermal and chemical stability is a notable choice.[32,33] Yuan et al. made use of silica in the production of encapsulated stearic acid (SA) PCM.[34] Besides, $Mg(NO_3)_2 \cdot 6H_2O$ was used by Zhang et al. because of its high melting point and high latent heat.[35] In comparison, organic–inorganic hybrid shells serve as a method to combine the collective advantages of organic and inorganic materials and neutralize the respective drawbacks present. The organic components contribute to structural flexibility while the inorganic parts help strengthen rigidity, thermal stability and conductivity.[29] Wang et al. created a PCM using a hybrid wall comprising melamine-urea-formaldehyde (MUF) modified with nanosilicon carbide (nano-SiC),[36] which displayed good crosslinking.[36]

Enhancement of the PCMs with the aid of encapsulation has significant improvements to its performance due to its enhanced thermal and mechanical properties and thus it has found application in areas such as photothermal storage, textiles and coatings for building components.[37] Su et al. noted that encapsulated PCMs prepared with nano-SiO$_2$ hydrosol as the surfactant were applicable in buildings as thermal energy storage due to their characteristic of enhanced shell integrity and core material content.[30] Wang et al. recorded a photothermal conversion rate of up to 74.4% for the synthesized microcapsule.[36] Encapsulated PCM application in textiles may be a result of improved elasticity and thermal storage properties, as demonstrated by Fu et al.[38] The microcapsule composite fabricated by Yu and He containing an octadecanol core and silicone elastomer wall can be used in thermal management for electronics, as well as thermoelectric energy harvesting.[39,40]

The preparation of PCM encapsulation can be categorized into mechanical, chemical and physical–chemical methods. Mechanical processes are purely physical and involve methods like drying and adhesion, with spray-drying and solvent evaporation.[11,41] Chemical microencapsulation methods commonly involve polymerization or condensation of the materials at an oil/water (O/W) emulsion. Polymerization can be further divided into in situ, interfacial, suspension and emulsion polymerization.[8,11] In situ polymerization is usually performed by adding the PCMs to an aqueous surfactant to form the O/W emulsion and then adding the prepolymer solution to afford the resulting product microcapsule or nanocapsule.[42] Wang et al. used in situ polymerization to successfully fabricate a PCM with a capric acid (CA) core and nano-SiC/MUF shell.[36]

For interfacial polymerization, the O/W emulsion is composed of the PCM and a hydrophobic monomer, and then the initiation of polymerization is achieved under the appropriate conditions.[11] Ma et al. successfully fabricated microcapsules through encapsulation.[43] Suspension polymerization is conducted by releasing the free radicals present in the oil-soluble initiator into the emulsion system created through suspending the PCMs, monomers and initiators in a continuous aqueous phase with surfactants and stirring.[44] Emulsion polymerization was successfully used by Zhang et al. in tandem with reverse micellization to synthesize a PCM with a sodium sulfate decahydrate core and silica shell.[45] With intense agitation and surfactants, suspension is obtained using the dispersed phase comprising PCMs and monomers in the continuous phase.[11] Physical–chemical techniques combine physical and chemical processes for encapsulation. The sol–gel method is an example of a physical–chemical process, in which the reactive materials, including the PCM, are evenly distributed in a continuous phase for hydrolysis to obtain a colloidal solution. Subsequently, a three-dimensional network gel system is formed via monomer condensation polymerization before the final formation of microcapsules.[46] As an example, the sol–gel process was utilized in Ref. 47 for the encapsulation of Tris(hydroxymethyl)methyl aminomethane (Tris) with a SiO_2.

Encapsulated PCMs were characterized in terms of Scanning Electron Microscopy (SEM), Transmission Electron Microscopy (TEM), Fourier-Transform Infrared Spectroscopy (FTIR) and X-ray Diffraction (XRD). Successful PCM encapsulation can be elucidated by surface morphology from SEM and TEM, together with physical evidence such as spherical structure with smooth and dense surface. FTIR is useful to verify the chemical composition of the core and shell present in the synthesized composite PCM.[48] For example, Sun et al. utilized FTIR to analyze a PCM composite comprising Pa with a silicon nitride (Si_3N_4) core. The characteristic stretching vibration peaks for Si_3N_4 were observed at 600–1100 cm^{-1} in the microcapsules, indicating complete encapsulation.[49] XRD is especially suited for characterization pertaining to inorganic components. For instance, XRD analysis showed the silica shell completely encapsulated the SA core with a lower sample crystallinity.[50] Thermal characterization methods play a crucial role in determining the performance of the encapsulated PCMs. Some common methods include Differential Scanning Calorimetry (DSC), Thermogravimetric Analysis (TGA) and thermal cycling.[11] In addition, mechanical properties of PCM composites including tensile strength, elongation at break and hardness were studied in some cases.[51]

To the best of our knowledge, despite the numerous literatures pertaining to PCM with silicon and other derivatives, there has not been a comprehensive review regarding the incorporation of silicon encapsulation for PCM. This review is categorized based on different Si-containing compounds used to encapsulate PCMs, silicone, siloxane, silicon rubber, Si_3N_4, SiC, silica and SiO_2, and other silicon derivatives. We hope that this review provides insights into silicon-containing encapsulation of PCMs for different applications such as thermal energy storage. We have also added in tables of summary (Tables 1-5) for convenience of reference.

2. Silicone Additives for Encapsulation

2.1. Siloxane and Silicone

Organic siloxane derivatives contain alternating Si–O backbone. Silicone has the ability to enhance the structural flexibility when encapsulated into PCMs, therefore improving its overall thermal and mechanical properties.[67,68] Silicone has great potential when applied as a microcapsule shell owing to its superior thermal and mechanical properties such as thermal conductivity and solvent resistance.[52,53] Its application[54-56] in PCMs can be attributed to its low glass transition temperature, high gas permeability, biocompatibility and elastic deformation.[38] It is a semi-inorganic polymer possessing high crosslinking with the main chain comprising alternating Si–O bonds while the organic groups constitute the side chains.

Such arrangement allows this material to have dual characteristics, with the ability to resolve the rigidity present in inorganic substances as well as provide structural integrity.[69,70] Its applications are evident in coatings and encapsulated electronic components.[57,58] The microcapsules in this section were fabricated using microfluidic techniques, *in situ* polymerization and cast molding.[39,52] Silicone can be in the form of oil which is used as a fluid in the process of synthesizing PCMs, and it can also be applied as a heat transfer fluid (HTF).[59,60] Akamatsu *et al.* employed photopolymerizable silicone oil for PCM encapsulation in a glass capillary through a microfluidic process, while Wang *et al.* utilized silicone oil in a vertical model as a HTF, aiming at enhancing PCM thermal conductivity with a nitrogen supply to prevent the silicone oil from oxidation.[59,60] The miscibility present between the PCM and the silicone oil resulted in unclear distinction between the inner and middle phases, but the corresponding immiscibility between the middle and outer face ensured the stable formation of the emulsion.[59] The most common silicone rubber (SR) possesses numerous advantages for it to be incorporated into PCMs, including its elastic, electrical insulating, thermally stable, chemically inert and durable properties.[61] For example, Wu *et al.* showed that methyl vinyl silicone rubber (MVQ) can be utilized in the preparation of PCM composite and it was found that the optimal amount of the additive gave the best overall thermal and mechanical performance.[62]

Li *et al.* applied single-step *in situ* polymerization to fabricate a novel organic silicon microencapsulated phase change material (OS-MPCMs), with n-octadecane (OD) as the core material and silicone as the shell.[52] The silicone shell was prepared with vinyl polysiloxane (VPS), poly(methylhydrosiloxane) (PMHS) with platinum (Pt) as catalyst. Samples with varying core-shell ratios were synthesized and labeled as OS-MPCMs-1–OS-MPCMs-9. The OS-MPCM morphology was determined by optical microscopy and SEM in which a spherical structure was observed when the core-shell ratio was between 1:2 and 1:1. The presence of obvious cavities indicated a complete core-shell arrangement. FTIR analysis showed

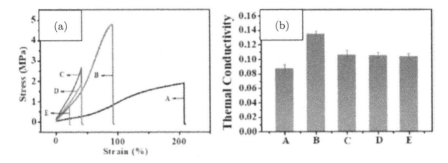

Fig. 1. (a) Stress–strain curves and (b) thermal conductivity histograms samples with different PMHS/VPS composite of (A) 5wt.%, (B) 9wt.%, (C), 15wt.%, (D) 20wt.%, (E) 30wt.%. Reprinted with permission from Ref. 52. Copyright 2022 Elsevier.

complete polymerization, as well as a strongly bonded organic silicon shell structure. XPS further proved the encapsulation of n-OD by organic silicone. From DSC, the melting enthalpy of OS-MPCMs could reach a value of 103.3 Jg^{-1}. The enthalpy values increased with the loading of n-OD. OS-MPCM encapsulation efficiency could be enhanced up to 86.45%. TGA and Derivative Thermogravimetry (DTG) were used to evaluate the thermal stability. Mechanical tests displayed optimal performance when 9wt.% PMHS was used due to a higher degree of crosslinking (Fig. 1(a)). Thermal conductivity was also the highest for this composite performing at 0.14 Wm^{-1}K^{-1} (Fig. 1(b)). Further DSC over 200 thermal cycles concluded good heat absorption and release ability for the OS-MPCMs, verifying its better cycle durability. It is worth noting that OS-MPCMs could be potentially used for energy-efficient buildings, thermos-regulated fabrics.[38,57]

Zhu et al. reported a novel flexible form-stable PCM (FSPCM) composite prepared using OD as the PCM and silicone as the supporting matrix for the encapsulation of the PCM.[27] The composites were characterized via FTIR, DSC, TGA and XRD. Interestingly, they found an upshift in OD melting temperature of 30.3 to between 34.4 and 37.8, attributing to the thermal insulation by silicone. The silicone PCM composite also possesses a relatively high

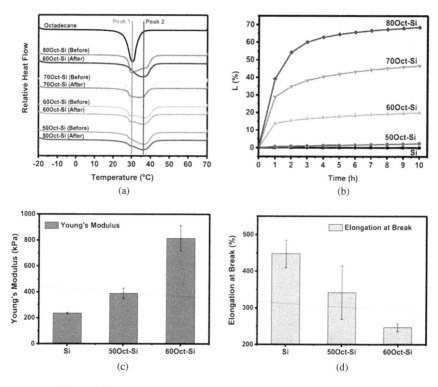

Fig. 2. (a) DSC (b) leakage test when heated to 40°C for 10 h, mechanical test of (c) strain and (d) elongation of samples with various OD loading. Reprinted with permission from Ref. 27. Copyright 2022 Elsevier.

latent heat of 103.8 Jg^{-1} and leakage was found to be only 2.44% when heated to 40°C for 10 h (Figs. 2(a) and 2(b), respectively). Furthermore, to reinforce the mechanical enhancements of the composite PCM, a mechanical test further confirmed that the silicone PCM composites had good flexibility and durability as it exhibited a Young's Modulus of 388.92 kPA and an elongation of 341.42% rendering its use in wearable devices and energy harvesting (Figs. 2(c) and 2(d), respectively).[71,72]

Li et al. designed a MEPCM consisting of elastic silicone/n-hexadecyl bromide microcapsules via a microfluidic method with co-flowing channels. Double oil1-in-oil2-in-water (O1/O2/W) droplets with a core-shell structure were prepared in the channels (Fig. 3(a)[38,64]). From the optical microscope images and SEM analysis

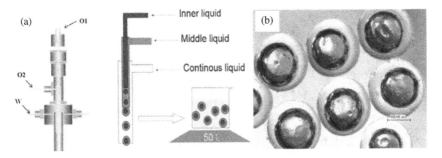

Fig. 3. (a) Schematic diagram of the microfluidic device and synthesis of double emulsions, (b) optical microscope images for MEPCMs. Reprinted with permission from Ref. 38. Copyright 2014 Elsevier.

in Fig. 3(b), it was observed that the microcapsule surfaces were smooth and the cross sections were compact. Subsequent observation through the optical microscope found that the silicone shell contributed to expansion owing to its elasticity. DSC analysis indicated good energy storage ability through the respective melting and freezing enthalpies of 76.35 and 78.67 Jg^{-1}. The microencapsulation ratio was found to be 49 wt.%. TGA concluded superior thermal stability through the 35% weight loss of a neat silicone elastomer at 400°C with slow decomposition. The MEPCMs had two weight loss stages up to 610°C. Silicone/n-hexadecyl bromide microcapsules have potential usage in thermal regulating textile and thermal insulation materials.

He et al. designed a shape-remodeled PCM macrocapsule with an octadecanol core encapsulated incorporating silicone elastomer using a cast molding technique.[39] The macrocapsule is capable of dynamic and repetitive restructuring to achieve a complex arrangement with large-scale deformation. Four PCM macrocapsules were prepared: octadecanol with silicone, octadecanol with silicone mixed with 80wt.% EBiInSn and 1wt.% extended graphite (EG), octadecanol and 2wt.% EG with silicone mixed with 80wt.% EBiInSn and 1wt.% EG, and octadecanol and 3wt.% EG with silicone mixed with 80wt.% EBiInSn and 1wt.% EG. SEM images of the macrocapsule core showed the complete impregnation and absorption of octadecanol into the EG micropores (Fig. 4(a)).

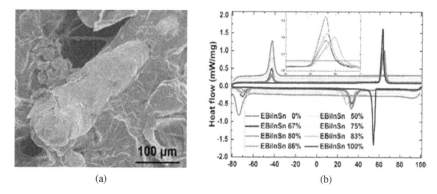

Fig. 4. (a) SEM image for the mixture of EG and octadecanol, (b) melting–freezing DSC curves for the EBiInSn-silicone composites. Reprinted with permission from Ref. 39. Copyright 2019 Elsevier.

DSC results concluded insignificant latent heat change in the macrocapsule core due to EG addition since the 5wt.% EG only caused a drop in latent heat of melting from 241.5 kJkg^{-1} to 234.1 kJkg^{-1}. TGA measurements indicated higher 5% and maximum weight loss temperatures occurring in EG(5wt.%)-octadecanol reaching up to 190°C and 259°C relative to that of pure octadecanol. The improved latent heat storage and heat conductivity were attributed to the embedding of high-concentration, low-melting Bi–In–Sn eutectic alloy microparticles (80wt.%) in the flexible shell. The DSC curves in Fig. 4(b) demonstrated the partial coincidence of the melting endothermic range of octadecanol with that of pure EBiInSn, which had an endothermic peak at 62.3°C, contributing to heat storage. TGA presented good thermal stability and leakage prevention in the EBiInSn-silicone composite, with only one degradation peak at up to 483°C for 50 wt.% EBiInSn addition. The shell retained its high stretchability with a strain at 432%. Thermal conductivity of the macrocapsule experienced a 428% increase in the macrocapsule core to 1.53 Wm^{-1}K^{-1} compared to pure octadecanol, while the shell thermal conductivity rose by 890% relative to pure silicone to 1.98 Wm^{-1}K^{-1}. The shape remodeling performance, energy storage capacity and heat charging and discharging rates of the macrocapsule were reported. The solid PCM core helped maintain the remodeled arrangement due to good encapsulation with the flexible shell. The macrocapsule was found to possess a high latent heat density value at 210.1 MJm^{-3}. It was observed from two thermal cycles that 3wt.% EG and the other additives in the PCM helped decrease the respective charging and discharging times to 53 min and 20 min, respectively. Further thermal cycles verified the thermal durability as there was no apparent weight loss and the charging and discharging times remained constant. The synthesized macrocapsule may be applied in thermal management for flexible electronic devices and thermal storage for thermoelectric energy harvesting.[65,66] Therefore, flexible PCM capsules may be useful in energy storage and thermal control engineering.

Akamatsu et al. synthesized an encapsulated n-tetradecane and n-hexadecane as PCMs in silicone shells via a simple microfluidic technique with a glass capillary device, in which PCM comprised double emulsion droplets acted as the inner fluid and photopolymerizable silicone oil as the middle fluid.[59] Due to the immiscibility between the PCM and the silicone oil, instant UV irradiation was applied for the formation of microcapsule shells through the solidification of the middle fluid. Three samples were prepared with a n-tetradecane core (T1, T2 and T3), while two samples (H1, H2) were composed of n-hexadecane cores. Optical microscopy with a temperature-regulated sample stage was conducted using H1 as example to evaluate the phase change performance as indicated in Figs. 5(a) and 5(b). TGA concluded that the microcapsules displayed superior thermal stability below 400°C, as the silicone microspheres maintained a weight of 97.4% under heating below this temperature, while it was found from DSC that the H2 microcapsules had the largest energy storage capacity at 96 Jg^{-1}. It was noted that the PCM encapsulation ratio could be adjusted by shifting the flow rates for the innermost and middle fluids producing the emulsion droplets in the glass capillary device.

Wang et al. suggested a gradient porosity metal foam to enhance heat transfer to mitigate low thermal conductivity commonly observed in PCMs hindering its practicality.[60] Latent thermal energy storage (LTES) has potential applications for efficient

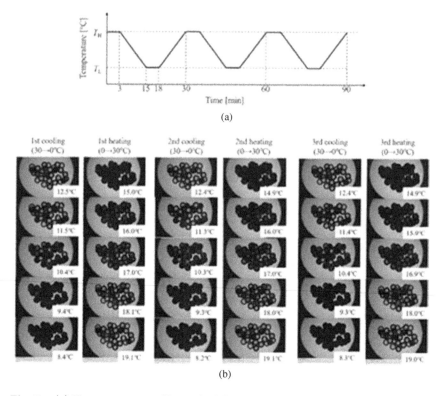

Fig. 5. (a) Temperature profile applied for PCM observation, (b) optical microscope images demonstrating the H1 microcapsule phase change for three-cycle cooling–heating program. Reprinted with permission from Ref. 59. Copyright 2019 Elsevier.

solar energy usage. Experiments and comparisons were conducted on the thermal behavior of a gradient copper foam (GCF) and widely used homogeneous copper foam (HCF) in a mid-temperature solar energy storage system. A153 was utilized as the PCM in the annulus of the two lab-scale shell-and-tube units built, with the GCF and HCF occupying one unit of each, respectively. Silicone oil acted as the HTF present in the inner tube. Charging processes demonstrated thermal reliability through the repeated heating and cooling from the experiment. Temperature evolution tests showed that lower porosity in large temperature gaps contributed to higher thermal conductivity (Figs. 6(a)–6(d)). The melting time

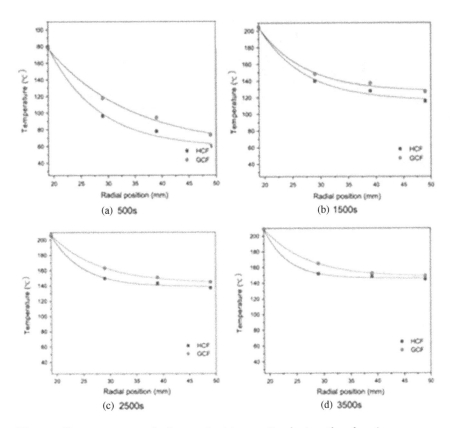

Fig. 6. Temperature evolution against two units during the charging process. Reprinted with permission from Ref. 60. Copyright 2020 Elsevier.

was decreased by 37.6%. An infrared photograph presented the superior heat transfer of GCF, as the gradient porosity arrangement enhanced the LTES system and thermal conductivity through the combination with temperature gradient, direction of heat flow and structure. Therefore, the gradient porosity metal foam was proven to improve energy storage system performance and can be applied accordingly, with additional usage for promoting solar energy use at medium temperature ranges.

Wu et al. fabricated a new silicone rubber (MVQ) based phase change composite, MVQ/OD/P(St-MMA), comprising n-octadecane (OD)/poly(styrene-methyl methacrylate) (St-MMA) microcapsule, where n-OD was the PCM core and P(St-MMA) was the shell.[62]

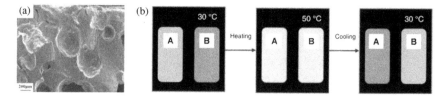

Fig. 7. (a) SEM micrograph for MVQ/OD/P(St-MMA) with 2phr microcapsules. (b) Infrared camera images for samples with (A) and without (B) microcapsules in a thermal cycle. Reprinted with permission from Ref. 62. Copyright 2018 Springer.

Four composite samples were prepared with 0 to 3 phr (per hundred rubber) microcapsules. The mechanical properties of the composite were evaluated to show better performance with relatively high tensile strength (2.57Mpa) and elongation (296.43%) after aging. SEM analysis on the morphology in Fig. 7(a) further showed that MVQ/OD/P(St-MMA) had neat fracture surfaces, the least damage in the microcapsules, in addition to even distribution of the microcapsules in the SR. The 2phr samples had well-embedded microcapsules uniformly distributed in the SR matrix. TGA investigated the effect of microcapsule content on the composite thermal properties with 2phr microcapsule content in the composite having the best thermal stability. The composites experienced a lower mass loss rate relative to pure SR below a temperature of 450°C. FTIR indicated that MVQ/OD/P(St-MMA) maintained all the properties of each component. From DSC, the energy-storage performance was proven to be improved, with an enthalpy value of 67.6 Jg^{-1}. Thermal response tests were applied with an infrared thermocamera, displaying a higher temperature in the sample containing microcapsules (Fig. 7(b)). In conclusion, RTV with 2phr content cured at room temperature gave superior thermal and mechanical properties in the composites.

Yang *et al.* designed a Pa@graphene/SR composite FSPCM, where MEPCM was first synthesized through the electrostatic self-assembly method, then it was added to the SR matrix for composite preparation.[63] Composite samples with different Pa@graphene MEPCM contents were prepared (MEPCM-5/SR, MEPCM-10/SR,

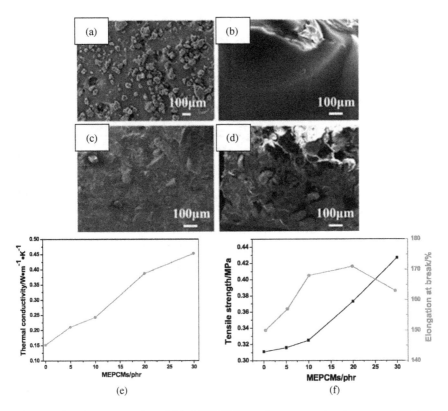

Fig. 8. SEM micrographs of (a) MEPCMs, and cross-section images of (b) SR, (c) MEPCM-5/SR, (d) MEPCM-30/SR, (e) thermal conductivity, (f) tensile strength test of MEPCMs. Reprinted with permission from Ref. 63. Copyright 2019 Taylor & Francis Online.

MEPCM-15/SR, MEPCM-20/SR, MEPCM-25/SR and MEPCM-30/SR). SEM was conducted for microstructure characterization as shown in Figs. 8(a)–8(d), in which a rough surface was observed in MEPCM-5/SR and MEPCM-30/SR due to the dispersion of the microcapsules in SR, with some graphene penetrating through the rubber. Composites synthesized through mechanical blending were favorable for heat conduction. FTIR verified the presence of MEPCM and SR in the composite. DSC and thermal conductivity analysis showed the highest phase change latent heat of 53.39 Jg^{-1} and thermal conductivity of 0.453 $Wm^{-1}K^{-1}$, respectively, in MEPCM-30/SR

(Fig. 8(e)). TGA recorded two thermal decomposition processes for the MEPCM/SR composites at 150–300°C and 350–600°C, respectively, and the MEPCMs have a higher thermal decomposition temperature than that of neat Pa. Pa@graphene content influenced the tensile strength of the material (Fig. 8(f)). The thermal and mechanical properties of SR were enhanced by the microcapsules, rendering them suitable for thermal energy storage.

Zhang *et al.* prepared a multi-responsive FSPCM through the encapsulation of crosslinkable bottlebrush crystalline polysiloxanes ($Si_{0.75}$-18-x), a very flexible supporting material, in Pa via self-assembly and 3D networks.[67] Superhydrophobic surfaces enhance the long-term use of solar energy and environmental compliance of multi-responsive FSPCMs.[73,74] This work reports the direct synthesis of the multi-responsive FSPCM through *in situ* curing after spray-coating $Si_{0.75}$-18-x/Pa solution with a multi-walled carbon nanotube (MWCNT) dispersion.[67,75] FTIR showed that MWCNT and spraying did not affect hydrosilylation. The Contact Angle (CA) was increased correspondingly to the increase in MWCNT content. SEM-Energy Dispersive Spectroscopy (SEM-EDS) results shown in Figs. 9(a)–9(c)

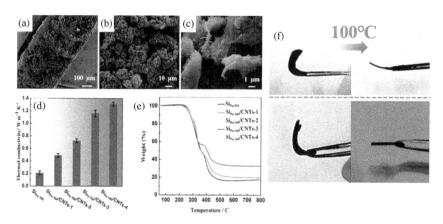

Fig. 9. (a)–(c) Cross-sectional micrographs for the Si_{Pa-160}/CNTs-3 film. (d) Thermal conductivity of the Si_{Pa-160}/CNTs-x films, respectively. (e) TGA curves of Si_{Pa-160}/CNTs-x. (f) (Top) Thermally actuated and (bottom) photon-actuated shape memory behaviors of the Si_{Pa-160}/CNTs-3 film (2 cm × 0.6 cm × 0.1 cm), respectively. Reprinted with permission from Ref. 67. Copyright 2020 Royal Society of Chemistry.

also indicated the even distribution of MWCNTs, Pa and $Si_{0.75}$-18-3/polymethylhydrosiloxane (PMHS) on the surface of Si_{Pa-160}/CNTs-3. XRD characterization further showed even heat dissipation, while subsequent DSC verified that latent heat loss experienced an increase as more MWCNTs were added. Si_{Pa-160}/CNTs-3 was optimal in combining higher latent heat with superhydrophobicity at 138 Jg^{-1}. Leakage prevention for Si_{Pa-160}/CNTs-3 was excellent at 0 wt.%. Further shape stability was observed through the insignificant surface tension shift after melting. Si_{Pa-160}/CNTs-3 had good thermal conductivity at 1.16 ± 0.05 $Wm^{-1}K^{-1}$ (Fig. 9(d)). The photothermal conversion spectrum for Si_{Pa-160}/CNTs-3 demonstrated that 1.0 sun gave a sunlight-to-heat storage efficiency of around 65%. After 50 thermal cycles, it could be inferred that the Si_{Pa-160}/CNTs-3 experienced insignificant melting/freezing temperature shifts and a latent heat drop of less than 1%. Irradiation cycling tests verified the superior photo-driven phase change reversibility. TGA presented high thermal stability in Si_{Pa-160}/CNTs-x films, with a maximum mass loss rate for Si_{Pa-160} occurring at 323°C (Fig. 9(e)). Under 100°C or near-infrared (NIR) light as indicated in Fig. 9(f), Si_{Pa-160}/CNTs-3 had good thermally actuated and photon-actuated shape variation ability, since the fast shape recovery was observed. The uncured Si_{Pa-160}/CNTs-3 can be sprayed layer-by-layer on different substrates for energy conversion and storage due to its excellent properties. Potential applications include wearable devices, intelligent buildings and electronic devices.[76-78]

Kim et al. synthesized a new PCM based on inorganic salt hydrates via a sol–gel method in combination with interfacial polymerization.[68] Calcium chloride hexahydrate (CCH) acted as the core material, while organoalkoxysilane contributed to the hybrid quality by mediating the hydrophilic core and hydrophobic shell. Eight samples with varying core-shell ratios were fabricated. FTIR verified the successful encapsulation of the CCH core with the siloxane and polyurea shell. SEM results in Fig. 10 showed that granular and spherical molecules could only be successfully prepared under a pH of 1.5–3.0, with 2.35 being the most desirable pH condition for complete encapsulation. Further morphology tests presented a nearly spherical structure and a well-defined core-shell arrangement.

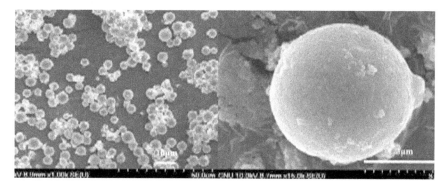

Fig. 10. SEM images for the microencapsulated CCH prepared at pH2.35 with 70/30 core shell ratio. Reprinted with permission from Ref. 68. Copyright 2017 Wiley.

TGA showed good thermal stability with four degradation steps for the microcapsules with CCH. The sample with a weight ratio of 65/35, i.e., Sample 4 had the least weight loss at 24.6%. DSC results indicated absorption of thermal energy with phase change during melting and significant supercooling during crystallization, with the microcapsules prepared with pH 2.35 and 2.55 having superior enthalpies of 95.83 and 88.20 Jg^{-1}, as well as higher encapsulation ratio and efficiency of up to 71% and 83%, respectively. Durability tests to gauge leakage of the siloxane polymer and polyurea present in the shell concluded a 5.0% leached PCM for Sample 4 synthesized under pH 2.35, demonstrating relatively good leakage prevention.

Zhou *et al.* reported the preparation of three waterborne polyurethane-acrylate (WPUA) hybrid emulsions with organic silicon.[51] The components consist of isophorone diisocyanate, polytetramethylene ether glycol, double hydroxyl terminated polydimethylsiloxane (PDMS) and acrylates. The hybrid WPUA dispersions consist of various arrangements, mainly, core-shell, inverse core-shell and semi-interpenetrating network (semi-IPN). They were labeled as WPUA1, WPUA2 and WPUA3, respectively. TEM in Figs. 11(a)–11(c) was employed for morphological analysis. WPUA1 presented a conventional core-shell shape, while WPUA2 had the inverse core-shell structure and WPUA3 a relatively irregularly shaped spherical structure due to the presence of IPN. WPUA1 had the smallest particle size and WPUA3 the largest, with a smaller particle size contributing to the increase in viscosity. FTIR verified the chemical composition and the incorporation of the hydroxyl groups in PDMS into the polyurethane chain. XPS showed the WPUA structure, which was found to influence silicon mobility, a factor subsequently affecting the film water resistance, determined by CA measurements. WPUA1 had the smallest CA (72.6°) and the highest water swelling absorption, while WPUA3 the largest CA (93.1°) and lowest water swelling absorption. The water swelling curves are indicated in Fig. 11(g). WPUA had the highest degree of silicon migration and water resistance. WPUA1 possessed optimal tensile strength at 18.3MPa as well as a high elongation at break. Thermal stability measurements through TGA proved the relatively superior thermal stability of WPUA, which recorded a relatively higher residual mass at the same decomposition temperature, for example a 5% weight loss at a higher temperature of 273.28°C compared to that of WPUA2 and WPUA3. SEM imaging demonstrated the highest phase separation degree in WPUA3, as shown in Figs. 11(e) and 11(f).

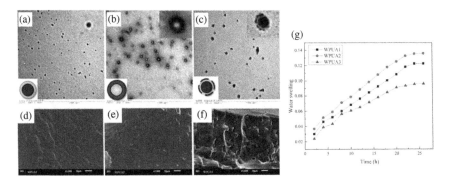

Fig. 11. TEM micrographs for (a) WPUA1, (b) WPUA2, (c) WPUA3. (g) Curves of water swelling against time of WPUA films. SEM micrographs of (d) WPUA1, (e) WPUA2, (f) WPUA3. Reprinted with permission from Ref. 51. Copyright 2019 Elsevier.

Table 1. Summary of siloxane and silicone-containing encapsulation and its parameters.

Additives	PCM	Method of preparation	Latent heat (Jg^{-1})	Melting point (°C)	Unique properties	Encapsulation efficiency	Ref.
—	n-OD	Single-step in situ polymerization	103.3	35.5	—	86.45	52
—	n-OD	Physical mixing	103.8	34.4–37.8	—	—	27
—	n-hexadecyl bromide	Microfluidic	78.67	16.7	—	—	38
$Bi_{31.6}In_{48.8}Sn_{19.6}$ alloy (EBiInSn), EG	Octadecanol	Vacuum filtration assisted drop-casting method	192.4	55.7	Can be dynamically and repeatably remodeled for large-scale formation	—	49
—	n-tetradecane, n-hexadecane	Microfluidic method with the PCM as inner fluid and photopolymerizable silicone oil as the middle fluid	96	~15	—	—	59
SiC foam, copper foam	A153	Two lab-and-scale shell-and-tube units built with PCM in the annulus	262.6	151–155	Mid-temperature LTES units	—	60

Silicon-Containing Additives in Encapsulation of PCMs 101

Shell	Core	Method			Property		Ref.
P(St-MMA)	n-OD	Mixing of the microcapsules	67.6	—	—	—	62
Graphene	Paraffin	Self-assembly	53.39	45.03	—	—	63
MWCNTs	Paraffin	Self-assembly and spray-coated	138	25–50	Superhydrophobicity	54.27	67
Polyurea	CCH	Sol–gel process and interfacial polymerization	95.83	29.5	Relatively good leakage properties	71	68
Isophorone diisocyanate, acrylates	Polytetramethylene ether glycol	Emulsions	—	—	—	—	51

2.2. Silicon nitride

Si_3N_4 is a widely applicable and economical ceramic powder with superior strength, wear resistance, thermal resistance, abradability and resistance to oxidation.[49,79] Tang et al. and Sun et al. highlighted its efficiency as a supplementary and filler material to enhance the thermal performance of PCMs. A novel desiccation technique and Pickering emulsion method were utilized on Si_3N_4 to circumvent the agglomeration occurring in the particles, an issue attributed to its high surface energy.[49,80] Incorporation of Si_3N_4 was performed to enhance the thermal properties of PCMs.

Tang et al. prepared a novel phase change microcapsule with a n-OD core and polymethylmethacrylate (PMMA) shell with modified Si_3N_4 powder supplement to improve the thermal properties of PCMs, since thermal performance is among the most vital criteria for phase change microcapsules in solar energy storage applications.[80] Five Si_3N_4-modified microcapsules were synthesized with Si_3N_4 weight percentages of 0%, 5%, 10%, 15%, 20% and 30%, labeled Sample 1–Sample 5. SEM micrographs displayed a uniform sphere with a well-defined core-shell arrangement as in Fig. 12(a).

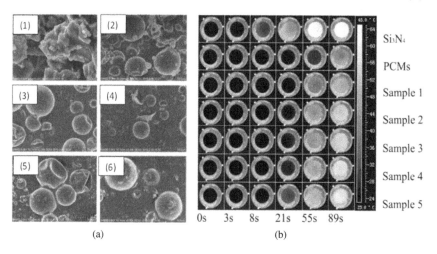

Fig. 12. (a) SEM images for the Si_3N_4 and microcapsules (1) Si_3N_4, (2) Sample 1, (3) Sample 2, (4) Sample 3, (5) Sample 4, (6) Sample 5, (b) phase change microcapsule composite materials. Reprinted with permission from Ref. 80. Copyright 2016 Elsevier.

Table 2. Summary of silicon nitride-containing encapsulation and its parameters.

Additives	PCM	Method of preparation	Latent heat (Jg^{-1})	Melting point (°C)	Unique properties	Encapsulation efficiency (%)	Ref.
PMMA	n-OD	*In situ* polymerization	110.6	25.48	Solar energy storage	—	80
Water-dispersible Si_3N_4 nanoparticles (nano-Si_3N_4) functionalized with amphiphilic polymer chains	Paraffin	Pickering emulsion	134.64	59.25	—	79.97	49

FTIR and EDS demonstrated effective cross-linking and polymerization of Si_3N_4 with the microcapsules after modifying the surface. The DSC curves indicated lower latent heat values for the Si_3N_4-enhanced microcapsules, with 110.6 Jg^{-1} for Sample 5. TGA showed two weight loss stages for the samples, beginning at around 230°C and a constant being achieved at a temperature of approximately 490°C. 500 thermal cycles with subsequent DSC analysis concluded insignificant shift in the melting and freezing temperatures as the heat storability values, quantified with the latent heat value of 108.1 Jg^{-1} for Sample 5, experienced only a 5% decrease. Forward-looking infra-red system was applied to investigate the microcapsule temperature fluctuation under unchanged environmental temperatures, as indicated in Fig. 12(b). The results showed a higher heating rate of the modified microcapsules corresponding with the amount of Si_3N_4 added, proving its superior thermal conductivity. The microcapsule thermal conductivity was heightened by 56.8% to 0.3630 $Wm^{-1}K^{-1}$ relative to that without addition of Si_3N_4.

Xiao et al. utilized an environmentally friendly Pickering emulsion process to synthesize a new composite PCM through the encapsulation of solid Pa with water-dispersible Si_3N_4 nanoparticles (nano-Si_3N_4) functionalized with amphiphilic polymer chains (Fig. 13(a)).[49,81] This technique was employed since the oil phase of melted Pa and the monomers could be effectively encapsulated and stabilized by the nano-Si_3N_4 Pickering stabilizer in an aqueous form. The formation of the PCM microcapsules embedded with nano-Si_3N_4 in the polymer shell was allowed by the subsequent polymerization and polymer-induced oil core phase separation. Five different samples were prepared, i.e., Pa wax microcapsules with 0%, 6%, 10%, 13% and 16% nano-Si_3N_4. FTIR characterized the composite chemical structure. SEM results determined the successful fabrication of the composite PCMs with a well-defined spherical shape as indicated in Fig. 13(b). It was observed that the PCMs with 10% nano-Si_3N_4 had the best performance, with an encapsulation efficiency of 79.97% and latent heats of melting and crystallization at 134.64 Jg^{-1} and 133.47 Jg^{-1}, respectively. TGA indicated a residual

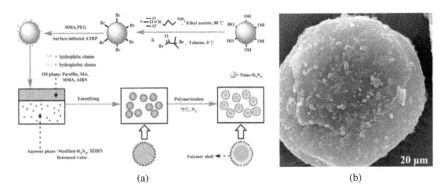

Fig. 13. (a) Schematic illustration for the preparation of MEPCM embedded with nano-Si$_3$N$_4$, (b) SEM image for Pa wax microcapsules with 10% nano-Si$_3$N$_4$. Reprinted with permission from Ref. 49. Copyright 2016 Springer.

weight of 36.10% for PCMs with 10% nano-Si$_3$N$_4$ and generally enhanced thermal stability. Thermal conductivity tests concluded an improved value of 0.32 Wm^{-1}K^{-1} due to nano-Si$_3$N$_4$.

2.3. Silicon carbide

SiC is known as a suitable additive to PCMs owing to its low cost, superior thermal conductivity, chemical stability and mechanical properties such as wear resistance.[82,83] SiC has found applications in photothermal areas due to its notable photothermal conversion and solar absorptivity at the NIR.[82,84] In most cases, modifications are made to incorporate SiC into the coreshell material to increase compatibility.[85] Therefore, SiC is sometimes incorporated into the PCM wall material or as nanowires to enhance the existing properties of the resulting product.[36,84,85] *In situ* polymerization is also a common method used in the preparation process while other common preparation methods include chemical vapor deposition (CVD).[36,84,85]

Shi *et al.* reported a corrosion-resistant carbon fiber reinforced SiC composite (C/C–SiC) that encapsulates the metallic PCM (mPCM), AlSi$_{12}$.[86] The eutectic alloy mPCM has a high meting

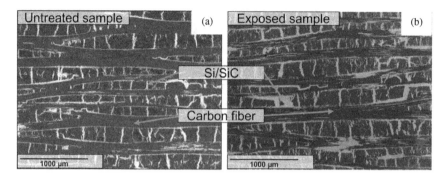

Fig. 14. SEM images of C/C–SiC samples: (a) without exposure and (b) after exposure to the corrosive MPCM. Reprinted with permission from Ref. 86. Copyright 2020 Wiley.

point of 557°C and a latent heat of fusion of 481 Jg^{-1}. The encapsulation is achieved via liquid silicon infiltration and the compatibility of the two components are assessed via CA measurements, showing non-wetting characteristic up to 1102°C. The results as seen in Fig. 14 also showed that C/C–SiC composite retains its mechanical strength and microstructure after exposure to the corrosive mPCM. The thermal output and input and the thermal storage capability of this material were also studied via a system calorimeter, indicating its potential in applications for containment of the corrosive mPCM.

Ryou et al. reported the use of PCM/SiC-based composite aggregates in concrete.[87] The concrete samples were prepared by replacing natural coarse aggregates with PCM/SiC-based composite aggregates and various ratios through impregnating and coating methods to encapsulate the PCM. Pa wax was used as the PCM and the physical properties of the composite aggregates in the concrete were assessed. The composite aggregate microstructures were also characterized using FTIR, XRD, SEM and TGA. It was found that the mechanical strength of the composite aggregate concrete was not compromised as observed in the compressive strength test, albeit the decrease in compressive strength upon addition of PCM/SiC aggregate, as they met the target designed strength of 30 Mpa as seen in Figs. 15(a) and 15(b). Furthermore, samples that contain

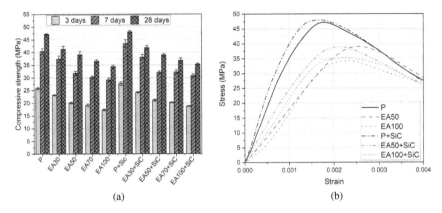

Fig. 15. (a) Compressive strength test and (b) stress–strain test of the different PCM/SiC composite aggregates. Reprinted with permission from Ref. 87. Copyright 2020 Elsevier.

SiC as sand replacement also produced enhanced microstructure as compared to those without SiC. These results suggest that the PCM/SiC-based composite aggregates can help in efficient use in partition walls of building components for thermal energy storage and reduction in energy demand.

Ma et al. reported a simple in situ synthesis method of SiC nanofibers on graphite felt as scaffold for improving performance of Pa-based PCM.[88] In this report, the Pa PCM was encapsulated in a composite material of SiC and graphite fibers via vacuum impregnation in hope of enhancing the thermal conductivity and reducing leakage in organic PCMs. A composite material was prepared via a vapor–solid reaction on industrially produced porous graphite felt as raw material. The composite PCM exhibited an increase in thermal conductivity to 1.29 $Wm^{-1}K^{-1}$ as compared to pure Pa which only exhibited 0.25 $Wm^{-1}K^{-1}$ as evident in Fig. 16(a). The enhanced composite PCM exhibited chemical and form stability as seen in Figs. 16(b) and 16(c), and the leakage test showed that the composite retained its shape while pure Pa melted and stained the filter paper when heated to 80°C for an hour. The composite PCM also exhibited phase change enthalpies of 180.5 Jg^{-1} and 176.4 Jg^{-1} during melting and freezing, respectively via DSC analysis.

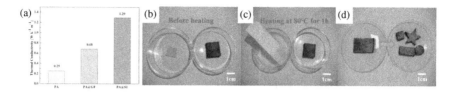

Fig. 16. (a) Thermal conductivity of PA and the composite PCM, (b) shape stability analysis and (c) leakage test of PA and composite PCM, (d) the shape design of the CPCM. Reprinted with permission from Ref. 88. Copyright 2022 The Royal Society of Chemistry.

Furthermore, the composite PCM is also very flexible in its shape design (Fig. 16(d)) and has excellent energy storage properties in terms of photo-thermal conversion rendering its potential and promising use in solar and building energy storage.

Kim et al. developed a novel thermal energy storage aggregate (TESA) system containing cement mortar and PCM/epoxy/SiC composite for improving the energy efficiency of buildings.[89] The composite fine aggregate was fabricated by first impregnating a zeolite scaffold with the Pa PCM, and thereafter, it was coated and encapsulated in epoxy resin, SiC and silica fume. The TESA was added into mortar by replacing sand content in the mortar and varying the ratio of zeolite to Pa (Z/P). Several specimens TESA30, TESA50, TESA70 and TESA100, Z/P50 and Z/P100, Z/PX or TESAX where X represents the percentage loading for the respective material were prepared. Their thermal and physical properties were studied. TESA 100 was highlighted as it exhibited desirable results with latent heat storage capacity of 19.76 Jg^{-1} and a melting point of 36.46°C obtained from DSC analysis. Furthermore, TGA analysis also showed the TESA was thermally stable up to 142°C. A thermal shock cycling test as seen in Fig. 17 revealed the long-term reliability of TESA, and a difference of 2.8°C and 6.3°C was observed in TESA100 and Z/P100 samples. Thermal conductivity was found to be 0.78 W/mK and thermal energy storage performance of the TESA mortar was also conducted in laboratory sized cell test. The results showed that the TESA100 was able to reduce the maximum temperature by 22% during the heating period when

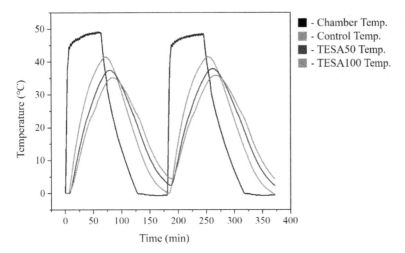

Fig. 17. Thermal energy storage performance results of TESA samples. Reprinted with permission from Ref. 89. Copyright 2021 Elsevier.

compared to control cell specimens. These results indicated the potential use of TESA for wall plastering cement mortar as it was able to reduce indoor temperature when heated and temperature fluctuations.

Wang *et al.* reported the *in situ* polymerization via ultrasonic dispersion used to prepare the microcapsule containing MEPCMs with a high photothermal conversion efficiency.[36] CA was the PCM and nano-SiC modified MUF resin as the wall material of the microcapsule.[36,90] The composite presented good cross-linking between nano-SiC and the MUF shell. Superior thermal conductivity and thermal storage properties were observed with the addition of nano-SiC in micro-PCMs. Various loadings of nano-SiC were prepared, namely MicroPCMs-2%, MicroPCMs-4%, MicroPCMs-6% and MicroPCMs-8%. DSC analysis indicated successful encapsulation and no chemical reaction between CA and the copolymer shell. From TGA study, the appropriate nano-SiC amount in the microcapsules had a higher maximum decomposition temperature of 415.2°C, indicating higher thermal stability. MicroPCM-6% had the best performance in micro-PCMs with 65.7% encapsulation efficiency and thermal conductivity of 0.2265 $Wm^{-1}K^{-1}$ which was increased by 59.2%. With a proper amount of SiC, there is more effective leakage prevention since the nano-SiC properly fills the tiny holes in the MUF shell, in addition to performing well in photothermal conversion. Under light irradiation, nano-SiC effectively captures and absorbs photons through internal molecular vibration.[91] MicroPCMs-6% and MicroPCMs-8% had exceptional photothermal conversion properties with rates of 74.4% and 71.1%, respectively, as calculated from the DSC energy storage values of 195.60 J and 184.12 J.[92] The illumination energy of the micro-PCMs from the transition phase is shown in Fig. 18, where it was observed that MicroPCMs-6% had the highest energy output and photothermal efficiency. Therefore, micro-PCMs may be potentially useful in solar energy utilization and thermal energy storage.

Fei *et al.* investigated how hydrophobic-SiC (H-SiC) dosage would affect the mechanical properties and thermal conductivity of H-SiC modified melamine-formaldehyde (MF) microcapsules.[85] *In situ* polymerization was used to prepare a novel MEPCMs, with a Pa core and H-SiC modified MF resin shell.[84,93] Characterization of the modified microcapsules was done with FTIR, SEM and EDS while their thermal properties were conducted using DSC, TGA

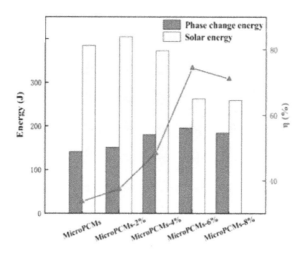

Fig. 18. Energy diagram and photothermal conversion diagram for different micro-PCMs. Reprinted with permission from Ref. 36. Copyright 2021 Elsevier.

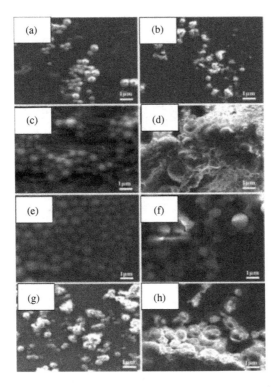

Fig. 19. SEM images of H-SiC modified MF microcapsules with varying H-SiC dosages: (a) 0%, (c) 1%, (e) 2%, (g) 3%, and SEM images of H-SiC modified MF microcapsules with varying H-SiC dosages after pressing: (b) 0%, (d) 1%, (f) 2%, (h) 3%. Reprinted with permission from Ref. 85. Copyright 2020 Elsevier.

and thermal conductivity meter. 0%, 1%, 2% and 3% H-SiC were used, in which 2% H-SiC was found to have the best performance. The results from SEM as shown in Fig. 19 indicated less occurrence of rupture under pressure since its spherical structure was maintained, with the different images indicating different H-SiC dosages. There was a 10% decrease in melting permeability after pressing due to even dispersion from the nano-toughening principle and stress dispersion, causing shell resistance to externally applied forces to be toughened. The thermal conductivity was increased by 55.82% and the melting enthalpy was found to be 93.21 Jg^{-1}

obtained from DSC. TGA and DTG proved the modified microcapsules had their excellent thermal stability, with the Pa volatilization temperature of 282°C and the decomposition temperature of 382°C. This can be attributed to the uniform distribution of H-SiC in the polymer shell, enhancing the physical and mechanical properties of the latter. FTIR and EDS showed successful incorporation and uniform distribution of the modified microcapsules on the microcapsule surfaces.

Zhao et al. reported the synthesis and characterization of a novel microcapsule with n-OD as the PCM and PMF/SiC shell with heightened thermal conductivity and light-heat performance to be applied in NIR light harvesting, light-heat conversion, thermal energy storage and heat-transfer enhancement.[84] Successful fabrication of hybrid microcapsules was achieved through in situ polymerization in an oil-in-water emulsion synergistically stabilized by sodium salt of styrene-maleic anhydride copolymer and SiC. 0%, 1%, 3%, 5% and 7% nano-SiC were used.[42] The microcapsules presented regular spherical morphology as evidenced by SEM. EDS-XRD, FTIR and X-ray powder diffraction confirmed the chemical compositions and crystalline structures of the resultant microcapsules, showing that SiC nanoparticles were embedded in the PMF shell.[94,95] The high thermal stability was observed by TGA, with a maximum decomposition temperature of 395.9°C for the material with 7% nano-SiC. DSC analysis showed that hybrid microcapsules could sustain effective core content and high encapsulation efficiency despite having reduced absolute phase change enthalpies at 168 Jg^{-1} for 7% nano-SiC compared to pure n-OD, which has a high phase change enthalpy of 250.6 Jg^{-1}. Nano-SiC incorporation inhibited supercooling crystallization because of the suitable nucleation sites from the hybrid microcapsule inner walls. The thermal transfer properties of the PMF/SiC hybrid microcapsules were enhanced as compared to non-hybrid microcapsules and the thermal conductivity experienced a 60.34% increase with 7% nano-SiC. Effective NIR light absorption and good photothermal conversion performance under light radiation were observed in the synthesized

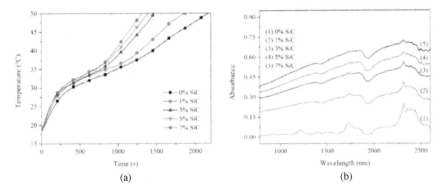

Fig. 20. The time-temperature curves (a) and the NIR absorption spectra (b) of the micro-PCMs with different nano-SiC levels. Reprinted with permission from Ref. 84. Copyright 2018 Elsevier.

hybrid microcapsule. Initially, the temperature for all samples increased quickly before the occurrence of a slow-temperature-change platform that indicated phase change thermal energy storage in the process of core material melting as seen in Fig. 20(a). The temperature of the hybrid samples then rose again, with the microcapsule containing 7% nano-SiC rising to 10.7°C under the same exposure time. The NIR light absorption for hybrid microcapsules was significantly higher than that of PMF micro-PCMs, and therefore nano-SiC incorporation heightens effective light harvesting (Fig. 20(b)).

Zhu *et al.* applied the spouted bed CVD to encapsulate mPCMs.[96] Despite being promising high-temperature PCMs with high heat storage density and high heat exchange rate for high-temperature heat storage, encapsulation of mPCMs is necessary due to its being highly corrosive to encapsulation materials, as well as its volume expansion during phase change.[97,98] With CVD, a novel heat storage capsule was developed utilizing an iron core encapsulated with porous pyrolytic carbon (PyC) layer, dense PyC layer and dense SiC layer in turn from inside to outside as shown in Fig. 21(a). The porous PyC layer accommodates the volume expansion of the iron core and the dense SiC layer acts as a package and oxidation resistance layer. The SiC/C-shells/Fe-core capsules have a working temperature up to 1100°C. A solid–solid phase change (48.85 Jg^{-1}, 686°C) and solid–liquid phase change (128 Jg^{-1}, 1136°C) were also observed from the DSC curves.[99] The results showed that this material possesses high thermal conductivity owing to its physical properties, in addition to the high thermal conductivity of PyC and SiC at increased temperatures relative to metallic package materials; this property contributes to the high thermal density in tandem with the low densities of the PyC and SiC layers.[100] SEM displays the iron capsule cross-section morphology in Figs. 21(b) and 21(c), where full encapsulation of the iron balls by the layers of SiC and PyC was present, as well as maintained density and integration observed with EDS owing to the superior oxidation resistance of SiC. The porous carbon was dissolved by the iron core during melting with increased cycling times, resulting in intermittent holes appearing at the iron core–carbon

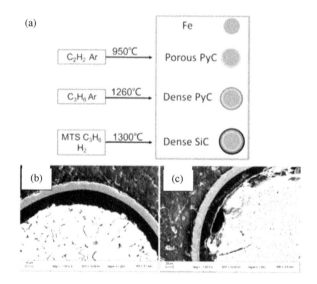

Fig. 21. (a) Schematic diagram of the encapsulation process. Change in cross-sectional morphology for capsules with differing charge–discharge cycles ranging from 1100 to 1200: (b) 100 cycles, (c) 1000 cycles. Reprinted with permission from Ref. 96. Copyright 2017 Elsevier.

shell interface.[101] These holes provide space for expansion in core volume for the subsequent cycles. Therefore, effective thermal cycling can be observed with an iron core packaged with porous PyC, dense PyC and dense SiC layers.

To resolve the drawback of form stability during phase transition and enhance the slow PEG[102] heat transfer rate, Li et al. prepared polyethylene glycol (PEG)-enwrapped SIC nanowires (SiC NWs) network/expanded vermiculite (EVM) form-stable composite PCMs (PSE FS-CPCMs).[103] PSE0, PSE1.05, PSE2.13 and PSE3.29 were prepared, in which PSE3.29, containing 3.29 wt.% SiC NW, displayed a maximum PEG adsorption ratio up to 73.12 wt.%. The SiC NWs filler could improve the heat transfer of PSE FS-CPCMs. The confinement effect of EVM nanoscale pore structures and the interactions between the EVM and SiC NW surfaces greatly contributed to the thermal energy storage performance of PSE FS-CPCMs. Figure 22 depicts SEM of the different PSE samples and EDS findings. SEM mapped the images of EVM, PSE0 and PSE3.29, showing that the melted PEG and the PEG-enveloped intertwining SiC NWs were evenly embedded and dispersed in the EVM pores and surfaces. EDS verified successful enwrapping of PEG due to the higher carbon level (69.9 wt.%). Synergy between the pore structures of EVM and SiC NW surfaces overcame flowability. Leakage of melted PEG and EVM was prevented due to the large surface area resulting in good PEG absorption by SiC NWs. The thermal conductivity of PSE3.29 was 8.8 times higher than PEG at 0.53 $Wm^{-1}K^{-1}$. TGA and DTG evaluated the PSE FS-CPCM thermal stability, with PSE3.29 showing a weight loss percentage of 73.91 wt.% that approximately corresponded with PEG weight loss of 99.92 wt.%, further indicating even distribution. PSE FS-CPCMs had desirable thermal stability below 110°C as there was no notable decomposition at this range. The PSE2.13 DSC curves showed that it possesses good thermal reliability within a minimum of 200 phase change cycles, and the dip in latent heat (4.14% and 3.77% for melting and solidification, respectively) was feasible for applications in thermal energy storage. PSE2.13

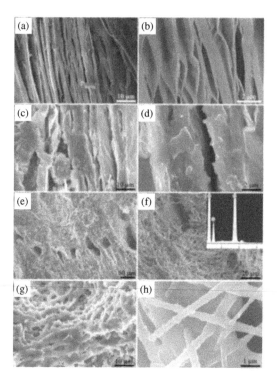

Fig. 22. SEM images for different magnifications of (a), (b) EVM, (c), (d) PSE0, (e)–(h) PSE3.29, and ((f) inset) the corresponding EDS spectrum. Reprinted with permission from Ref. 103. Copyright 2018 Elsevier.

presented good chemical compatibility after 200 cycles since the absorption peaks were maintained.

2.4. Silica/silicon dioxide

Unlike other organic polymers containing carbon bones,[104–106] SiO_2 is an important linking agent. There are numerous reports on silica or SiO_2 as a component in PCM encapsulation. Hassabo et al. noted the inertness, stability and biocompatibility of silica, as well as its nontoxic quality, therefore deducing it as a superior solid matrix for other applications.[107] The silica shell was reported to possess high thermal and chemical stability, as well as the ability to decrease any present flammability from organic components,[32,33,108] Furthermore,

Table 3. Summary of silicone carbide-containing encapsulation and its parameters.

Additive	PCM	Methods of preparation	Latent heat (Jg^{-1})	Melting point (°C)	Unique application	Efficacy (%)	Ref.
—	AlSi$_{12}$	Liquid silicon infiltration	481	557	Battery electric vehicles	—	86
Concrete	Paraffin	Impregnation and coating methods	—	—	Energy storing concrete aggregates	—	87
Graphite	Paraffin	Vacuum impregnation	180.5	51	Flexible shape design	—	88
Epoxy, silica fume	Paraffin wax	Impregnation and coating	19.76	36.46	Reduce indoor temperature fluctuations	—	89
MUF	CA	Polymerization via ultrasonic dispersion	97.80	33.16	Photothermal conversion	—	36
MF	Paraffin	In situ polymerization	93.21	56.44	—	—	85
PMF	n-OD	In situ polymerization in an oil-in-water emulsion	168	10–29.5	Light-heat performance	—	84
Encapsulated with porous PyC	Iron alloy balls (Q195)	CVD	48.85, 128	686, 1136	—	—	96
EVM	PEG	Stirring and impregnation	81.88	57.1	—	—	103

silica has also found its uses in curbing liquid-phase micro-PCM dissolution at a dispersion in organic HTF with the capability to withstand degradation under high temperatures.[109] Different methods such as polymerization, Pickering emulsion, sol–gel technique and more were employed to fabricate the encapsulation shell.

Hassabo et al. synthesized pure inorganic silica capsules containing inorganic hydrated salt by crosslinking with poly(aminoorganosiloxane)s (PAOS) using a water-in-oil emulsion method based on Pickering emulsion.[107] Six different inorganic hydrated salts were utilized for evaluation in the silica-based microcapsules: calcium nitrate tetrahydrate ($Ca(NO_3)_2 \cdot 4H_2O$), calcium chloride hexahydrate ($CaCl_2 \cdot 6H_2O$), sodium sulphate decahydrate ($Na_2SO_4 \cdot 10H_2O$), disodium hydrogen phosphate dodecahydrate ($Na_2HPO_4 \cdot 12H_2O$), ferric nitrate nonahydrate ($Fe(NO_3)_3 \cdot 9H_2O$) and manganese (II) nitrate hexahydrate ($Mn(NO_3)_2 \cdot 6H_2O$). Optimal results were observed for sodium sulphate decahydrate and disodium hydrogen phosphate dodecahydrate due to their respective melting temperatures of 32°C and 34.5°C, as well as their higher enthalpies of fusion at 78.20 kJmol^{-1} and 94.9 kJmol^{-1}, respectively.[110] Silica was a useful material due to it being less reactive, stable, biocompatible and relatively safe. The microcapsules displayed superior phase change properties; DSC analysis indicated lower latent heat in both inorganic salts when incorporated with silica compared to those of the pure salts, with sodium sulphate decahydrate and silica at 38.6 Jg^{-1} and 35.4 Jg^{-1} for the first and second heating, respectively, while disodium hydrogen phosphate dodecahydrate and silica recorded values of 69.3 Jg^{-1} and 70.1 Jg^{-1} for the two heatings. SEM in Fig. 23(b) verified that the particle size fits the micrometer range. Such inorganic PCMs can be applied in textiles and 5% PCM-silica capsules were embedded in a polypropylene (PP) film, a strongly flexible polymer commonly used in interior textiles, to form an inorganic PCM polymer composite. The main mechanical properties for the novel material are similar to those of pure PP. Further DSC analysis demonstrated a small effect for the PCM capsule-sodium sulphate decahydrate endothermic peak at phase transition, i.e., melting at approximately 53°C. From SEM

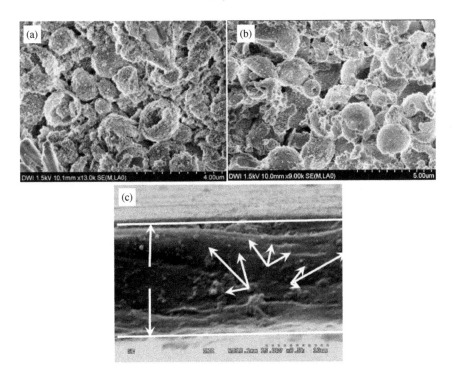

Fig. 23. (a), (b) SEM image for encapsulation of inorganic PCM in silica, (c) SEM image of cross-section of PP film with 5% PCM capsules. Reprinted with permission from Ref. 107. Copyright 2015 Trade Science Inc.

in Fig. 23(c), the microcapsules were homogeneously distributed in the PP film, and therefore, there is satisfactory compatibility between the polymer matrix and the additive.

Ram et al. reported the use of dimethyl terephthalate (DMT) as a PCM to improve the performance of a synthetic oil in an estimated temperature interval of 100°C to 170°C.[109] Micro-PCMs are suggested to enhance thermal conductivity and thermal energy storage ability of a HTF. SiO_2 was the microencapsulant owing to its superior containment capabilities and thermal stability.[111,112] The resulting SiO_2-coated DMT micro-PCM was examined and studies were conducted to investigate its appropriateness for HTF performance. SEM findings in Fig. 24(a) illustrated the morphology, in which 2-propanol and water were chosen as the mixture for the

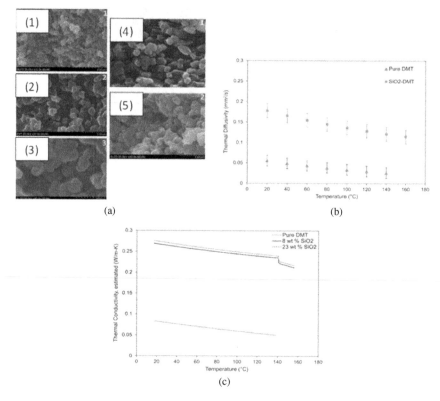

Fig. 24. (a) (1)–(3) SEM images for SiO$_2$-coated DMT micro-PCMs prepared in 2-propanol and water at different magnifications and (4), (5) SEM images for various magnifications of SiO$_2$-coated DMT micro-PCMs prepared in 2-propanol/water mixture. (b) Thermal diffusivity against temperature for pure and SiO$_2$-coated DMT, with extrapolated fit for SiO$_2$-coated DMT shown in dashes. (c) Thermal conductivity estimates based on thermal diffusivity values obtained and known thermophysical properties of DMT and SiO$_2$. Reprinted with permission from Ref. 109. Copyright 2016 Wiley.

fabrication of SiO$_2$-coated DMT micro-PCMs due to the smaller particle sizes obtained. The high thermal diffusivity values in Fig. 24(b) via laser flash concluded a higher heat transfer in the material for SiO$_2$-coated DMT compared to pure DMT. TGA indicated that roughly 8–23 wt.% of the MEPCM was occupied by the SiO$_2$, after which subsequent thermal conductivity estimates for 8% and 23% SiO$_2$ in Fig. 24(c) demonstrated only minimal shift due to

SiO$_2$ proportion change; encapsulated PCM generally had higher thermal conductivity than DMT. The values were derived from the thermal diffusivity values and known thermophysical properties of DMT and SiO$_2$.[113,114] XRD measurements presented the diminishing of the sharp peak intensity of DMT because of SiO$_2$ coating, with the X-ray pattern similar to amorphous SiO$_2$ particles.[115] From DSC, the coated micro-PCMs had lower latent heat than pure DMT, at 96.2 Jg^{-1} and 118 Jg^{-1} compared to the literature value for DMT of 163 Jg^{-1}.[113] DSC proved the ability of the particles to enhance HTF at 100–170°C, even beyond this range when using pressurized pipes or vessels. The results showed that thermal conductivity and the thermal energy storage heat capacity for the HTF would be enhanced by addition of micro-PCM. The PCM latent heat will significantly enhance the fluid thermal energy storage capacity. Full dispersion was observed for SiO$_2$-coated DMT micro-PCM in the synthetic oil, Therminol SP, silicone oil at and beyond 100°C.

Lee *et al.* utilized the sol–gel method to carry out the microencapsulation of SA, a PCM applied in energy storage.[50] Encapsulation was done using SA of varying amounts (5, 10, 15, 20, 30 and 50 g) taken against 10 mL tetraethyl orthosilicate (TEOS), labeled as SATEOS1 to SATEOS6, respectively. FTIR, XRD, XPS and SEM analyses were performed to characterize the synthesized MEPCM. FTIR inference and XRD analysis indicated the complete encapsulation of the SA core with the SiO$_2$ shell, with the latter demonstrating that sample crystallinity experienced subsequent decrease.[116,117] SEM in Fig. 25 illustrates the MEPCM morphology; observation of SATEOS1 indicated the deposition of SiO$_2$ particles and agglomeration, with the shell providing mechanical strength to the SA core and leakage prevention, suggesting possible application in energy storage. SA content variation as in SATEOS1, SATEOS2 and SATEOS3 was found to influence the particle size distribution, efficiency and shape with constant surfactant amount, with SATEOS6 having proper microcapsule formation compared to a MEPCM with a lower SA content. EDS showed that the Si amount decreased from SATEOS1 (12.34%) to SATEOS6 (2.68%) with

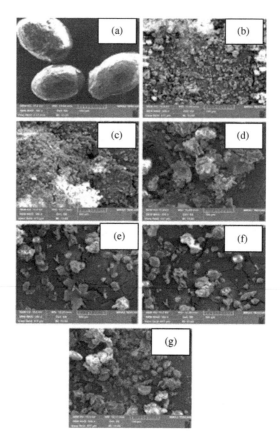

Fig. 25. SEM images for (a) SA (at 100), (b) SATEOS1, (c) SATEOS2, (d) SATEOS3, (e) SATEOS4, (f) SATEOS5, (g) SATEOS6. Reprinted with permission from Ref. 50. Copyright 2020 Scientific Reports.

increase of the SA amount, and therefore this increase resulted in less SiO_2 deposition on the SA surface. SATEOS6 had the highest melting and solidifying temperatures of 70.37°C and 64.27°C, respectively. The temperature interval of 6.10–8.37°C between the melting and solidification processes showed that MEPCM was potentially useful in energy storage because of the superior thermal conductivity of the SiO_2 shell.[118] SATEOS6 had the highest enthalpy among all MEPCMs; this can be attributed to best encapsulation, with the overall highest encapsulation ratio (90.86%) and efficiency (86.68%). TGA displayed the improved thermal stability through the formation of SiO_2 layers from SA decomposition at 415°C. DSC of 30 cycles demonstrated that SATEOS6 possessed the highest latent heats of melting and solidification at 182.53 Jg^{-1} and 160.12 Jg^{-1}, respectively, as well as significantly lower fluctuation relative to bulk SA.[119,120]

Luo et al. applied polydopamine (PDA)-SiO_2 hybrid shell material and n-OD as the PCM to synthesize a novel NEPCM, i.e., n-OD@SiO_2-PDA (NEPCM-50-4), in which the PDA layer had a very tunable thickness and was prepared from oxidative dopamine (DA) self-polymerization in Tris-HCl buffer solution on preformed NEPCMs with a SiO_2 shell.[121,122] XPS analysis was performed and the results indicate successful encapsulation, as well as the presence of an oxidative self-polymerization mechanism. DSC thermograms reported the phase change properties, in which PCM supercooling was found to be undesirable thus, a nucleating agent (n-octacosane) was added.[123,124] The increase in PDA layer thickness results in the decline in phase change enthalpies of NEPCMs with PDA-SiO_2 hybrid shell. Thermal stability of the NEPCMs was notably enhanced compared to that of the unencapsulated n-OD, as indicated by TGA. SEM images illustrated good dispersion in the morphology of thermo-regulating rigid polyurethane (RPU) foams with NEPCMs, in addition, n-OD@SiO_2-PDA exhibited better mechanical strength. Subsequent DSC confirmed that RPU foam with n-OD@SiO_2-PDA had higher phase change enthalpy than that containing only n-OD@SiO_2 due to better leakage prevention properties; the maximal melting and solidifying enthalpies of the composite after the addition of 20.37 wt.% of n-OD@SiO_2-PDA was 23.24 Jg^{-1} and 24.77 Jg^{-1}, respectively. The mechanical properties of the composites were examined and the compressive strength (σ) and compressive modulus (E) displayed a small drop with the increase of NEPCM content from 7.87 wt.% to 20.37 wt.% shown in Fig. 26(a). Likewise, the compressive stress–strain curves in Fig. 26(b) demonstrate the enhanced ductility and compressive properties in composite foams with 14.59 wt.% and 20.37 wt.% n-OD@SiO_2-PDA. Therefore, incorporation of the NEPCMs containing PDA-SiO_2 into thermoregulating RPU with good mechanical properties is feasible

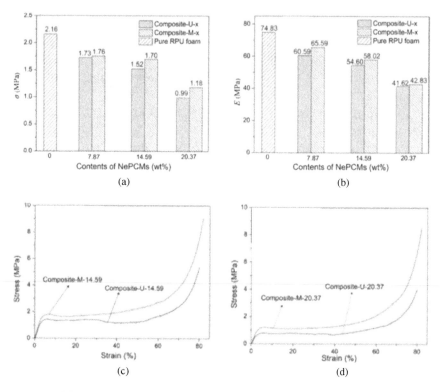

Fig. 26. (a) Compressive strength, (b) compressive modulus for the thermoregulating RPU foam with different NEPCMs. Compressive stress–strain curves of composite foams with (c) 14.59wt.% and (d) 20.37wt.% n-OD@SiO$_2$-PDA. Reprinted with permission from Ref. 122. Copyright 2019 Elsevier.

because of enhanced interfacial compatibility and interactions between the PDA layer and polyurethane matrix.[125]

Liang et al. reported the synthesis of a new polymer-SiO$_2$ hybrid shelled NEPCMs in one pot with n-OD core through the sequential execution of interfacial hydrolysis-polycondensation of alkoxy silanes and the radical polymerization of vinyl monomers.[126,127] The organic–inorganic hybrid materials are potentially useful for PCM encapsulation to achieve exemplary capsule properties.[128,129] Three samples of NEPCMs with polystyrene (PS)-SiO$_2$ shell and two NEPCMs with poly(hydroxylethyl methacrylate) (PHEMA)-SiO$_2$ were prepared. FTIR verified the nanocapsule chemical composition and effective synthesis of organic–inorganic hybrid shell materials; XRD showed complete n-OD encapsulation and the identical triclinic crystal structure in both pure n-OD and nanocapsules.[130] Observations from TGA indicated that increasing vinyl monomer addition resulted in a gradual rise in nanocapsule polymer content that corresponded with the decrease of n-OD and SiO_2 content. DSC concluded that supercooling performance is significantly influenced by the polymer type. The NEPCM phase change enthalpies experience decline with increased vinyl monomer usage. Thermal cycling tests revealed that the n-OD@SiO_2 melting and solidifying enthalpies decreased by respective values of 9.0% and 8.7%, yet NEPCMs with incorporated hybrid materials, experienced a less intense drop, proving the thermal energy storage capability of the latter. Additional thermal cycles heightened heterogeneous nucleation due to the nanocapsule inner shells. Polymer-enhanced NEPCMs also possessed better thermal conductivity, with 5 mL styrene (St) or hydroxylethyl methacrylate (HEMA) providing the maximal increase of 15.4% and 14.7%, respectively. Leakage prevention was also improved, with PHEMA-SiO_2 being marginally inferior to PS-SiO_2 as seen in Fig. 27(a) from its higher relative weight loss at heating. The compressive load obtained from Figs. 27(b) and 27(c) yielded impressive results of >34.6 μN for NEPCMs with PS-SiO_2 shell, and 65 μN for that with PHEMA-SiO_2 shell, compared to the low value of 14.7 μN for that with merely a SiO_2 shell.

Su et al. investigated the synthesis and characterization of fabricated MEPCMs using PMF as the shell, with nanosilicon dioxide hydrosol as a surfactant for n-OD core encapsulation due to its notable thermal stability and superior combination properties with organic and inorganic PCMs.[30] Unlike SiO_2, PMF is a type of polymer made from carbon bones, similar to many other polymers.[131–134] This is an alternative to current MEPCM which possesses low thermal conductivity and low mechanical strength. Nanosilicon dioxide hydrosol was employed as a surfactant to prepare oil-in-water (O/W) emulsion and for thermal enhancement during PCM encapsulation. Three samples, i.e., MF-1, MF-2 and MF-3 were prepared. Results demonstrated better particle dispersion and

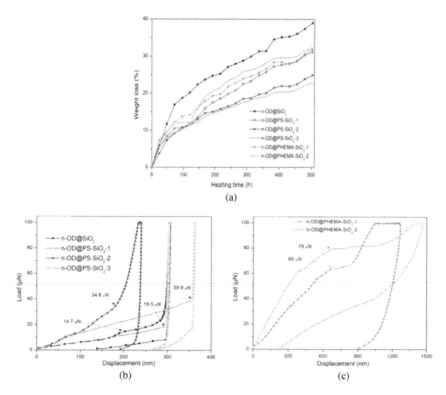

Fig. 27. (a) Leakage prevention of NEPCMs with different shell materials (weight loss against heating time). Load–displacement curves for NEPCMs with different shell materials: (b) n-OD@SiO$_2$ and n-OD@PS-SiO$_2$, (c) n-OD@PHEMA-SiO$_2$. Reprinted with permission from Ref. 126. Copyright 2018 Elsevier.

shell integrity, in addition to successful encapsulation for MF-1 and MF-2 as shown in the SEM images in Figs. 28(a) and 28(b), while MF-3 showed some deformation in Fig. 28(c). Particle size distribution and base material content were affected.[135,136] Nucleating agents, such as ammonium chloride, influenced morphology. EDS verified the elements present and successful retainment of nanosilicon dioxide particles, which contributed to the significantly improved capsule thermal conductivity of 1.4 Wm^{-1}K^{-1} compared to those of n-OD and PMF.[137] MF-2 exhibited the best performance, with a notable thermal stability temperature rise of

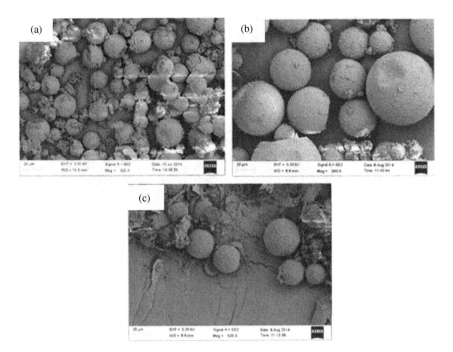

Fig. 28. SEM images for (a) MF-1, (b) MF-2, (c) MF-3. Reprinted with permission from Ref. 30. Copyright 2018 Advance Access Publication.

around 78°C from TGA, as well as relatively higher core material content of 8–25%.

He *et al.* used two sol–gel assisted methods to synthesize a new shape-stabilized phase change material (SSPCM), labeled as SSPCM1 and SSPCM2 based on temperature-assisted and $CaCl_2$-assisted techniques respectively.[138] Polyethylene glycol/SiO_2 (PEG/SiO_2) composites were prepared. The composite with 80 wt.% PEG did not exhibit leakage and remained as a solid above the PEG melting point. XRD analysis inferred that SiO_2 and $CaCl_2$ negatively impacted ideal PEG crystallization as it resulted in crystal structure transformation and significant PEG phase change. From FTIR, PEG and SiO_2 exhibited physical interactions. The PEG free movement was restrained due to the coordinate bonds formed during PEG and $CaCl_2$ chemical action. Polarizing optical microscopy

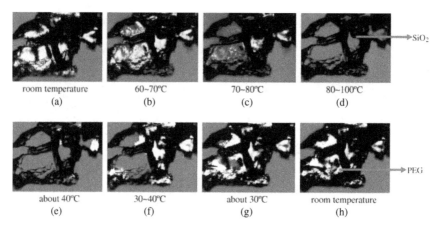

Fig. 29. Phase change behaviors of SSPCM2: (a)–(d) the heating process; (e), (f) the cooling process. Reprinted with permission from Ref. 138. Copyright 2014 Elsevier.

(POM) showed PEG dispersion into the mesoporous silica structure and the formation of stable core-shell structures during the heating (~100°C) (Figs. 29(a) –29(d)) and cooling process (Figs. 29(e) –29(f)). The two synthesis methods were then compared in terms of their impact on phase change characteristics, in which the temperature-assisted technique proved to be superior. DSC presented a higher phase change enthalpy with the temperature-assisted method relative to $CaCl_2$ assistance, recording values of 122.0 Jg^{-1} and 118.3 Jg^{-1}, respectively, for the melting and crystallizing enthalpies. TGA presented better thermal stability for the SSPCM fabricated through the temperature-assisted method, with the final SSPCM2 degradation temperature at 412.6°C, higher than that of PEG, and therefore SSPCM2 possesses potential application in a wide temperature range.

Lee et al. prepared a PCM microcapsule by utilizing lauric acid (LA) as the core material and TEOS as the precursor for the SiO_2 shell formation.[139] They prepared LA/SiO_2 micro-PCM samples with varying amounts of LA, i.e., 5, 10, 15, 20, 30 and 50 g, which were labeled as LAPC-1–LAPC-6. The pH was maintained at 2.5 for the microencapsulated LA composition as agglomeration

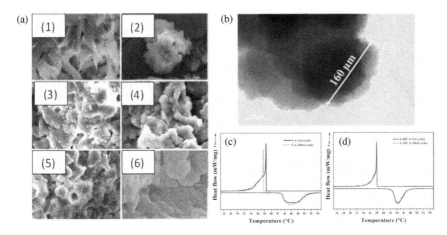

Fig. 30. (a) SEM images for (1) LAPC-1, (2) LAPC-2, (3) LAPC-3, (4) LAPC-4, (5) LAPC-5, (6) LAPC-6, (b) TEM image for LAPC-6. Thermal cycles for (c) LA, (d) LAPC-6 at the 1st and 30th heating and cooling cycles. Reprinted with permission from Ref. 139. Copyright 2021 Elsevier.

from Van der Waals forces caused uniform dense SiO_2 layer formation on the LA surface.[140,141] FTIR, XRD, XPS, SEM and TEM analyses were used to characterize the LA/SiO_2 samples. As shown in Fig. 30(a) and 30(b), SEM and TEM confirmed complete SiO_2 encapsulation in LA and agglomeration in samples with low LA amount.[42] Smooth and compact morphology at low pH contributed to the PCM mechanical strength and leakage proof properties at higher temperatures.[142] LAPC-6 demonstrated microencapsulation with good definition and exhibited the highest melting temperature of 44.295°C and latent heat of melting of 160.91 Jg^{-1}. It also has a solidifying temperature and latent heat of solidification of 39.009°C and 152.82 Jg^{-1}, respectively, showing its potential as a high latent heat storage material.[143] DSC showed that LAPC-6 had optimal performance in terms of encapsulation efficiency at 96.50% and encapsulation ratio at 96.15%. TGA presented superior thermal stability of LA/SiO_2 compared to pure LA, and LAPC-6 experienced the relative greatest weight loss percentage of 82.80% at approximately 288°C. As illustrated in Fig. 30(c), thermal cycling tests indicated insignificant and negligible changes in the thermal

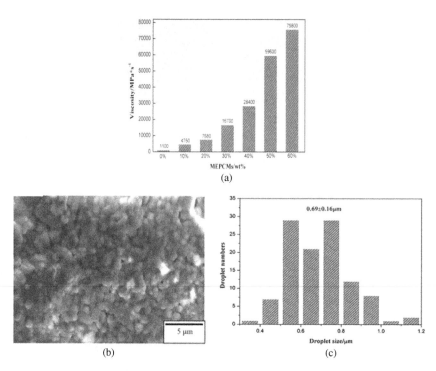

Fig. 31. (a) MEPCM wt.% effect on PUR-MEPCM viscosity. (b) The SEM at 5000 times of MEPCM and (c) the mean droplet size of MEPCM. Reprinted with permission from Ref. 144. Copyright 2019 Wiley.

properties like melting and solidifying temperatures, in addition to latent heat, proving its thermal reliability.

Li *et al.* reported *in situ* foaming of PU prepolymer blended with MEPCM to synthesize polyurethane rigid foam composites (PUR-MEPCM).[144] The MECPCM is composed of a n-OD core and SiO_2 shell using chemical precipitation. Six samples were prepared, i.e., pure PU, 10 wt.% PUR-MEPCM to 50 wt.% PUR-MEPCM. FTIR characterization demonstrated the absence of chemical interactions between the MEPCM core and shell, while XRD inference found OD segment motion limitation causing a decrease in crystal diffraction intensity. MEPCM addition to PU increased its viscosity by a significant amount as shown in Fig. 31(a). SEM images in Fig. 31(b) showed successful encapsulation for the MEPCM and that the average diameter is 0.69m (Fig. 31(c)). DSC recorded a lower melting and crystallization enthalpy of 105.4 Jg^{-1} and 106.8 Jg^{-1} in the MEPCM relative to n-OD, in addition to a calculated encapsulation rate of 47.5%. TGA presented a 20°C rise in decomposition temperature for encapsulated OD, confirming that the silica shell plays a protective role to enhance MEPCM core-shell thermal stability. Subsequent DSC performed on PUR-MEPCM indicated that a higher amount of MEPCM in the composite would result in higher latent heat. A heating and/or cooling device verified the thermal control capacity of PUR-MEPCM, specifically 50 wt.% PUR-MEPCM, which possessed a melting enthalpy and crystallized enthalpy value of 56.3 Jg^{-1} and 53.1 Jg^{-1}, respectively, besides a stable temperature interval of 26–28°C. Therefore, PUR-MEPCMs are potentially applicable in temperature regulation.

Bai et al. investigated SA nanocomposite PCMs with silica coating synthesized with TEOS and SA using the sol–gel technique.[34] Composite PCMs are widely applicable in processes involving energy saving and energy efficiency, particularly, NEPCMs which can be tailored to recover or store waste heat. FTIR showed the absence of any chemical reaction between SiO_2 and SA. TEM and field emission SEM (FE-SEM), shown in Figs. 32(a) and 32(b), verified the formation of monodisperse microspheres containing a

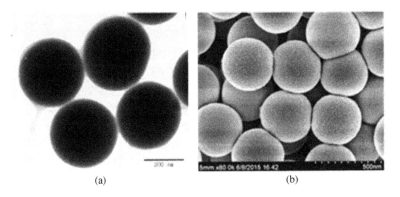

Fig. 32. (a) TEM image, (b) SEM photograph (80kX), of the NEPCM. Reprinted with permission from Ref. 34. Copyright 2016 Springer.

stable core shell arrangement along with SA encapsulation in the SiO_2 shell. The silica coating served to enhance the mechanical strength of the composite. DSC results demonstrated favorable heat capacity in the composite PCM, with the NEPCMs possessing 33% SA encapsulation ratio having a respective lower melting temperature and latent heat of 67.25°C and 66.4 kJ^{-1} compared to pure SA, indicating lower phase change temperature and better thermal conductivity. It can be inferred that SiO_2 enhances thermal conductivity in composite PCMs.

Xu et al. synthesized colored microcapsules and investigated their applications in architectural interior wall coating. Preparation of the Pa@SiO_2 microcapsules was achieved through reactive dye-grafted SiO_2 shell to encapsulate Pa via interfacial polymerization and chemical grafting.[43,145] SEM analysis in Fig. 33(a) demonstrated regular spherical and precise core-shell structure. FTIR verified that the amorphous SiO_2 shell and reactive dyes were present, along with successful grafting of the dye on the microcapsule surface. XRD showed high crystallinity in Pa. The absence of chemical interactions between Pa and SiO_2 ensured the thermal stability of pristine Pa. The Pa@SiO_2 colored microcapsules had higher thermal conductivity relative to pure Pa by 62% at 0.7011 $W^{-1}K^{-1}$. DSC analysis presented high encapsulation efficiency of 30% and good latent heat storage capacity in the colored microcapsules, with lower latent crystallization and melting enthalpies of 60.49 Jg^{-1} and 59.03 Jg^{-1}. Solar energy thermal storage ability tests in Fig. 33(b) showed that the phase change end temperatures of Pa@SiO_2 colored microcapsules recorded a value of 27.9°C, and the latent heat release during crystallization resulted in the plateau in the 28–22°C temperature interval (Fig. 33(c)). From UV–Vis on the Pa@SiO_2 colored microcapsules, superior UV protection properties were observed as displayed in Fig. 33(d). A decomposition temperature of 148°C was found for the colored microcapsule PCM, indicating better thermal stability. Leakage proof properties were also present in the microcapsule samples. Infrared imaging showed that the latex paint had notable temperature-controlled properties. The temperature experienced an approximate decrease of 2.5°C compared to

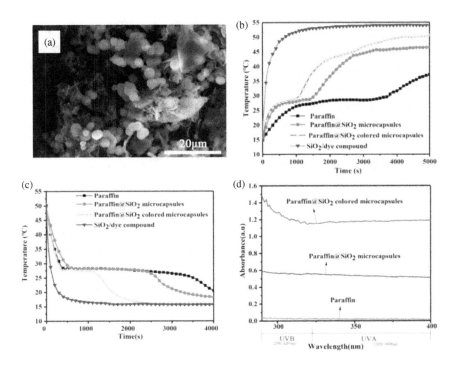

Fig. 33. (a) SEM micrographs for Pa@SiO$_2$ colored microcapsules. Temperature/time curves (b) under simulated solar irradiance of 500 Wm^{-1} and (c) natural cooling process at 15°C for pristine Pa, Pa@SiO$_2$ microcapsules and Pa@SiO$_2$ colored microcapsules and SiO$_2$ nanoparticles, (d) UV–Vis spectra of pristine Pa, Pa@SiO$_2$ microcapsules and Pa@SiO$_2$ colored microcapsules with the same mass fraction. Reprinted with permission from Ref. 43. Copyright 2021 MDPI.

normal interior wall coatings. The Pa@SiO$_2$ colored microcapsules exhibited good performance in solar thermal energy storage, temperature regulation, as well as enhancing the colored fixing capabilities in colored latex paint coating.

Wu et al. applied the sol–gel process to prepare encapsulated Tris(hydroxymethyl)methyl aminomethane (Tris) with a SiO$_2$ shell.[47] Polyalcohol is a notable medium range temperature PCM for thermal energy storage, with solid–solid PCMs free of liquid being more favorable despite some sublimation when crystals enter the "plastic phase".[146,147] The Tris@SiO$_2$ microcapsules were synthesized using the hydrolysis and polycondensation of the bicomponent

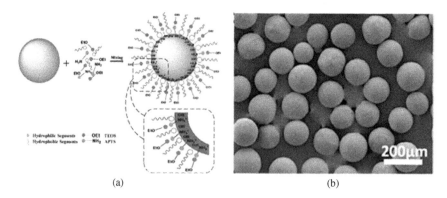

Fig. 34. (a) Schematic illustration for Tris@SiO$_2$ microcapsule preparation, (b) SEM image of the capsule with 18m:7mL TEOS to APTS ratio under 7h reaction time. Reprinted with permission from Ref. 47. Copyright 2015 Elsevier.

silicon precursors added to a mixture tetraethoxysilane (TEOS) and 3-aminopropyl triethoxysilane (APTS) as shown in Fig. 34(a). The SEM micrographs in Fig. 34(b) of products with varying TEOS to APTS ratio highlighted the importance of a proper ratio required to prepare Tris@SiO$_2$, with 20 mL:5 mL of TEOS to APTS giving uniform distribution and higher encapsulation ability. The excellent sealing properties contributed to Tris leakage prevention as evident by SEM results in Fig. 34(b). FTIR and XRD were utilized to characterize the microcapsule. The thermal conductivity of the silica-encapsulated PCM was doubled to 0.478 Wm^{-1}K^{-1}. TGA presented a 67.9% weight ratio for Tris in Tris@SiO$_2$. DSC analysis presented reversible phase change from 110–155 for the Tris@SiO$_2$. The latent heat was observed to have stabilized at around 146 Jg^{-1} at 70 thermal cycles. The Tris@SiO$_2$ capsules are potentially useful for applications in medium temperature thermal energy storage.

Bai et al. fabricated LA/SiO$_2$ NEPCMs of different particle sizes to investigate their effects on heat transfer and long-term usage.[148] NEPCMs possessing high thermal storage density play vital roles in solar energy systems.[149,150] Several samples were prepared, with varying ratios of ammonia-to-TEOS for M1–M5, ethanol-to-TEOS for M6–M8 and water-to-TEOS for M9–M11 for the SiO$_2$ shell material, while LA acted as the core. FTIR and XRD indicated successful LA encapsulation with SiO_2. The latent heat of the nanocapsules was measured to be 165.6 Jg^{-1} and an 85.9% encapsulation ratio. FESEM analysis on the samples showed the general presence of spherical structures in the NEPCMs. TG observation indicated that melted LA degradation was curbed by the silica shell. From DSC results, the ideal nanocapsules presented high thermal storage capabilities, experiencing a latent heat decrease of lower than 10% after 1200 thermal cycles, demonstrating superior thermal reliability with 0.8% damping decrease. Analysis was performed under synthetic conditions to investigate the sample size change, in which the nanocapsules were mainly affected by OH^- concentration as illustrated in Fig. 35 from pH testing, and the encapsulation ratio was impacted by the cumulative percentage of emulsion droplets and shell thickness. Therefore, LA/SiO_2 size control mechanism analysis concluded that the optimal particle size is 340nm, which corresponds to the maximum latent heat of the nanocapsules.

Qin et al. synthesized a novel composite hygroscopic PCM using MEPCM and diatomite, i.e., MPCM/diatomite (CMPCM) material in which the sol–gel method was applied on methyl triethoxysilane (MTES) to fabricate the shell, and the core was composed of an alkane mixture.[151] Diatomite acted as the hygroscopic material. SEM analysis in Fig. 36(a) showed the microcapsule and diatomite morphology, in which the SiO_2 prevented leakage occurrence. DSC analysis presented the thermal properties of the MPCM, showing encapsulation of the microcapsules in the SiO_2 shell and lower latent heats of melting and solidifying for the MPCMs relative to PCM and MPCM, at 19.0 Jg^{-1} and 18.4 Jg^{-1}, respectively. TGA indicated superior MPCM and CMPCM thermal stability because they have a higher starting temperature value for a maximum weight loss at 175°C. From moisture transfer coefficient and moisture buffer value (MBV) analysis on diatomite, CMPCM, gypsum board and wood, it was observed that CMPCM possessed optimal hygroscopic behavior, with a moisture transfer coefficient at 510^{-1} kg/ms%RH contributing to the increased porosity. The CMPCM MBV reflected in Fig. 36(b) recorded a significantly higher value of 1.57 g/m^2%RH

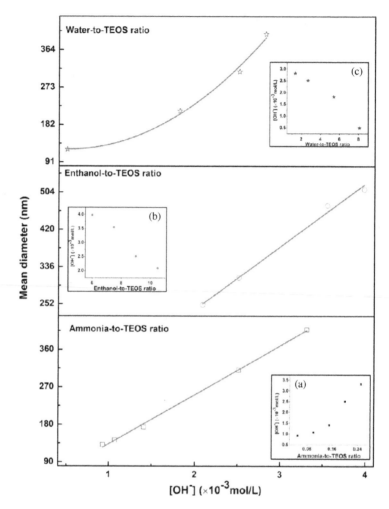

Fig. 35. Diagram of particle size relation to OH⁻ concentration: (a) ammonia-to-TEOS ratio, (b) ethanol-to-TEOS ratio, (c) water-to-TEOS ratio. Reprinted with permission from Ref. 148. Copyright 2019 Elsevier.

compared to the other analytes. Sorption and desorption isotherms in Fig. 36(c) demonstrated almost linear progression in the standard hygroscopic range of 20–85% RH. The hygroscopic PCM can be applied in indoor temperature and moisture moderation, with a calculated energy saving rate of 17.7%.

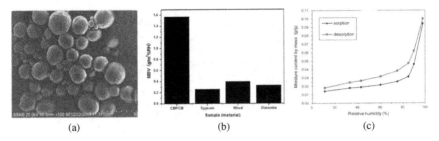

Fig. 36. (a) SEM image for CMPCM, (b) moisture transfer coefficient for CMPCM, diatomite, wood and gypsum, (c) sorption and desorption isotherms for the CMPCM. Reprinted with permission from Ref. 151. Copyright 2015 Elsevier.

Fang et al. utilized the sol–gel technique to fabricate a composite thermal energy storage material through microencapsulation of SA PCM core with a SiO_2 shell, in which the latter prevented leakage of melted SA.[152] SA, MPCM1, MPCM2 and MPCM3 samples were prepared. FTIR and SEM were conducted for microstructure determination of the microencapsulated SA with SiO_2 shell. The SEM images of MPCM1, MPCM2 and MPCM3 in Figs. 37(a)–37(b), 37(c)–37(d) and 37(e)–37(f), respectively, displayed complete encapsulation of SA in the SiO_2 shell. DSC and TGA reported the thermal properties and thermal stability. From DSC, the composite possessed solidifying and melting temperatures of 52.6°C and 53.5°C respectively, while the values for latent heat of solidification and melting were recorded at 162.0 kJkg^{-1} and 171.0 kJkg^{-1}. The SA encapsulation ratio was 90.7% for MPCM3. TGA verified thermal stability improvement with the presence of SiO_2 shell.

Zhu et al. applied a new sol–gel method to synthesize polyethylene glycol/SiO_2 shape-stabilized PCMs (PEG/SiO_2 SSPCMs) of different PEG mass fractions, i.e., 30–90% PEG/SiO_2, in which the PEG was the PCM and silica gel retained a stable composite shape.[153] Instead of coagulant addition, gelatinization was done through temperature adjustment. Stable core-shell structures were observed in the composites through PEG impregnation into multi-mesoporous silica gel.[154,155] The solid state of the composite was maintained even when exceeding the melting temperature of PEG.

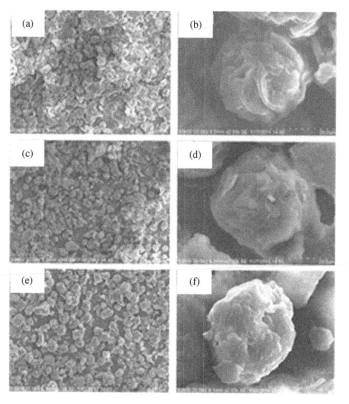

Fig. 37. SEM images for (a) MPCM1 (200X), (b) MPCM1 (2500X), (c) MPCM2 (200X), (d) MPCM2 (2500X), (e) MPCM3 (200X), (f) MPCM3 (2500X). Reprinted with permission from Ref. 152. Copyright 2013 Elsevier.

Physical adsorption occurred between PEG and silica gel with no effect on the PEG crystal structure as concluded by XRD and FTIR, therefore superior phase change was maintained by the PEG and good compatibility was observed. From DSC analysis, the composite enthalpies ranged from 63.4 Jg^{-1} to 128.4 Jg^{-1} (50–80% PEG/SiO_2), proportional to PEG content. Thermal energy capacity was stable as the SSPCM enthalpies experienced negligible change under repeated heating cycles. The composites demonstrated great thermal stabilities indicated by TGA as the decomposition temperatures proved excellent thermal durability of 330–350°C. Thermal conductivity recorded a rise to 0.558 $Wm^{-1}K^{-1}$ through

2.7% mass addition of graphite. Therefore, the composites have potential for thermal energy storage in building envelopes.

Jiang et al. prepared a FSPCM of polyurethane rubber (PU)/n-octadecane (n-OD) @silicon dioxide (SiO_2)-polymethyl acrylate (PHEMA), i.e., n-OD@SiO_2-PHEMA microcapsules using one-pot interfacial hydrolysis polycondensation of alkoxysilanes and HEMA radical polymerization.[156,157] Four composites were prepared: PU/n-OD@SiO_2-PHEMA-40, 60, 80 and 100. The PU/n-OD@SiO_2-PHEMA composites were retrieved through the addition of different wt.% of n-OD@SiO_2-PHEMA to the PU matrix. The SiO_2-PHEMA shell helps to enhance microcapsule/PU interfacial bonding and provides mechanical strength to the composite. FTIR verified the microcapsule functional groups present, while XRD presented the n-OD core encapsulation with a similar triclinic phase structure. SEM and TEM showed full encapsulation of SiO_2 in the core, in addition to successful synthesis of n-OD@SiO_2 and n-OD@SiO_2-PHEMA. SEM topography for the different composites of n-OD@SiO_2-PHEMA content indicated that PU/n-OD@SiO_2-PHEMA-100 possessed higher density, enhancing leakage prevention through the dual encapsulation of the PU with the microcapsules. The melting enthalpy of PU/n-OD@SiO_2-PHEMA-100 was 85.1 Jg^{-1}. This property is also contributed to significantly by the 33.7% n-OD content, observed in the PU/n-OD@SiO_2-PHEMA-100 composite. Dynamic mechanical analysis (DMA) in Figs. 38(a) and 38(b) concluded a lower composite glass transition temperature than that of neat PU by around 5°C. Besides, the fillers were found to enhance the dynamic mechanical properties of the composites in the elastic region due to the higher storage and loss moduli stability as seen in Fig. 38(c) and 38(d), respectively. From TGA, PU/n-OD@SiO_2-PHEMA had two weight loss temperatures and excellent thermal. The PU/n-OD@SiO_2-PHEMA-100 composite had improved tensile strength at 12.98 MPa. Along with tests for elongation at break and hardness, the composites were concluded to have better mechanical properties.

Fang et al. prepared a PCM for thermal energy storage by applying a new technique of reverse micellization and emulsion

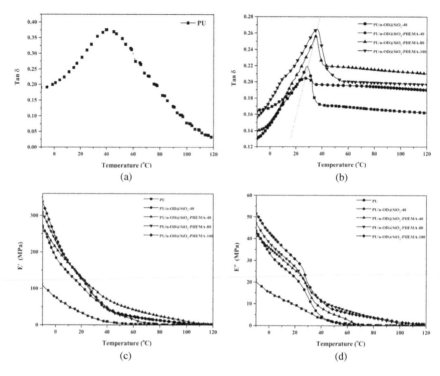

Fig. 38. DMA curves for (a) neat PU, (b) PU/n-OD@SiO$_2$-40, PU/n-OD@SiO$_2$-PHEMA-40, PU/n-OD@SiO$_2$-PHEMA-80, PU/n-OD@SiO$_2$-PHEMA-100, (c) storage modulus curves, (d) loss modulus curves. Reprinted with permission from Ref. 156. Copyright 2021 Springer.

polymerization.[45] Sodium sulfate decahydrate was the core material microencapsulated within a silica shell synthesized through silicon precursors of tetraethoxysilane and 3-aminopropyl-triethoxysilane. Five samples were prepared with varying Triton-X-100 surfactant amounts and labeled as MPCM1–MPCM5, with MPCM3 having optimal performance. FTIR verified the microcapsule sample composition and XRD low crystallinity in the silica shell. MPCM3 containing 1.0 mL surfactant possessed the best morphology when compared with MPCM4 and MPCM5, as presented in the SEM images in Fig. 39(a). It was observed that reduction of surfactant amount, used as the stabilizer, affected the microcapsule size,

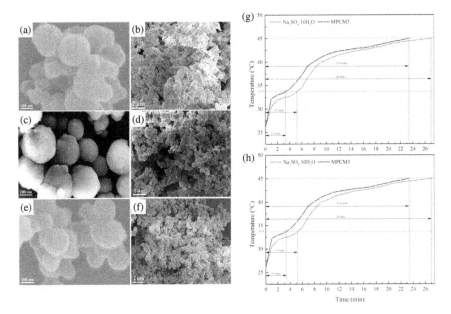

Fig. 39. SEM images for (a), (b) MPCM4, (c), (d) MPCM3, (e), (f) MPCM5. Temperature curves of Na$_2$SO$_4 \cdot$ 10H$_2$O and MPCM3 in the (g) melting and (h) solidifying processes, respectively. Reprinted with permission from Ref. 45. Copyright 2019 Elsevier.

reflected in the 500 nm to 28 μm regulation. DSC analysis concluded that the MPCM3 microcapsule melting and solidifying temperatures were at 33.6°C and 6.0°C, respectively. The corresponding latent heat values were 125.6 kJkg^{-1} and 74.0 kJkg^{-1}, respectively. Silica mesopores contributed to the inhibition of the phase segregations of different hydrate salts through its confining effect.[158] TGA displayed three weight loss stages of up to 650°C for the removal of aminopropyl groups on the silica surface.[159] The heat storage properties for the hydrate salts were enhanced by the SiO$_2$ matrix, as evidenced in the T-history method utilized, the results of which in Fig. 39(b) temperature curves indicated lower charging and discharging time at 23.4 min and 12 min for MPCM3.[160] The MEPCM is thus a suitable candidate for thermal energy storage material.

Bai et al. employed a sol–gel technique to fabricate SA/SiO$_2$ PCMs, with SA as the PCM and SiO$_2$ the shell material.[161] PCMs

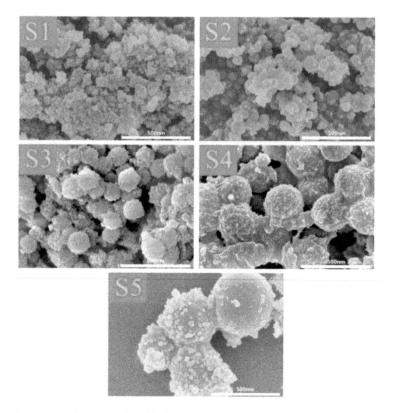

Fig. 40. FESEM images for SA/SiO$_2$ nanocapsules with varying ammonia-to-ethanol ratios, (S1) 1.2/90, (S2) 1.2/90, (S3) 2.0/90, (S4) 2.80.4/90, (S5) 3.6/90. Reprinted with permission from Ref. 161. Copyright 2018 Elsevier.

with large thermal storage capacities are applicable in solar energy storage systems.[162,163] Five samples were synthesized with varying ammonia-to-ethanol ratios, labeled S1–S5, with S3 possessing the best overall performance. FESEM results in Fig. 40 showed an even spherical structure for the nanocapsules, with the diameter rise corresponding to an increasing ammonia-to-ethanol ratio. FTIR found that S3 was the ideal sample due to its closest peak area to pure SA. The composition of the PCM was verified, containing SA and amorphous silica encapsulated together through physical interactions. From DSC analysis, it can be inferred that the nanocapsules possess superior energy storage ability, with latent heats of up to

169.4 Jg^{-1} only experiencing a 5.2 Jg^{-1} decrease after 3000 thermal cycles. The solidifying and melting temperatures were 63.9°C and 64.1°C, respectively. S3 exhibited the highest encapsulation ratio at 73%. TGA results indicated higher residual weights for the SA/SiO$_2$ samples at 298°C compared to that of SA, which was observed to be close to zero. The SA/SiO$_2$ nanocapsules indeed have significant potential in solar energy storage and can be applied in solar heating systems and other thermal management.[164–167]

Yang et al. reported the synthesis of a SR/Pa@SiO$_2$ composite PCM. PA was the PCM core, while the SiO$_2$ shell was based on tetraethoxysilane (TEOS) and γ-aminopropyl triethoxysilane (APTES) used as the precursors.[157,168] The Pa@SiO$_2$ microcapsules were subsequently embedded in SR to form the composites, then their properties were compared with SR/Pa and SR/Pa/SiO$_2$. Different compositions were prepared for the composite PCM, i.e., SR/Pa@SiO$_2$_40, SR/Pa@SiO$_2$_60, SR/Pa@SiO$_2$_80 and SR/Pa@SiO$_2$_100. The presence of SiO$_2$ as shell material contributed as a barrier to prevent leakage and reinforced the SR as an inorganic filler.[169,170] In situ SiO$_2$ shell modification with APTES containing amino groups was conducive toward avoiding agglomeration and increasing SR/Pa@SiO$_2$ interfacial bonding, as indicated by SEM analysis in Figs. 41(a)–41(f). Good dispersibility was also indicated. FTIR analysis proved the complete modification of amine groups on the SiO$_2$ surface. XRD demonstrated that Pa retained its crystallinity in the amorphous SR matrix structure. DSC was used to characterize the PCM thermal properties and the composite phase change latent heat achieved up to 92.39 Jg^{-1} for SR/Pa@SiO$_2$_100, with the value rising corresponding to Pa@SiO$_2$ content. The encapsulated efficiency was 76.05%. After 100 heating–cooling cycles, the general consistency in the phase change properties for the Pa@SiO$_2$ and SR/Pa@SiO$_2$_100 showed only slight changes in the phase change latent heats as seen in Figs. 41(g)–41(j). Low leakage rates were observed through leakage tests. SR/Pa@SiO$_2$_100 had the lowest leakage rate at 1.14%. Mechanical tests were conducted, with good results of 1.041 MPa tensile strength in the SR/Pa@SiO$_2$_100 composite, compared to that of pure SR. Hardness tests to gauge

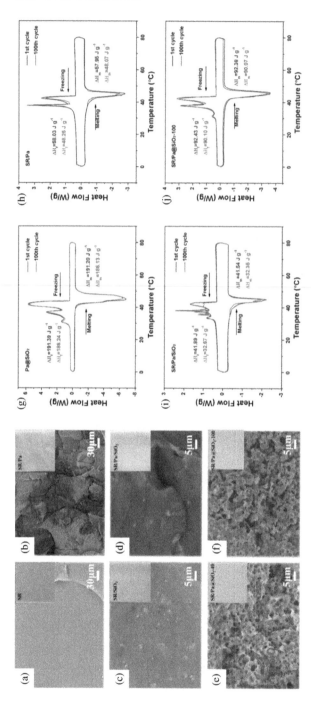

Fig. 41. SEM cross-sectional images for (a) neat SR, (b) SR/Pa, (c) SR/SiO$_2$, (d) SR/Pa/SiO$_2$, (e) SR/Pa@SiO$_2$-40, (f) SR/Pa@SiO$_2$-100. (The inset image in the top right corner is the corresponding rubber sheet.) DSC curves of (g) Pa@SiO$_2$, (h) SR/Pa, (i) SR/Pa/SiO2 and (j) SR/Pa@SiO$_2$-100 after experiencing 1 and 100 heating–cooling cycles. Reprinted with permission from Ref. 157. Copyright 2019 Elsevier.

compressibility found that SR/Pa@SiO$_2$ composites had increased hardness. SR/SiO$_2$ achieved a 71.43% improvement, proving that SiO$_2$ contributed hardness to SR. Thermal conductivity values for the composites were the highest compared to the other samples, notably SR/Pa@SiO$_2$_100 reaching a 146.7% enhancement relative to neat SR, recording a value of 0.375 Wm^{-1}K^{-1}. It was concluded that for both properties, SR/Pa@SiO$_2$ composites had better performance than that of SR/Pa and SR/Pa/SiO$_2$.

Yang et al. synthesized a SR/Pa (Pa)@silicon dioxide (SiO$_2$)@ polydopamine (PDA) double-shelled PCM through dopamine (DA) oxidative self-polymerization and interfacial polycondensation in a Tris-HCl buffer solution.[171,172] Four microcapsule samples, i.e., Pa@SiO$_2$@PDA-0.5, Pa@SiO$_2$@PDA-1, Pa@SiO$_2$@PDA-1.5 and Pa@SiO$_2$@PDA-2 were first prepared with different DA content, then SR matrix was incorporated into the samples and labeled as SR/Pa@SiO$_2$@PDA-0.5 to SR/Pa@SiO$_2$@PDA-2. SEM performed on the microcapsules and composites in Figs. 42(a)–4(d) showed respective successful encapsulation of PDA in the Pa@SiO$_2$ microcapsules and the relatively strong bonding between SR and Pa@SiO$_2$@PDA. FTIR and XPS characterization further verified PDA modification.

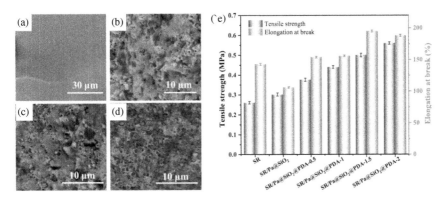

Fig. 42. Cross-sectional SEM images for (a) SR, (b) SR/Pa@SiO$_2$, (c) SR/Pa@SiO$_2$@PDA-0.5, (d) SR/Pa@SiO$_2$@PDA-2, (e) mechanical properties of neat SR, SR/Pa@SiO$_2$ and the composites. Reprinted with permission from Ref. 172. Copyright 2021 American Chemical Society.

Table 4. Summary of silicone dioxide/silica-containing encapsulation and its parameters.

Additive	PCM	Methods of preparation	Latent heat (Jg^{-1})	Melting point (°C)	Unique application	Efficacy (%)	Ref.
—	$Na_2SO_4 \cdot 10H_2O$, $Na_2HPO_4 \cdot 12H_2O$	Pickering emulsion	38.6, 35.4	32, 34.5	Possible application in PP textiles	—	107
—	DMT	Coating through deposition	96.2–118	134–136	Improve performance of synthetic oil	—	109
—	SA	Sol–gel method	182.53	70.37°C	Energy storage	86.68	50
PDA	n-OD	Oxidative self-polymerization	23.24	28	Good compressive properties in composite foam	—	122
PS	n-OD	Radical polymerization	76.7–104.3	27.5–28.7	—	38–51.6	126
PMF	n-OD	Oil-in-water emulsion	179	23.55	—	—	30
—	PEG, $CaCl_2$	Sol–gel method	122	55–57	Display crystallization properties	—	138
—	LA	Agglomeration from Van der Waals forces	160.91	44.295	—	96.15	139
Polyurethane foam	n-OD	Chemical deposition	56.3	32.1	—	47.5	144
—	SA	Sol–gel method	66.4	67.25	—	—	34

Reactive blue dye (Reactive Blue 4(X-Br))	Paraffin	Interfacial polymerization and chemical grafting	59.03	27.9	Architectural wall coating	30	43
—	Tris(hydroxymethyl) methyl aminomethane	Sol–gel method	146	110–155	—	—	47
—	LA	Sol–gel method	165.6	35.1–37.9	Possible application in solar energy application	85.9	148
Diatomite	Alkane	Sol–gel method	19	27	Building applications	—	151
—	SA	Sol–gel method	171.0	53.5	—	90.7	152
—	PEG	Sol–gel method impregnation	128.4	57.4	—	—	153
PS, PHEMA, polyurethane	n-OD	Radical polymerization	85.1	30.7	—	—	156
Triton-X-100	Sodium sulfate decahydrate	Reverse micellization, emulsion polymerization	125.6	33.6	—	—	45
Sodium dodecyl sulfate	SA	Sol–gel method	169.4	63.9	—	68	161
SR	Paraffin	Sol–gel method	92.39	45.67	—	76.05	157
SR, PDA	Paraffin	Oxidative self-polymerization	171.0	45.26	—	—	172

SiO$_2$, PDA and SR matrix were observed to contribute to the SR/Pa@SiO$_2$@PDA composite protection, with a notably minimal leakage rate of 0.45% for SR/Pa@SiO$_2$@PDA-2, indicating good stability. DSC analysis concluded a reduction in phase change enthalpies for the Pa@SiO$_2$@PDA microcapsules and SR/Pa@SiO$_2$@PDA composites with corresponding DA content increase, with SR/Pa@SiO$_2$@PDA-2 having a latent heat of melting and crystallization of 57.8 Jg^{-1} and 56.3 Jg^{-1}, respectively. Heightened mechanical strength was recorded in Fig. 42(e) as shown by the tensile strength in SR/Pa@SiO$_2$@PDA-2 of 0.560MPa, 1.8 times higher than that of pure SR/Pa@SiO$_2$. Elongation at break for the composite was 188% and the hardness was found to be 46 shore A, a relevant scale for hardness. This is attributed to the hydrogen bonding occurring between the numerous groups on the PDA surface and the matrix rendering a physical interlocking present on the rough surface of the PDA-modified composites thus, improving the interface interaction.

2.5. *Silicon with other additives*

Due to the cross-linking nature of polymers,[173–175] other silicon compounds may be potentially useful for PCM encapsulation as well. As indicated by Jia *et al.*,[176] low-cost and common materials are more feasible for industrial production. The preparation of multiphase composites via the inclusion of different components with superior electronic conductivity is another useful way for heightened silicon electrochemical performance. Another example can be seen in Han *et al.*, where small-grain mullite was selected as the PCM shell due to its notable resistance and durability attributed to its tetrahedral arrangement partially overlaid with oxygen vacancies and Al and Si atoms.[177] Its applications to enhance the relevant qualities in ceramics and coatings can be incorporated in a similar way in PCMs through relieving the thermal stress experienced by PCMs during rapid temperature changes.[178] Using encapsulation methods incorporating thin and tough shell helps with sustaining good thermal cycling performance and lowering loss of latent heat.[178]

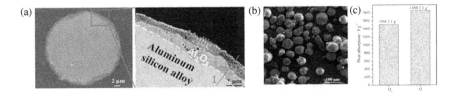

Fig. 43. SEM micrographs of (a) cross-sectional microcapsule, (b) microcapsules. (c) Calculated values of heat absorption. Reprinted with permission from Ref. 178. Copyright 2020 American Chemical Society.

Gu et al. prepared a double shell microcapsule PCM for heat storage and thermal cycling performance investigation with aluminum silicon alloy acting as the core, while Al_2O_3 and mullite were the inner and outer shells, respectively.[178] Encapsulation is a useful way of preventing molten PCM leakage.[179] The product was synthesized via steam corrosion before silica sol immersion and high-temperature calcination. SEM images in Fig. 43(a) showed a rough thickness of 1.5 μm for the microcapsule, and near-perfect microcapsule formation with its shell having a network arrangement in Fig. 43(b), in which some Al_2O_3 crystal clusters were embedded. Further SEM after thermal cycles demonstrated the leakage proof properties in the microcapsules as the structures were mostly intact. From TGA, it was observed that the microcapsules experienced a rapid mass rise at above 850°C, where it remained constant below 800°C. However, thermal stability was assured since the total mass increase was insignificant after the thermal cycle performances (TCPs). Calculations using studies from TCP results concluded latent heat storage and release values of 367.1 Jg^{-1} and 298.4 Jg^{-1}, respectively, with 367.1 Jg^{-1} being 78% of the aluminum silicon alloy latent heat. Notably, these values did not have significant changes after 3000 thermal cycles at 90.4% and 80.0%. The latent heat storage and release rates were derived, at 72.0 $Jg^{-1}min^{-1}$ and 20.3 $Jg^{-1}min^{-1}$. The microcapsule total heat absorption increased by 22.7% relative to that without undergoing phase change, achieving a value of up to 1850.3 Jg^{-1} when the temperature was adjusted from 40°C to 950°C (Fig. 43(c)). Therefore, thermal energy storage systems may be an application for this double shell microcapsule.

Zou et al. suggested a method of eutectic Al–Si PCM self-encapsulation without a container by using Si-rich Si-poor cladding ([SiC&Si-rich] E_{Al-Si}) to prepare a 3D-honeycomb-Sic reinforced Al matrix to resolve corrosion affecting eutectic Al–Si PCM encapsulated in an iron-based container with a rotating magnetic field (RMF), shown in Fig. 44(a).[180,181] This is a notable problem in high temperature waste heat recovery for the metallurgical industry. The addition resulted in significantly improved structural stability and energy storage capacity. The images in Fig. 44(b) confirmed the self-encapsulation of the eutectic Al–Si alloy, while XRD and Electron Probe Microanalysis (EPMA) revealed that the 3D-honeycomb SiC ceramic acted as an appropriate supporting system for the Si-rich shell to encapsulate eutectic the Al–Si PCM well. Heat treatment over 12 h found that the [SiC&Si-rich] E_{Al-Si} retained its original structure over temperature rise without any leakage. TGA indicated no mass loss, highlighting the thermal stability in the cladding. The 3D-honeycomb SiC incorporation also resulted in superior mechanical strength of [SiC&Si-rich] E_{Al-Si} as the Si-rich layer had a strength of 16.8/13.2 MPa under a high temperature of 650°C. The maximum explosion pressure was investigated, recording a higher value of 4.3 MPa corresponding with the increase in height of [SiC&Si-rich] E_{Al-Si} to 3 m. The thermal

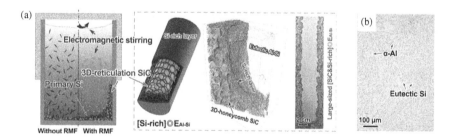

Fig. 44. (a) Schematic diagram illustrating eutectic Al–Si self-encapsulation through the synthesis of [SiC&Si-rich] E_{Al-Si}, (b) cross-sectional image of the Al–Si alloy eutectic structure indicating self-encapsulation. Reprinted with permission from Ref. 180. Copyright 2021 Elsevier.

conductivity value rose by approximately 10% after 1000 cycles to 96–62 Wm^{-1}K^{-1}. The discharge energy density achieved a value of 490 Jg^{-1}. After 1000 thermal cycles, the capacity was well maintained at 98.9%. The [SiC&Si-rich] E$_{Al-Si}$ discharge energy density was observed to be heightened by a range of 19.3–105.7% relative to that of the Al@Si@Al$_2$O$_3$. ([SiC&Si-rich] E$_{Al-Si}$ also presented 10.7% higher charge efficiency and 16.2% higher discharge efficiency. From DSC, it was found that the eutectic Al-Si PCM with [SiC&Si-rich] E$_{Al-Si}$ had stable performance, as well as melting and solidification latent heats of 490.63 Jg^{-1} and 468.44 Jg^{-1}, respectively. The eutectic Al–Si PCM can be applied in metallurgical industries since this preparation method may be performed in large scale [SiC&Si-rich] E$_{Al-Si}$ production using electromagnetic continuous separation technology.[182]

Zhang et al. demonstrated a two-step encapsulation resulting in the production of a composite phase change material (CPCM) comprised of a shape-stabilized Mg(NO$_3$)$_2$·6H$_2$O.[35] Owing to its melting point and high latent heat, Mg(NO$_3$)$_2$·6H$_2$O has potential in thermal energy storage applications if its poor thermal stability, attributed to dehydration, is resolved. Graphitic carbon nitride (g-C$_3$N$_4$) was used as the microhousing when Mg(NO$_3$)$_2$·6H$_2$O was infiltrated within as a means for liquid leakage prevention, thereafter, the composite was shaped into a cylinder. An epoxy resin structural adhesive with silicon sealant was employed for the macroencapsulation of the cylinder to mitigate the dehydration occurring in hydrated salt. Samples with 75 wt.%, 80 wt.% and 85 wt.% Mg(NO$_3$)$_2$·6H$_2$O/ g-C$_3$N$_4$ were prepared. From SEM imaging in Fig. 45, the Mg(NO$_3$)$_2$·6H$_2$O was found to have been evenly absorbed into the pores, along with the presence of g-C$_3$N$_4$. FTIR and XRD showed the absence of chemical interactions between Mg(NO$_3$)$_2$·6H$_2$O and g-C$_3$N$_4$. It was observed from DSC that 80 wt.% Mg(NO$_3$)$_2$·6H$_2$O/g-C$_3$N$_4$ composite recorded a 87.0°C phase change temperature and 112.30 kJkg^{-1} specific phase change enthalpy. The thermal conductivity of the composite was 0.4 Wm^{-1}K^{-1}. The 80 wt.% Mg(NO$_3$)$_2$·6H$_2$O/g-C$_3$N$_4$ left no residue in leakage testing. From TGA, the Mg(NO$_3$)$_2$·6H$_2$O/ g-C$_3$N$_4$ composite experienced a 32.70% weight loss at 300°C, with

Fig. 45. SEM image of the CPCM. Reprinted with permission from Ref. 35. Copyright 2019 Elsevier.

the $Mg(NO_3)_2 \cdot 6H_2O$ composite losing 2.31% less weight than the pure inorganic salt itself, proving its thermal stability. Thermal cycling test concluded a mere respective 0.84% and 6.25% weight loss over one hundred cycles in the CPCM when epoxy resin and silicon sealant coatings were used. The specific phase change enthalpy did not experience significant change and remained at up to values of 120.90 kJkg^{-1} and 117.60 kJkg^{-1}. Therefore, the usage of g-C_3N_4 and the sealants created a positive effect enhancing $Mg(NO_3)_2 \cdot 6H_2O$ thermal stability and reliability.

3. Outlook and Conclusion

In this review, we have summarized recent advances in the development of silicon containing encapsulation of PCM to enhance its performances and application, these silicon-containing compounds used are, namely, commercial silicone, siloxanes, Si_3N_4, SiC, SiO_2 and other additives. Siloxane, silicone and silica may act as the base of the shell material due to their polymeric characteristic and may impart their desired thermal and mechanical properties to the core shell while Si_3N_4, SiC and other additives serve to enhance the shell material that is based on the polymeric shell material such as polyurethane or PMMA. Collectively, the applications of these silicon

Table 5. Summary of silicone with other additives encapsulation and its parameters.

Additives	PCM	Method of preparation	Latent heat (Jg^{-1})	Melting point (°C)	Unique properties	Encapsulation efficiency (%)	Ref.
Al_2O_3 and mullite	Aluminum silicon alloy	Steam corrosion before silica sol immersion	298.4		—	—	178
Si-rich Si-poor cladding Al_2O_3	Eutectic aluminum silicon alloy	Pickering emulsion			Applications in metallurgical industry and can be performed in large scale	79.97	180
Graphitic carbon nitride	$Mg(NO_3)_2 \cdot 6H_2O$	Infiltration	112.30	87.0	—	—	35

compounds, especially in the thermal and mechanical properties, are able to push the limits of its performance in the encapsulated systems and ensure the efficient use of energy, pushing their limits as thermal energy storage materials.

Fortunately, there has been increasing research of these silicon-containing encapsulation additives for thermal storage application in the recent year. For example, the use of solar thermal energy conversion by making use of the electronic and thermal properties in silicon compounds is an excellent method to harness solar and thermal energy as it allows absorption at NIR region of the spectrum. While a myriad of accounts of silica containing encapsulation promises strong capability and great versatility in fabrication procedure, this process may compromise compatibility of the composite shell material and cause problems such as solubility issues. Therefore, advantages for each compound should be fully taken into account so as to minimize the drawback of it. In addition, to push for a large-scale application, the issues of material cost, safety and long-term stability should be fully addressed. To reduce material cost, perhaps further research and development can be invested to reduce the loadings of relatively expensive nanoadditives, such as Si_3N_4, or to discover cheaper alternatives that can also offer comparable properties and performance. Many types of nanoparticles are still generally toxic to health and therefore cause the safety issue. Likewise, given these coating materials will likely be exposed to different external environments such as acidic or basic conditions, high heat or under mechanical stress for a prolonged period, long-term chemical and physical stabilities of the materials should be improved to ensure optimal performance over time. Meanwhile, research and development should also continue on the discovery and development of novel materials suitable for functional compounds for encapsulation applications. These are not limited to inorganic nanomaterials but also organic polymer materials that generally offer lower cost and toxicity.

In conclusion, in a move toward a greener use of energy for greater sustainable living in the urban built environment, thermal energy storage materials play integral roles to achieve efficient use of resources while keeping the integrity of its performance as compared to conventional methods for occupants. Nonetheless, in addition to the types of silicon compounds discussed herein, other functional additives compounds may also be equally relevant and important in the grand domain of sustainable energy, such as carbon-based nanoparticles for the purpose of increasing thermal conductivity and polymer-based encapsulation to increase mechanical strength and stability of the encapsulation. It is envisaged that continual research work in the sustainable materials domain will spur the development of different functional encapsulation materials with desired properties and performance, and better adoptability for large-scale applications.

Acknowledgments

The authors acknowledge financial support from the Individual Research Grant (Grant reference No.: A20E7c0109) of the Agency for Science, Technology and Research of Singapore (A*STAR). Johnathan Joo Cheng Lee and Natalie Jia Xin Lim contributed equally to this work.

References

1. M. S. Dresselhaus and I. Thomas, *Nature*, 2001, **414**, 332–337.
2. Y. Lin, Y. Jia, G. Alva and G. Fang, *Renew. Sustain. Energy Rev.*, 2018, **82**, 2730–2742.
3. M. Waterson, *Energy Policy*, 2017, **104**, 466–473.
4. J. Cao, Y. Sim, X. Y. Tan, J. Zheng, S. W. Chien, N. Jia, K. Chen, Y. B. Tay, J.-F. Dong, L. Yang, H. K. Ng, H. Liu, C. K. I. Tan, G. Xie, Q. Zhu, Z. Li, G. Zhang, L. Hu, Y. Zheng, J. Xu, Q. Yan, X. J. Loh, N. Mathews, J. Wu and A. Suwardi, *Adv. Mater.*, 2022, **34**, 2110518.
5. Q. Zhu, S. Wang, X. Wang, A. Suwardi, M. H. Chua, X. Y. D. Soo and J. Xu, *Nano-Micro Lett.*, 2021, **13**, 119.
6. E. Yidirim, Q. Zhu, G. Wu, T. L. Tan, J. W. Xu and S. W. Yang, *J. Phys. Chem. C*, 2019, **123**, 9745–9755.

7. H. Zhou, M. H. Chua, Q. Zhu and J. W. Xu, *Compos. Commun.*, 2021, **27**, 100877.
8. S. D. Sharma and K. Sagara, *Int. J. Green Energy*, 2005, **2**, 1–56.
9. W. Su, J. Darkwa and G. Kokogiannakis, *Renew. Sustain. Energy Rev.*, 2015, **48**, 373–391.
10. D. Chandra, R. Chellappa and W.-M. Chien, *J. Phys. Chem. Solids*, 2005, **66**, 235–240.
11. G. Peng, G. Dou, Y. Hu, Y. Sun and Z. Chen, *Adv. Polym. Technol.*, 2020, **2020**, 1–20.
12. H. Nazir, M. Batool, F. J. B. Osorio, M. Isaza-Ruiz, X. Xu, K. Vignarooban, P. Phelan and A. M. Kannan, *Int. J. Heat Mass Transf.*, 2019, **129**, 491–523.
13. W. Wang, X. Yang, Y. Fang and J. Ding, *Appl. Energy*, 2009, **86**, 170–174.
14. H. Ji, D. P. Sellan, M. T. Pettes, X. Kong, J. Ji, L. Shi and R. S. Ruoff, *Energy Environ. Sci.*, 2014, **7**, 1185–1192.
15. Z. M. Png, X. Y. D. Soo, M. H. Chua, P. J. Ong, J. W. Xu and Q. Zhu, *J. Mater. Chem. A*, 2022, **10**, 3633–3641.
16. Z. M. Png, X. Y. D. Soo, M. H. Chua, P. J. Ong, A. Suwardi, C. K. I. Tan, J. W. Xu and Q. Zhu, *Solar Energy*, 2022, **231**, 115–128.
17. N. Zhang, Y. Yuan, X. Cao, Y. Du, Z. Zhang and Y. Gui, *Adv. Eng. Mater.*, 2018, **20**, 1700753.
18. L. Wang and D. Meng, *Appl. Energy*, 2010, **87**, 2660–2665.
19. J.-L. Zeng, Y.-H. Chen, L. Shu, L.-P. Yu, L. Zhu, L.-B. Song, Z. Cao and L.-X. Sun, *Solar Energy Mater. Solar Cells*, 2018, **178**, 84–90.
20. J.-W. Kook, K. Hwang and J.-Y. Lee, *Materials*, 2021, **14**, 3157.
21. P. J. Ong, Z. M. Png, X. Y. Debbie Soo, X. Wang, A. Suwardi, M. H. Chua, J. W. Xu and Q. Zhu, *Mater. Chem. Phys.*, 2022, **277**, 125438.
22. F. Gao, X. Wang and D. Wu, *Solar Energy Mater. Solar Cells*, 2017, **168**, 146–164.
23. H. Liu, X. Wang and D. Wu, *Appl. Therm. Eng.*, 2018, **134**, 603–614.
24. X. Geng, W. Li, Y. Wang, J. Lu, J. Wang, N. Wang, J. Li and X. Zhang, *Appl. Energy*, 2018, **217**, 281–294.
25. L. Yang, X. Jin, Y. Zhang and K. Du, *J. Clean. Prod.*, 2021, **287**, 124432.
26. A. A. Abuelnuor, A. A. Omara, K. M. Saqr and I. H. Elhag, *Int. J. Energy Res.*, 2018, **42**, 2084–2103.

27. X. Y. D. Soo, Z. M. Png, M. H. Chua, J. C. C. Yeo, P. J. Ong, S. Wang, X. Wang, A. Suwardi, J. Cao, Y. Chen, Q. Yan, X. J. Loh, J. Xu and Q. Zhu, *Mater. Today Adv.*, 2022, **14**, 100227.
28. R. Al-Shannaq, J. Kurdi, S. Al-Muhtaseb and M. Farid, *Solar Energy*, 2016, **129**, 54–64.
29. M. M. Umair, Y. Zhang, K. Iqbal, S. Zhang and B. Tang, *Appl. Energy*, 2019, **235**, 846–873.
30. W. Su, J. Darkwa and G. Kokogiannakis, *Int. J. Low-Carbon Technol.*, 2018, **13**, 301–310.
31. E. Katoueizadeh, S. Zebarjad, K. Janghorban and A. Ghafarinazari, *J. Polym. Res.*, 2018, **25**, 1–9.
32. G. Alva, X. Huang, L. Liu and G. Fang, *Appl. Energy*, 2017, **203**, 677–685.
33. Y. Lin, C. Zhu, G. Alva and G. Fang, *Appl. Energy*, 2018, **231**, 494–501.
34. H. Yuan, H. Bai and Y. Wang, *Energy Technology 2016*, Springer, 2016, pp. 99–106.
35. W. Zhang, Y. Zhang, Z. Ling, X. Fang and Z. Zhang, *Appl. Energy*, 2019, **253**, 113540.
36. X. Wang, C. Zhang, K. Wang, Y. Huang and Z. Chen, *J. Colloid Interface Sci.*, 2021, **582**, 30–40.
37. M. Aguayo, S. Das, A. Maroli, N. Kabay, J. C. Mertens, S. D. Rajan, G. Sant, N. Chawla and N. Neithalath, *Cem. Concr. Compos.*, 2016, **73**, 29–41.
38. Z. Fu, L. Su, J. Li, R. Yang, Z. Zhang, M. Liu, J. Li and B. Li, *Thermochim. Acta*, 2014, **590**, 24–29.
39. D.-H. Yu and Z.-Z. He, *Appl. Energy*, 2019, **247**, 503–516.
40. Y. E. Milián, A. Gutiérrez, M. Grágeda and S. Ushak, *Renew. Sustain. Energy Rev.*, 2017, **73**, 983–999.
41. T. Y. Wang and J. Huang, *J. Appl. Polym. Sci.*, 2013, **130**, 1516–1523.
42. H. Zhang and X. Wang, *Colloids Surf. A: Physicochem. Eng. Asp.*, 2009, **332**, 129–138.
43. E. Ma, Z. Wei, C. Lian, Y. Zhou, S. Gan and B. Xu, *Materials*, 2021, **14**, 4012.
44. G. Alva, Y. Lin, L. Liu and G. Fang, *Energy Build.*, 2017, **144**, 276–294.
45. Z. Zhang, Y. Lian, X. Xu, X. Xu, G. Fang and M. Gu, *Appl. Energy*, 2019, **255**, 113830.

46. A. Arshad, M. Jabbal, Y. Yan and J. Darkwa, *Int. J. Energy Res.*, 2019, **43**, 5572–5620.
47. C.-B. Wu, G. Wu, X. Yang, Y.-J. Liu, T. Liang, W.-F. Fu, M. Wang and H.-Z. Chen, *Appl. Energy*, 2015, **154**, 361–368.
48. Y. Wang, J. Chen, J. Xiang, H. Li, Y. Shen, X. Gao and Y. Liang, *React. Funct. Polym.*, 2009, **69**, 393–399.
49. N. Sun and Z. Xiao, *J. Mater. Sci.*, 2016, **51**, 8550–8561.
50. S. Ishak, S. Mandal, H.-S. Lee and J. K. Singh, *Sci. Rep.*, 2020, **10**, 15023.
51. R. Lin, Y. Zhang, H. Li, Y. Shi and C. Zhou, *Progr. Org. Coat.*, 2019, **135**, 65–73.
52. Y. Guo, Y. Gao, N. Han, X. Zhang and W. Li, *Solar Energy Mater. Solar Cells*, 2022, **240**, 111718.
53. S. S. F. Duran, D. W. Zhang, W. Y. S. Lim, J. Cao, H. F. Liu, Q. Zhu, C. K. I. Tan, J. W. Xu, X. J. Loh and A. Suwardi, *Crystals*, 2022, **12**, 307.
54. Q. Zhu, M. H. Chua, P. J. Ong, J. J. Cheng Lee, K. Le Osmund Chin, S. Wang, D. Kai, R. Ji, J. Kong, Z. Dong, J. Xu and X. J. Loh, *Mater. Today Adv.*, 2022, **15**, 100270.
55. K. W. Shah, P. J. Ong, M. H. Chua, S. H. G. Toh, J. J. C. Lee, X. Y. D. Soo, Z. M. Png, R. Ji, J. W. Xu and Q. Zhu, *Energy Build.*, 2022, **262**, 112018.
56. V. P. Bui, H. Z. Liu, Y. Y. Low, T. Tang, Q. Zhu, K. W. Shah, E. Shidoji, Y. M. Lim and W. S. Koh, *Energy Build.*, 2017, **157**, 195–203.
57. D. Lin, X. Xu, J. Wang, T. Zhang, F. Xie, L. Gong, J. Chen, T. Shi, J. Shi, P. Liu and W. Xie, *ACS Appl. Mater. Interfaces*, 2021, **13**, 8138–8146.
58. B. A. Kamino and T. P. Bender, *Chem. Soc. Rev.*, 2013, **42**, 5119–5130.
59. K. Akamatsu, M. Ogawa, R. Katayama, K. Yonemura and S.-i. Nakao, *Colloid Surf. A: Physicochem. Eng. Asp.*, 2019, **567**, 297–303.
60. Z. Wang, J. Wu, D. Lei, H. Liu, J. Li and Z. Wu, *App. Energy*, 2020, **261**, 114472.
61. H. T. Chiu, T. Sukachonmakul, M. T. Kuo, Y. H. Wang and K. Wattanakul, *Appl. Surf. Sci.*, 2014, **292**, 928–936.
62. W. Wu and Z. Chen, *J. Therm. Anal. Calorim.*, 2018, **133**, 1365–1370.

63. H. Deng, Y. Guo, F. He, Z. Yang, J. Fan, R. He, K. Zhang and W. Yang, *Fuller. Nanotub. Carbon Nanostructures*, 2019, **27**, 626–631.
64. M. Liu, S. Chen, Qi, X. b.; B. Li, R. Shi, Y. Liu, Y. Chen and Z. Zhang, *Chem. Eng. J.*, 2014, **241**, 466–476.
65. H. Xu, A. Romagnoli, J. Y. Sze and X. Py, *Appl. Energy*, 2017, **187**, 281–290.
66. S. K. Sahoo, M. K. Das and P. Rath, *Renew. Sustain. Energy Rev.*, 2016, **59**, 550–582.
67. J. Ma, T. Ma, W. Duan, W. Wang, J. Cheng and J. Zhang, *J. Mater. Chem. A*, 2020, **8**, 22315–22326.
68. J. M. Yang and J. S. Kim, *J. Appl. Polym. Sci.*, 2018, **135**, 45821.
69. G. Zu, K. Kanamori, A. Maeno, H. Kaji and K. Nakanishi, *Angew. Chem.*, 2018, **130**, 9870–9875.
70. G. Zu, K. Kanamori, K. Nakanishi and J. Huang, *Chem. Mater.*, 2019, **31**, 6276–6285.
71. M. M. Rahman, I. Lee, H. H. Chun, H. D. Kim and H. Park, *J. Appl. Polym. Sci.*, 2014, **131**, 1–7.
72. H. Zhang, S. Sun, X. Wang and D. Wu, *Colloids Surf. A: Physicochem. Eng. Asp.*, 2011, **389**, 104–117.
73. G. Li, G. Hong, D. Dong, W. Song and X. Zhang, *Adv. Mater.*, 2018, **30**, 1801754.
74. J. Lyu, G. Li, M. Liu and X. Zhang, *Langmuir*, 2019, **35**, 943–949.
75. Y. Zhang, J. Xiu, B. Tang, R. Lu and S. Zhang, *AIChE J.*, 2018, **64**, 688–696.
76. M. M. Umair, Y. Zhang, S. Zhang, X. Jin and B. Tang, *J. Mater. Chem. A.*, 2019, **7**, 26385–26392.
77. X. Chen, H. Gao, G. Hai, D. Jia, L. Xing, S. Chen, P. Cheng, M. Han, W. Dong and G. Wang, *Energy Storage Mater.*, 2020, **26**, 129–137.
78. S. Wu, T. Li, Z. Tong, J. Chao, T. Zhai, J. Xu, T. Yan, M. Wu, Z. Xu and H. Bao, *Adv. Mater.* 2019, **31**, 1905099.
79. Y. Ge, W. Cui, Q. Wang, Y. Zou, Z. Xie and K. Chen, *J. Amer. Ceram. Soc.*, 2015, **98**, 263–268.
80. Y. Yang, J. Kuang, H. Wang, G. Song, Y. Liu and G. Tang, *Solar Energy Mater. Solar Cells*, 2016, **151**, 89–95.
81. N. Sun, X. Meng and Z. Xiao, *Ceram. Int.*, 2015, **41**, 13830–13835.
82. W. Yuan, X. Yang, G. Zhang and X. Li, *Appl. Therm. Eng.*, 2018, **144**, 551–557.

83. N. Chisholm, H. Mahfuz, V. K. Rangari, A. Ashfaq and S. Jeelani, *Compos. Struct.*, 2005, **67**, 115–124.
84. X. Wang, C. Li and T. Zhao, *Solar Energy Mater. Solar Cells*, 2018, **183**, 82–91.
85. B. Zhang, S. Li, X. Fei, H. Zhao and X. Lou, *Colloids Surf. A: Physicochem. Eng. Asp.*, 2020, **603**, 125219.
86. V. Stahl, Y. Shi, W. Kraft, T. Lanz, P. Vetter, R. Jemmali, F. Kessel and D. Koch, *Int. J. Appl. Ceram. Technol.*, 2020, **17**, 2040–2050.
87. H. G. Kim, A. Qudoos, I. K. Jeon and J. S. Ryou, *Construct. Build. Mater.*, 2020, **262**, 120088.
88. X. Li, H. Wang, X. Yang, X. Zhang and B. Ma, *RSC Adv.*, 2022, **12**, 878–887.
89. D. H. Yoo, I. K. Jeon, B. H. Woo and H. G. Kim, *Appl. Therm. Eng.*, 2021, **198**, 117445.
90. S. Han, S. Lyu, Z. Chen, F. Fu and S. Wang, *Carbohydr. Polym.*, 2020, **234**, 115923.
91. F. Gomez-Garcia, J. Gonzalez-Aguilar, G. Olalde and M. Romero, *Renew. Sustain. Energy Rev.* 2016, **57**, 648–658.
92. W. Aftab, X. Huang, W. Wu, Z. Liang, A. Mahmood and R. Zou, *Energy Environ. Sci.*, 2018, **11**, 1392–1424.
93. X. Hu, Z. Huang and Y. Zhang, *Carbohydr. Polym.*, 2014, **101**, 83–88.
94. C. Vix-Guterl, I. Alix, P. Gibot and P. Ehrburger, *Appl. Surf. Sci.*, 2003, **210**, 329–337.
95. A. Salinas, M. Lizcano and K. Lozano, *J. Ceram.* 2015, **2015**, 1–5.
96. F. Zhang, Y. Zhong, X. Yang, J. Lin and Z. Zhu, *Solar Energy Mater. Solar Cells*, 2017, **170**, 137–142.
97. M. Liu, W. Saman and F. Bruno, *Renew. Sustain. Energy Rev.*, 2012, **16**, 2118–2132.
98. R. Fukahori, T. Nomura, C. Zhu, N. Sheng, N. Okinaka and T. Akiyama, *Appl. Energy* 2016, **163**, 1–8.
99. H. Okamoto, *J. Phase Equilibria*, 1992, **13**, 543–565.
100. J. Li, L. Porter and S. Yip, *J. Nucl. Mater.* 1998, **255**, 139–152.
101. R. Liu, B. Liu, K. Zhang, M. Liu, Y. Shao and C. Tang, *J. Nucl. Mater.* 2014, **453**, 107–114.
102. X. Y. D. Soo, Z. M. Png, X. Z. Wang, M. H. Chua, P. J. Ong, S. X. Wang, Z. B. Li, D. Z. Chi, J. W. Xu, X. J. Loh and Q. Zhu, *ACS Appl. Polym. Mater.* 2022, **4**, 2747–2756.

103. Y. Deng, J. Li and H. Nian, *Solar Energy Mater. Solar Cells*, 2018, **174**, 283–291.
104. H. J. Zhu, J. M. J. Qiang, C. G. Wang, C. Y. Chan, Q. Zhu, E. Y. Ye, Z. B. Li and X. J. Loh, *Bioact. Mater.*, 2022, **18**, 471–491.
105. M. H. Chua, S. H. G. Toh, P. J. Ong, Z. M. Png, Q. Zhu, S. X. Xiong and J. W. Xu, *Polym. Chem.* 2022, **13**, 967–981.
106. S. Sugiarto, R. R. Pong, Y. C. Tan, Y. Leow, T. Sathasivam, Q. Zhu, X. J. Loh and D. Kai, *Mater. Today Chem.* 2022, **26**, 101022.
107. A. Hassabo, A. Mohamed, H. Wang, C. Popescu and M. Moller, *Inorg. Chem.: Indian J.*, 2015, **10**, 59–65.
108. Q. A. Zhu and Y. X. Lu, *Org. Lett.*, 2010, **12**, 4156–4159.
109. M. Ram, J. P. Myers, C. Jotshi, E. Stefanakos, K. Arvanitis, E. Papanicolaou and V. Belessiotis, *Int. J. Energy Res.*, 2016, **41**, 252–262.
110. D. Vargas-Florencia, O. Petrov and I. Furó, *J. Phys. Chem. B*, 2006, **110**, 3867–3870.
111. S. Pendyala, *Macroencapsulation of Phase Change Materials for Thermal Energy Storage*, University of South Florida, 2012.
112. M. K. Ram, C. K. Jotshi, E. K. Stefanakos and D. Y. Goswami, *Method of Encapsulating a Phase Change Material with a Metal Oxide*, Google Patents, 2016.
113. J. H. Elliott and M. D. Chris, *J. Chem. Eng. Data*, 1968, **13**, 475–479.
114. C. W. Bale, E. Bélisle, P. Chartrand, S. A. Decterov, G. Eriksson, A. E. Gheribi, K. Hack, I. H. Jung, Y. B. Kang, J. Melançon, A. D. Pelton, S. Petersen, C. Robelin, J. Sangster, P. Spencer and M. A. Van Ende, *Calphad*, 2016, **54**, 35–53.
115. G. Nallathambi, T. Ramachandran, V. Rajendran and R. Palanivelu, *Mater. Res.*, 2011, **14**, 552–559.
116. Q. Zhu, E. Yildirim, X. Z. Wang, X. Y. D. Soo, Y. Zheng, T. L. Tan, G. Wu, S. W. Yang and J. W. Xu, *Front. Chem.*, 2019, **7**, 1–11.
117. Q. Zhu, E. Yildirim, X. Z. Wang, A. K. K. Kyaw, T. Tang, X. Y. D. Soo, Z. M. Wong, G. Wu, S. W. Yang and J. W. Xu, *Mol. Syst. Des. Eng.*, 2020, **5**, 976–984.
118. D. G. Cahill and T. H. Allen, *Appl. Phys. Lett.*, 1994, **65**, 309–311.
119. Y. Fang, L. Huang, X. Liang, S. Wang, H. Wei, X. Gao and Z. Zhang, *Solar Energy Mater. Solar Cells*, 2020, **206**, 110257.

120. B. Li, T. Liu, L. Hu, Y. Wang and L. Gao, *ACS Sustain. Chem. Eng.*, 2013, **1**, 374–380.
121. Y. Zhu, Y. Chi, S. Liang, X. Luo, K. Chen, C. Tian, J. Wang and L. Zhang, *Solar Energy Mater. Solar Cells*, 2018, **176**, 212–221.
122. Y. Zhu, Y. Qin, S. Liang, K. Chen, C. Wei, C. Tian, J. Wang, X. Luo and L. Zhang, *Thermochim. Acta*, 2019, **676**, 104–114.
123. R. Al-Shannaq, J. Kurdi, S. Al-Muhtaseb, M. Dickinson and M. Farid, *Energy*, 2015, **87**, 654–662.
124. Y. Wang, S. Li, T. Zhang, D. Zhang and H. Ji, *Solar Energy Mater. Solar Cells*, 2017, **171**, 60–71.
125. M. Hao, W. Zhao, R. Li, H. Zou, M. Tian, L. Zhang and W. Wang, *Ind. Eng. Chem. Res.* 2018, **57**, 7486–7494.
126. Y. Zhu, Y. Qin, C. Wei, S. Liang, X. Luo, J. Wang and L. Zhang, *Energy Convers. Manag.*, 2018, **164**, 83–92.
127. W. Li, G. Song, S. Li, Y. Yao and G. Tang, *Energy*, 2014, **70**, 298–306.
128. K. Yuan, J. Liu, X. Fang and Z. Zhang, *J. Mater. Chem. A*, 2018, **6**, 4535–4543.
129. L. Zhang, W. Yang, Z. Jiang, F. He, K. Zhang, J. Fan and J. Wu, *Appl. Energy*, 2017, **197**, 354–363.
130. F. He, X. Wang and D. Wu, *Energy*, 2014, **67**, 223–233.
131. J. Zheng, S. F. D. Solco, C. J. E. Wong, S. A. Sia, X. Y. Tan, J. Cao, J. C. C. Yeo, W. Yan, Q. Zhu and Q. Yan, *J. Mater. Chem. A*, 2022, **10**, 19787–19796.
132. S. Wang, P. J. Ong, S. Liu, W. Thitsartarn, M. Tan, A. Suwardi, Q. Zhu and X. J. Loh, *Chem. Asian J.*, 2022, **17**, e202200608.
133. H. Lambert, A. C. Bonillo, Q. Zhu, Y. W. Zhang and T. C. Lee, *npj Comput. Mater.*, 2022, **8**, 21.
134. M. T. Liu, Y. Chen, Q. Zhu, J. J. Tao, C. M. Tang, H. J. Ruan, Y. L. Wu and X. J. Loh, *Chem. — Asian J.*, 2022, **17**, e20220039.
135. J. K. Choi, J. G. Lee, J.-H Kim and H. S. Yang, *J. Ind. Eng. Chem.*, 2001, **7**, 358–362.
136. C. Y. Gao and S. Piao, *Colloid Polym. Sci.* 2017, **295**, 1–8.
137. M. G. Burzo, P. L. Komarov and P. E. Raad, *IEEE Trans. Compon. Pack. Technol.*, 2003, **26**, 80–88.
138. L. He, J. Li, C. Zhou, H. Zhu, X. Cao and B. Tang, *Solar Energy*, 2014, **103**, 448–455.

139. S. Ishak, S. Mandal, H.-S. Lee and J. K. Singh, *Sci. Rep.* 2021, **11**, 15012.
140. S.-H . Wu, C.-Y . Mou and H.-P . Lin, *Chem. Soc. Rev.*, 2013, **42**, 3862–3875.
141. P. Heaney, *Geochem. Mater. Appl.*, 1994, **29**, 1–40.
142. Z. Zhong-Qing, G. Qiang, Y. Jing-Min, W. Yan, Y. Jing-Yi, Z. Xue, F. Guo-Zhong, C. Zhi-Qiang, W. Shao-Jie and S. Hong-Ge, *Sci. Total Environ.* 2020, **704**, 135394.
143. G. Fang, H. Li and X. Liu, *Mater. Chem. Phys.*, 2010, **122**, 533–536.
144. H. Xiang, J. An, X. Zeng, X. Liu, Y. Li, C. Yang and X. Xia, *Polym. Compos.* 2020, **41**, 1662–1672.
145. C. Liang, X. Lingling, S. Hongbo and Z. Zhibin, *Energy Convers. Manag.* 2009, **50**, 723–729.
146. W. Wang, S. He, S. Guo, J. Yan and J. Ding, *Energy Convers. Manag.*, 2014, **83**, 306–313.
147. A. Shukla, D. Buddhi and R. L. Sawhney, *Renew. Energy*, 2008, **33**, 2606–2614.
148. H. Yuan, B. Hao, X. Lu, X. Zhang, J. Zhang, Z. Zhang and L. Yang, *Solar Energy Mater. Solar Cells*, 2019, **191**, 243–257.
149. N. Stathopoulos, M. El Mankibi and M. Santamouris, *Appl. Therm. Eng.*, 2017, **114**, 1064–1072.
150. J. Sarwar and B. Mansoor, *Energy Convers. Manag.*, 2016, **120**, 247–256.
151. Z. Chen, M. Qin and J. Yang, *Energy Build.*, 2015, **106**, 175–182.
152. Z. Chen, L. Cao, F. Shan and G. Fang, *Energy Build.*, 2013, **62**, 469–474.
153. J. Li, L. He, T. Liu, X. Cao and H. Zhu, *Solar Energy Mater. Solar Cells*, 2013, **118**, 48–53.
154. S. Karaman, A. Karaipekli, A. Sarı and A. Biçer, *Solar Energy Mater. Solar Cells*, 2011, **95**, 1647–1653.
155. J. Zhang, X. Zhang, Y. Wan, D. Mei and B. Zhang, *Solar Energy*, 2012, **86**, 1142–1148.
156. X. Lin, Y. Chen, J. Jiang, J. Li, Y. Jiang, H. Zhang and H. Liu, *Polym. Bull.* 2022, **79**, 3867–3889.
157. Y. Guo, W. Yang, Z. Jiang, F. He, K. Zhang, R. He, J. Wu and J. Fan, *Solar Energy Mater. Solar Cells*, 2019, **196**, 16–24.

158. J. Zhang, S. S. Wang, S. D. Zhang, Q. H. Tao, L. Pan, Z. Y. Wang, Z. P. Zhang, Y. Lei, S. K. Yang and H. P. Zhao, *J. Phys. Chem. C*, 2011, **115**, 20061–20066.
159. J. Wang, S. Zheng, Y. Shao, J. Liu, Z. Xu and D. Zhu, *J. Colloi Interf. Sci.*, 2010, **34**, 293–299.
160. A. Solé, L. Miró, C. Barreneche, I. Martorell and L. F. Cabeza, *Rene.Sustain. Energy Rev.* 2013, **26**, 425–436.
161. H. Yuan, H. Bai, X. Zhang, J. Zhang, Z. Zhang and L. Yang, *Solar Energy* 2018, **173**, 42–52.
162. S. Tahan Latibari, M. Mehrali, M. Mehrali, T. M. Indra Mahlia, Cornelis Metselaar, H. S., *Energy*, 2013, **61**, 664–672.
163. H. Wang, J. Luo, Y. Yang, L. Zhao, G. Song and G. Tang, *Solar Energy*, 2016, **139**, 591–598.
164. X. Yong, G. Wu, W. Shi, Z. M. Wong, T. Q. Deng, Q. Zhu, X. P. Yang, J. S. Wang, J. W. Xu and S. W. Yang, *J. Mater. Chem. A*, 2020, **8**, 21852–21861.
165. J. Cao, X. Y. Tan, N. Jia, J. Zheng, S. W. Chien, H. K. Ng, C. K. I. Tan, H. F. Liu, Q. Zhu, S. X. Wang, G. Zhang, K. W. Chen, Z. B. Li, L. Zhang, J. W. Xu, L. Hu, Q. Y. Yan, J. Wu and A. Suwardi, *Nano Energy*, 2022, **96**, 107147.
166. Y. Shi, C. Lee, X. Y. Tan, L. Yang, Q. Zhu, X. J. Loh, J. W. Xu and Q. Y. Yan, *Small Struct.*, 2022, **25**, 2100185
167. J. Cao, J. Zheng, H. F. Liu, C. K. I. Tan, X. Z. Wang, W. D. Wang, Q. Zhu, Z. B. Li, G. Zhang, J. Wu, L. Zhang, J. W. Xu and A. Suwardi, *Mater. Today Energy*, 2022, **25**, 100165.
168. C.-B. Wu, G. Wu, X. Yang, Y.-J. Liu, C.-X. Gao, Q.-H. Ji, M. Wang and H.-Z. Chen, *Colloids Surf. A: Physicochem. Eng. Asp.*, 2014, **457**, 487–494.
169. R. Jacob and F. Bruno, *Renew. Sustain. Energy Rev.*, 2015, **48**, 79–87.
170. T. Khadiran, M. Z. Hussein, Z. Zainal and R. Rusli, *Solar Energy Mater. Solar Cells*, 2015, **143**, 78–98.
171. H. Deng, Y. Yang, X. Tang, Y. Li, F. He, Q. Zhang, Z. Huang, Z. Yang and W. Yang, *ACS Appl. Mater. Interf.*, 2021, **13**, 39394–39403.
172. W. Xia, J. Zhou, T. Hu, P. Ren, G. Zhu, Y. Yin, J. Li and Z. Zhang, *Compos. Part A: Appl. Sci. Manuf.*, 2020, **131**, 105805.

173. W. T. Neo, Q. Ye, M. H. Chua, Q. Zhu and J. W. Xu, *Macromol. Rapid Commun.*, 2020, **41**, 200156.
174. J. K. Muiruri, J. C. C. Yeo, Q. Zhu, E. Y. Ye, X. J. Loh and Z. B. Li, *ACS Sustain. Chem. Eng.*, 2022, **10**, 3387–3406.
175. X. Y. D. Soo, S. X. Wang, C. C. J. Yeo, J. W. Li, X. P. Ni, L. Jiang, K. Xue, Z. B. Li, X. C. Fei, Q. Zhu and X. J. Loh, *Sci. Total Environ.*, 2022, **807**, 200156.
176. H. Jia, C. Stock, R. Kloepsch, X. He, J. P. Badillo, O. Fromm, B. Vortmann, M. Winter and T. Placke, *ACS Appl. Mater. Interfaces*, 2015, **7**, 1508–1515.
177. T. i. Mah and K. Mazdiyasni, *J. Amer. Ceram. Soc.*, 1983, **66**, 699–703.
178. C. Han, H. Gu, M. Zhang, A. Huang, Y. Zhang and Y. Wang, *Solar Energy Mater. Solar Cells*, 2020, **217**, 110697.
179. P. Moreno, L. Miró, A. Solé, C. Barreneche, C. Solé, I. Martorell and L. F. Cabeza, *Appl. Energy*, 2014, **125**, 238–245.
180. Q. Zou, Z. Dong, X. Yang, J. Jie, X. An, N. Han and T. Li, *Chem. Eng. J.*, 2021, **425**, 131664.
181. R. Jacob, A. Sibley, M. Belusko, M. Liu, J. Quinton and G. Andersson, *J. Energy Storage*, 2018, **17**, 249–260.
182. J. C. Jie, Q. C. Zou, J. L. Sun, Y. P. Lu, T. M. Wang and T. J. Li, *Acta Mater.*, 2014, **72**, 57–66

Comparison of the Properties of Additively Manufactured 316L Stainless Steel for Orthopedic Applications: A Review*

M. Nabeel[†], A. Farooq[†], S. Miraj[‡], U. Yahya[†], K. Hamad[§] and K. M. Deen[†,¶,∥]

[†]Corrosion Control Research Cell, Institute of Metallurgy
and Materials Engineering,
University of the Punjab, Lahore 54590, Pakistan
[‡]Punjab University College of Pharmacy, University
of the Punjab, Allama Iqbal Campus,
Lahore 54590, Pakistan
[§]School of Advanced Materials Science & Engineering,
Sungkyunkwan University,
Suwon, 16419, Republic of Korea
[¶]Department of Materials Engineering, The University
of British Columbia,
Vancouver Campus, Vancouver, BC, Canada V6T 1Z4
[∥]kashifmairaj.deen@ubc.ca

Owing to the low cost, ease of fabricability, good mechanical properties, corrosion resistance and biocompatibility of the 316L stainless steel (SS), this material is considered a suitable choice for orthopedic applications. Based on its properties and large utilization in orthopedics, this review focuses on the importance of additively manufactured (AM) 316L stainless steel. Owing to the large flexibility of the additive manufacturing process, the microstructure of the 316L SS can be easily tuned to modify the mechanical, corrosion and biological properties.

**Corresponding authors.
*To cite this article, please refer to its earlier version published in the *World Scientific Annual Review of Functional Materials*, Volume 1, 2230001 (2023), DOI: 10.1142/S281092282230001X.

To elucidate the benefits of additively manufactured 316L stainless steel, the properties of the selective laser melted (SLM) 316L stainless steel and wrought 316L stainless steel are compared. Particularly, the unique features of the SLM 316L stainless steel have been discussed in detail. The existing challenges associated with the additive manufacturing processes and implications of their widespread application are also highlighted. A brief overview of the biological properties and reactions sequence of the host immune system, i.e. tissue response, the activation of acute and chronic inflammatory processes and immunological reactions, is also provided to understand the reasons for implant failure or rejection by the body.

Keywords: Selective laser melting; 316L stainless steel; orthopedic implants; corrosion; biocompatibility.

1. Introduction

1.1. *A brief overview of the metals/alloys used in orthopedic applications*

Due to the increasing level of human activities, a significant surge in physical injuries and chronic orthopedic disorders has been experienced recently. Provoked by this issue, a considerable increase in the demand for orthopedic implants has been observed in the past two decades. There is a strict criterion in the selection of metallic materials for *in-vivo* use in orthopedic applications. In general, for orthopedic implants, high-performance materials with sufficient mechanical strength, high corrosion resistance, osteointegration and excellent biocompatibility are desirable. The world population is increasing day by day and the socio-economic growth of any country depends on various factors, i.e. health, education, transportation systems, physical and recreational facilities and sports. The increasing density of road traffic and physical activities also increase the chances of accidents that may cause loss of life, bone fracture, bone diseases and/or bone damage. To recover from bone injury and to bring back life to normal, implants are inserted by following the surgical protocols. Specifically, orthopedic implants are made up of a variety of materials but the use of various metal alloys has key importance due to their superior electrochemical, mechanical and biological properties. Mostly, metal implants are used for spinal fusion, joint replacement, bone trauma repair, etc. A substitute for the dysfunctional organ of the human anatomy is known as bioimplants or most commonly *implants*. Implants need to be custom-made and their geometries should be very specific.[1] Our skeleton is a framework of bones that supports us and provides protection to our tissues and organs in their smooth movement and working. The placements of 3D printed implants and the common alloys used to construct these are labeled in Fig. 1. The metallic implants have a wide range of applications in rehabilitation and retaining the functionality of the bone structure as shown in Figs. 1(a)–1(h). The types of various metals and alloys used in different parts of the human body are also labeled in Fig. 1(i). The 316L stainless steel (SS) has been widely used as an implant material including for the rehabilitation of maxillofacial, cranium and dental structures. Similarly, the most load-bearing parts that fail or malfunction in the orthopedic system (i.e. hip knee joints, femoral head, acetabular cup, etc.) are generally replaced with 316L SS implants due to their low cost, ease of manufacturing, good mechanical properties and acceptable biocompatibility.

Natural bones have an excellent ability to regenerate and an excellent self-healing power. The regenerative tendency of the bones also decreases and becomes fragile with ageing. Also, due to accidental damage, any serious trauma or systematic disease, the support of implants becomes necessary for the bone to recover its complete functioning. The need for bone replacement by an orthopedic implant in the last two decades has increased significantly. Patients with malfunctioned bones want quick and nonrepetitive treatments to perform the activities of life with the same quality.[4] The mechanical properties of the common bones that encounter serious injuries or failure are given in Table 1. The cortical bone has the highest elastic modulus (~15 GPa) and compressive yield strength (> 90 MPa) compared to the vertebra and femoral. However, sudden shock and accidental impact may cause the fracture of these bones and regenerative surgery may be required in these cases.

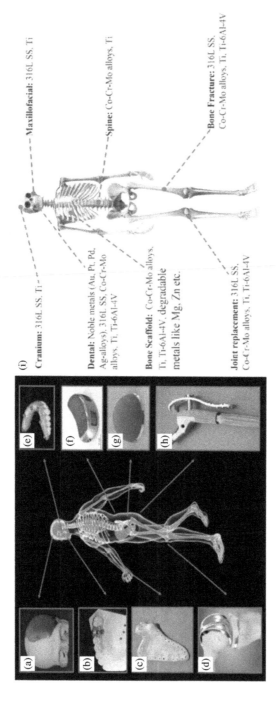

Fig. 1. The biomedical applications of 3D printing of biometals include (a) cranial prosthesis, (b) surgical guide, (c) scapula prosthesis, (d) knee prosthesis, (e) dental implants, (f) inter-body fusion cage, (g) acetabular cup and (h) hip prosthesis; and (i) details of different alloys used as an implant in the human body.[2,3] Figures reprinted with permission from Elsevier[TM].

Table 1. Mechanical properties of the bones in human anatomy.[5]

Human anatomy	Elastic modulus (GPa)	Ultimate tensile strength (UTS) (MPa)	Compressive yield strength (MPa)	Ultimate strain (%)
Cortical bone	15	60–130	90–200	1–5
Femoral head	2.9	135	68	—
Femoral condyle	4.9	—	32	—
Vertebra	1.5	—	4.1	—

1.2. Standard classification of orthopedic implants

To oversee the medical devices, the Federal Drug Association (FDA) has the Center for Devices and Radiological Health (CDRH). There are three criteria of the FDA for defining a medical device: (1) How a medical device affects the structure or function of the human body? (2) How it cures, treats, prevents, lessens or diagnoses the disease? (3) The main function is not achieved through any chemical action. All existing orthopedic implants are approved as medical devices. Medical devices are classified into three classes as listed in Table 2.

Some standards are related to the characterization and specifications of orthopedic implants as given in Table 3.

There are some standards separately related to the stainless steel, cobalt–chromium and titanium alloys for surgical implants and they are given in Table 4.

1.3. Current challenges

Metallic orthopedic implants have been used for more than 120 years. At the beginning of the development of orthopedic implants, poor corrosion resistance and inferior mechanical properties of the

Table 2. CDRH classification of medical devices.[6]

Classification	Risk level	Percentage devices	Controls	Clinical treatment	Examples
Class I	Low risk	55%	General	Not needed	Adhesive bandages, wheelchairs, tongue depressors
Class II	Moderate risk	40%	General and special	May be needed	Catheters, needles, contact lenses
Class III	High risk	5%	General and special	Always needed	Pace-makers, coronary stents, orthopedic implants

Table 3. International standards for orthopedic implants recognized by FDA.[7]

Standard	Description
ISO 10993-1	Biological evaluation of medical devices — Part 1: Evaluation and testing within a risk management process
ISO 10993-11	Biological evaluation of medical devices — Part 11: Tests for systemic toxicity
ISO 10993-6	Biological evaluation of medical devices — Part 6: Tests for local effects after implantation
ASTM F1983	Standard practice for assessment of the compatibility of absorbable/resorbable biomaterials for implant applications
ASTM F763	Standard practice for the short-term screening of implant materials
ASTM F981	Standard practice for assessment of the compatibility of biomaterials for surgical implants with respect to the effect of materials on muscle and bone
ISO/TR 37137	Cardiovascular biological evaluation of medical devices — Guidance for absorbable implants

implant materials were the main challenges. In 1920, 316L SS was introduced having good mechanical properties and sufficient corrosion resistance.[9] With time, many efforts have been made to reduce the harmful effects of Ni leaching from SS. For instance, Ni-free

Table 4. ASTM standards for surgical implants.[8]

ASTM standard	Description
ASTM F138	Standard specification for wrought 18 chromium–14 nickel–2.5 molybdenum stainless steel bar and wire for surgical implants
ASTM F139	Standard specification for wrought 18 chromium–14 nickel–2.5 molybdenum stainless steel sheet and strip for surgical implants
ASTM F745	Standard specification for wrought 18 chromium–12.5 nickel–2.5 molybdenum stainless steel for cast and solution-annealed surgical implant applications
ASTM F899	Standard specification for wrought stainless steels for surgical instruments
ASTM F1314	Standard specification for wrought nitrogen strengthened 22 chromium–13 nickel–5 manganese–2.5 molybdenum stainless steel alloy bar and wire for surgical implants

metallic materials with acceptable mechanical properties and excellent corrosion resistance are also being used in the construction of orthopedic implants. In addition to the metallic implants, various other materials, i.e. polymer, ceramic, glass, etc., with exceptionally good corrosion and wear resistance properties are also receiving attention.[10] To ensure osseointegration, the implant materials should specifically exhibit excellent biocompatibility and nontoxicity in addition to exceptionally good tensile strength, modulus of elasticity and corrosion resistance in the biological environment. Biomechanical properties and biodegradability for temporary implant materials and nontoxicity are the important aspects to be considered for the development of new metallic biomaterials. The permanent implant materials must possess excellent resistance to dissolution when these interact with the tissues, blood and body fluids. The biological-grade 316L stainless steel is the most common material used for the fabrications of internal fixtures, i.e. wire, pin, rods, screws, plates, intramedullary nails, etc., due to its low price, acceptable biocompatibility, good mechanical strength and convenient fabricability. However, the main challenges in arthroplasty surgeries are the need for customized surgical instruments and the design of hip and knee implants depending on the patient's age, height, body location and functionality.[11–15] Owing to the complexities in the implant design, the major issue is the rapid fabrication of the desired implant without involving any post-treatment and joining steps. By using the conventional manufacturing routes, the customized production of implants having an intricate design is still a big challenge faced by surgeons. For instance, via the conventional manufacturing processes (casting, turning, rolling, milling and forging), it is difficult to produce an implant of complex structural design, i.e. interconnected porous triply periodic minimal surfaces (TPMS) scaffold. Orthopedic implants should be structurally integral with excellent properties so that repetitive surgeries can be avoided.

The 316L stainless steel has been used in orthopedic implants because it provides good mechanical properties, biocompatibility and reliable corrosion resistance at a low cost.[16] But the use of 316L is limited due to its susceptibility to localized corrosion in saline solutions. Almost 90% of the failure of the 316L SS implant is due to localized corrosion.[17] Owing to the aggressive body environment, the corrosion-induced failure and mechanical fracture of the 316L SS constitute the major portion of the existing issues as shown in Fig. 2(a). Similarly, wear damage and adverse tissue reactions are the other main causes of implant failure reported in the literature.

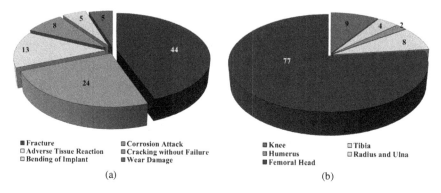

Fig. 2. History of the (a) possible reasons for the stainless steel implant malfunctioning and (b) failure of stainless steel implants at different anatomical positions.[5,18] Figures reproduced with permission from Elsevier™.

Based on the anatomical position of the implant, the quantitative representation of the implant failure was associated with the locations that experience extreme mechanical forces. For instance, ~86% of the failures were related to the femoral head and knee only as shown in Fig. 2(b).[18] From this data, it can be evaluated that the main drawback of 316L SS is its poor corrosion resistance, which could also deleteriously affect the mechanical performance of the implant. The mechanical, chemical and physical properties of any material greatly depend on its intrinsic microstructural features. For instance, the mechanical properties of 316L SS could be improved by post-thermomechanical and/or thermal treatment.[19] On the other hand, the microstructural features can also be tuned by optimizing the operating parameters of the additive manufacturing process to modify the mechanical and corrosion-resistant properties of 316L SS, which is the focus of this study.

1.4. Use of metal alloys as bioimplants

Orthopedic implants are used to maintain life routine, therefore the properties of the orthopedic implants should be compatible with those of the natural human anatomy. To improve the osteointegration of the implant materials, their properties should be comparable to the natural bones. In this regard, the most important properties of orthopedic implants are biocompatibility, high corrosion resistance and comparable mechanical properties. The first two issues are discussed rigorously in the following. However, a brief overview of the mechanical properties of common alloys with a particular focus on additively manufactured (AM) 316L SS is also given here.

Many metallic materials are used to manufacture fixtures and body implants for both *in-vitro* and *in-vivo* applications due to their good mechanical strength, high fracture toughness, sufficient corrosion resistance and biocompatibility. As a rule of thumb, mechanical failure could be avoided if the elastic modulus of the implant matches that of natural bones. In comparison to the characteristics of the human bone, the mechanical properties of the metallic and nonmetallic materials are different. As shown in Fig. 3(a), most

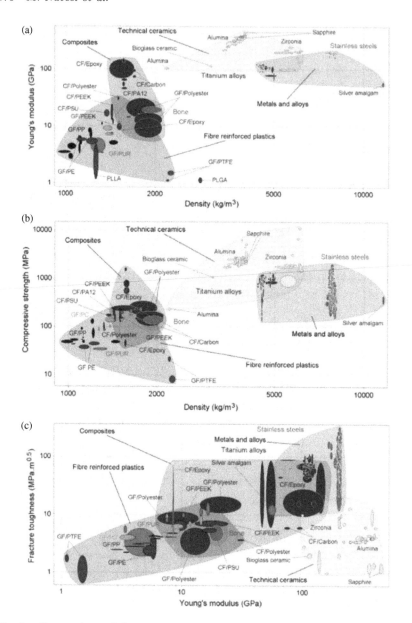

Fig. 3. Comparison of the mechanical properties of different biomedical materials (i.e. metals, ceramics, composites, polymers, etc.) with the human bone.[20] Figures reproduced with permission from Wiley[TM].

metals and alloys exhibit relatively high elastic modulus compared to bone. However, many composite and polymeric materials present comparable compressive strength and fracture toughness, which are essential requirements for orthopedic applications. As shown in Figs. 3(b) and 3(c), significantly large variations in the compressive strength (0.2–2 GPa) and fracture toughness (20–350 MPa·$m^{0.5}$) of stainless steel were associated with the variations in the microstructural features. Many other nonmetallic materials, i.e. ceramics, polymers and composites, used as orthopedic implants exhibit a wide range of mechanical properties depending on their processing parameters and intrinsic microstructures. For orthopedic applications, these materials are selected based on their intended application and desired functionality in the human body. For example, compared to stainless steel, ceramic materials exhibit sufficiently large compressive strength, excellent wear and corrosion resistance but a poor tensile strength. Similarly, the density of stainless steel is almost four times higher than that of human bone. Despite having several drawbacks, metallic implants are currently being used due to their superior mechanical properties. The physicochemical properties of metallic materials depend on many factors such as crystal structure, chemical composition, manufacturing route and post-processing thermal or thermomechanical treatments. Based on these characteristics, these metallic materials are selected for specific biomedical applications as enlisted in Table 5.

The most common metal alloys used as orthopedic implants and their mechanical properties are given in Table 6. Stainless steel, pure Ti and its alloys are the most commonly used materials in orthopedic implants due to their sufficiently large yield and tensile strengths. But, due to its intrinsic microstructural features, the SS registers a high modulus of elasticity but poor tensile strength compared to the Ti–6Al–4V alloy. Similarly, the Co–Cr–Mo alloy's yield and tensile strengths are comparable to SS. In contrast, pure Au, Pd, Mg and Zn metals have poor mechanical properties but the selection of these materials is related to their ultimate function. For instance, the Mg and Zn metals are used for a temporary fixture

Table 5. List of various metals and alloys used as bioimplants.[3]

Metal/alloy	Physiochemical properties	Material degradability	Applications
Pure Ti	Superior biocompatibility, good corrosion resistance and relatively inert	Nondegradable	Joint replacement, bone defect repair, spinal fusion, bone scaffold
Ti–6Al–4V alloy	Good biocompatibility, good corrosion resistance and relatively inert	Nondegradable	Joint replacement, bone defect repair, bone scaffold
Co–Cr–Mo alloy	Little biological toxicity, excellent corrosion and wear resistance	Nondegradable	Joint replacement, dental implant, bone scaffold
316L stainless steel	Acceptable biocompatibility, good corrosion resistance and bioinert material	Nondegradable	Joint replacement, dental implant, bone defect repair
Tantalum metal	Nontoxic, excellent biocompatibility and relatively inert	Nondegradable	Joint replacement
Gold	Excellent biocompatibility, superior corrosion resistance and completely unreactive	Nondegradable	Dental implant
Platinum metal	Excellent biocompatibility and superior corrosion resistance	Nondegradable	Dental implant, stent
Palladium metal	Excellent biocompatibility and superior corrosion resistance	Nondegradable	Dental implant
Silver	Good biocompatibility and good corrosion resistance	Nondegradable	Dental implant

Table 5. (Continued)

Metal/alloy	Physiochemical properties	Material degradability	Applications
Magnesium metal	Outstanding biocompatibility, promotes bone tissue regeneration, active nature and the corrosion product is nontoxic	Degradable with a fast degradation rate	Bone scaffold
Iron metal	Appropriate biocompatibility, less active properties and slower corrosion dissolution	Degradable with a slow degradation rate	Bone scaffold
Zinc metal	Good biocompatibility, active nature and dissolved species in the body are not poisonous	Degradable with a moderate degradation rate	Bone scaffold

Table 6. Mechanical properties of various metallic implants.[3,21]

Metallic implant	Yield strength (MPa)	Tensile strength (MPa)	Young's modulus (GPa)
316L SS	290–365	550–580	190–200
Selective laser melted (SLM) 316L SS*	450–500	550–640	165–185
Iron metal	50	540	200
Pure Ti	695	785	105
Ti–6Al–4V alloy	850–900	960–970	110
Tantalum metal	138–345	207–517	186
Co–Cr–Mo alloy	200–823	430–1028	230
Gold	25	130	80
Platinum metal	150	240	147

(Continued)

Table 6. (*Continued*)

Metallic implant	Yield strength (MPa)	Tensile strength (MPa)	Young's modulus (GPa)
Palladium metal	50	190	112
Silver	28	150	76
Magnesium metal	162	250	44.2
Zinc metal	30	37	96.5

Note: *Highly dense 316L SS having a porosity of 0.028 ± 0.021%, produced at 136-W laser power and a scanning velocity of 92.8 cm/s. The estimated quantitative information is given in this table, and these values may vary and depend on a large number of processing parameters.

for a short time and are biodegradable. On the other hand, SS and Ti alloys are selected for long-time fixation to maintain the overall integrity of the bone structure. Similarly, other properties such as strength-to-weight ratio, corrosion resistance and biocompatibility are equally important in the selection of implant material. The current focus is on the AM 316L stainless steel, which exhibits sufficiently large yield strength and comparable tensile strength and Young's modulus relative to the commercially available 316L SS. However, the superior corrosion resistance and biocompatibility of the AM 316L SS are related to its unique microstructural features as discussed in the following.

These properties affect the overall performance of any implant materials and are highly dependent on their microstructural features. However, the development of a specific microstructure is greatly influenced by the manufacturing and processing routes. In other words, the microstructural features could be tuned by adopting a suitable post-thermal, mechanical and/or thermomechanical processing of the implant material. Similarly, by changing the manufacturing route, the properties of the materials can be modified.

Additive manufacturing is a recently developed process to produce complex shapes and intricate designs of metallic structures with superior properties. For instance, via the additive manufacturing process, the customization of the implants is very easy and any shape/design can be produced in less time. Also, the whole implant can be made as a single unit in one step and does not require any post-assembly or joining procedures as associated with the conventionally manufactured parts. Briefly, the additive manufacturing process can be divided into two major categories, i.e. electron beam melting and selective laser melting. The selective laser melting process is further classified into the selective laser sintering (SLS), laser beam powder bed fusion (LB-PBF) and direct metal laser sintering (DMLS) processes. Considering these stimuli and comparing the resultant functional properties of the orthopedic implants, a comparison between the conventionally produced stainless steel and additively manufactured stainless steel is covered in this review. Recently, a detailed review on the additive manufacturing processes and properties, i.e. the microstructure, mechanical and corrosion behaviors of the common metallic alloys including AM 316L SS, describes the importance of additive manufacturing process parameters selection that could induce a significant impact on their overall performance.[22]

Kong et al.[23] studied the effects of post-heat treatment on the microhardness and tensile properties of the SLM 316L SS as shown in Fig. 4. The properties were compared with the commercially available wrought 316L SS. The Vickers hardness of the SLM 316L SS (without heat treatment) was significantly high (280 ± 08 HV) compared to the wrought 316L SS (200 ± 10 HV) as shown in Fig. 4(a) and was considered as due to the high density of dislocations in the as-produced SLM 316L SS. However, with the increase in time for solution treatment, a decrease in hardness was observed. A significant decay in the hardness of the SLM 316L SS was evident at a high solutionizing temperature (1200°C) with an increase in heat treatment time. The hardness becomes lower (< 200 HV) than that of wrought 316L SS by extending the solutionizing beyond 30 min at 1200°C. On the other hand, the solution treatment conducted at a relatively low temperature (at 1050°C) was found to be the least aggressive in terms of hardness deterioration. Similarly, the tensile properties of the as-quenched and solution-treated (at 1050°C and 1200°C for 2 h followed by quenching in water)

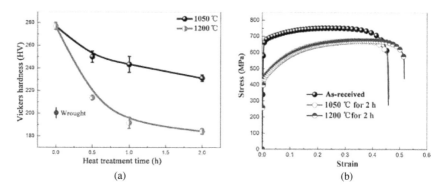

Fig. 4. Effects of heat treatment on (a) Vickers hardness and (b) tensile properties of the SLM 316L SS post-solutionized at 1050°C and 1200°C for 2 h followed by quenching in water.[23] Figures reprinted with permission from Elsevier^{TM}.

SLM 316L SS samples were calculated from the stress–strain curves as shown in Fig. 4(b).

As-received SLM 316L SS presented relatively higher tensile strength (750 MPa) and yield strength (640 MPa) compared to its wrought counterpart, which exhibited approximately 600 MPa and 300 MPa of tensile and yield strengths, respectively. On the other hand, a very minor difference in the elongation percentages at fracture (~40 ± 03%) of the as-received and low-temperature heat-treated (1050°C) SLM 316L SS samples was observed. However, at a high solution treatment temperature (1200°C), a significant increase in elongation percentage (~51%) was registered by SLM 316L SS. Compared to the as-received SLM 316L SS, a significant decrease in yield strength (~35%) was observed after solution treatment at a high temperature (1200°C for 2 h) attributing to the decrease in the dislocation density.

Ni et al.[24] also reported a relatively higher hardness (~50 HV) of SLM 316L SS compared to the wrought counterpart. This was attributed to the rapid solidification rate (~10^3–10^8 K/s) during the selective laser melting process and increased concentration of dislocation in the vicinity of subgrain boundaries. The effects of different laser scanning speeds (800, 1083, 1200 and 1400 mm/s) on the Vickers

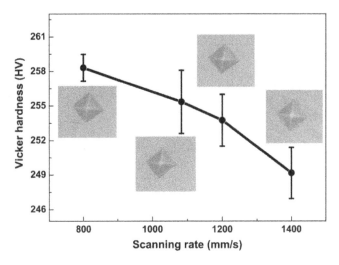

Fig. 5. Variation of Vickers hardness of the SLM 316L SS manufactured at different laser scanning speeds; results shown here are the averages of five test values.[24] Figure reprinted with permission from SpringerLink™.

hardness (HV) of SLM 316L SS samples were also studied. The Vickers hardness test is performed five times on each sample and the results are shown in Fig. 5. It was evident that an increase in laser scanning speed induced deleterious effects on the Vickers hardness due to the increase in the number and size of the defects that could form at high laser scanning speed. For instance, the hardness at 800-mm/s scan rate (~259 HV) decreased to ~250 HV at 1400 mm/s associated with the internal defects formed in the microstructure at high scanning speed.

In another study, the effects of laser power on the mechanical properties of the SLM 316L SS samples were investigated and compared with the quenched wrought 316L SS.[25] The SLM 316L SS samples were produced at 120-W and 195-W laser powers and the mechanical properties were determined before and after 288 h of immersion into the Simulated body fluid (SBF) solution as given in Table 7. The engineering stress–strain curves of the as-quenched wrought and SLM 316L SS samples produced at various laser powers are shown in Fig. 6. As evident from the results, the ultimate

Table 7. Mechanical properties of the wrought quenched and SLM 316L SS samples produced at 120-W and 195-W laser powers.[25]

Immersion time (h)	Material	UTS (MPa)	Yield strength (MPa)	e_f (%)
0	Quenched	620.6 ± 2.5	299 ± 4.8	49.7 ± 2.1
	SLM 120 W	610.1 ± 7.7	473.1 ± 8.4	35.5 ± 3.4
	SLM 195 W	618.6 ± 4.3	487.1 ± 7.2	55.7 ± 2.5
288	Quenched	618.1 ± 2.9	295.4 ± 2.7	44.8 ± 1.8
	SLM 120 W	602.9 ± 6.2	477.9 ± 4.1	17.9 ± 3.2
	SLM 195 W	614.3 ± 3.6	470.8 ± 6.8	47.3 ± 3.3

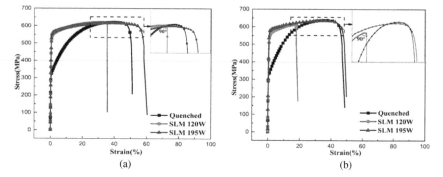

Fig. 6. Stress–strain curves of the wrought quenched and SLM 316L SS samples produced at different laser powers (120 W and 195 W) (a) before (0 h) and (b) after 288 h of immersion into SBF solution.[25] Figures reprinted with permission from Elsevier^{TM}.

tensile strengths of the wrought and SLM 316L SS samples (independent of the laser power) were comparable and there was found to be a negligible influence of immersion into SBF. However, a significantly higher yield strength of the SLM 316L SS sample compared to the wrought counterpart was quantified. Also, the yield strength was found to be increased from 473.1 MPa to 487.1 MPa with an increase in laser power from 120 W to 195 W. Similarly, a significantly higher elongation percentage (at fracture) (e_f) of 55.7% was registered by the SLM 316L SS produced at 195-W laser power compared to 35.5% at low (120-W) laser power. Compared to the nonimmersed samples, after 288 h of immersion into SBF, there was relatively a negligible influence on the UTS values of the wrought and SLM 316L SS samples produced at different laser powers. However, a considerable effect on the yield strength and e_f was observed after immersion. For instance, the yield strength of SLM 316L SS produced at 195 W decreased from 487.1 ± 7.2 MPa to 470.8 ± 6.8 MPa. In general, a decrease in the e_f values of all the samples was observed after immersion into SBF as quantified in Table 7. An appreciable decrease in e_f (from 35.5% to 17.9%) of the SLM 316L SS sample produced at 120-W laser power was associated with the occurrence of severe localized corrosion in SBF. The effect of laser power on the surface finish revealed a decrease in surface roughness with an increase in laser power and a decrease in the laser scan speed. It has also been reported that the yield strength, tensile strength and elastic modulus of the as-produced SLM 316L SS were comparable to the post-print heat-treated counterpart (soaking at 900°C for 45 min followed by slow cooling to room temperature in 6 h). However, a significant decrease in the elongation percentage was observed after post-print heat treatment.[26]

2. The 316L Stainless Steel for Orthopedic Implant Applications: Effect of the Manufacturing Route

In this section, the 316L stainless steel orthopedic implants manufactured by the conventional method and additive manufacturing process have been discussed. The 316L stainless steel is being used in orthopedic implants and prostheses owing to its good biocompatibility, superior corrosion resistance and better mechanical strength as discussed rigorously in the literature.[27-30] Recently, 3D printing or additive manufacturing has emerged that provides superior advantages over the conventional manufacturing processes. This technology allows flexibility in the design and manufacturing of intricate shapes with superior properties. Additively manufactured parts made of 316L SS result in the formation of the nonconventional subgranular structure due to its high cooling rate (10^3–10^8 K/s) and the autogenous heat treatment of the pre-deposited layers.[31,32] Additive manufacturing is an emerging method to produce customized orthopedic metallic implants of varying properties. The properties can be tuned by manipulating the manufacturing parameters. Similarly, additive manufacturing provides flexibility to design implant structures with desirable porosity and morphologies. This review compares the microstructural features, mechanical and electrochemical properties of the wrought and AM 316L SS samples specifically for their use in biomedical applications.

2.1. *Comparison of the microstructural features of wrought and AM 316L SS samples*

The microstructures of wrought and additively manufactured 316L SS samples are shown in Fig. 7. The wrought manufactured

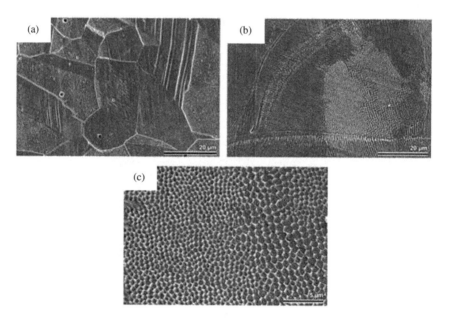

Fig. 7. The microstructures of 316L SS: (a) wrought sample, (b) AM sample and (c) magnified view of AM microstructure.[33] Figures reprinted with permission from Elsevier^{TM}.

316L SS shows equiaxed coarse grain structure and sharp grain boundaries [Fig. 7(a)]. On the other hand, fine cellular and subgrain columnar-shaped structures were exhibited by the AM sample [Fig. 7(b)]. These specific microstructural features of AM 316L SS were associated with the layer-by-layer deposition of molten metal and high cooling rates during solidification.

The wrought manufactured sample showed typical polygonal-shaped coarse austenitic grains, whereas the SLM sample depicted fine interconnected subgrain structure as shown in Fig. 8(a). These subgrains were decorated with the colonies of fine cellular and columnar subgrain structure [Fig. 8(b)] that was mainly associated with the layer-by-layer deposition of molten metal and high cooling rate. The formation of 400–800-nm polygonal-shaped equiaxed

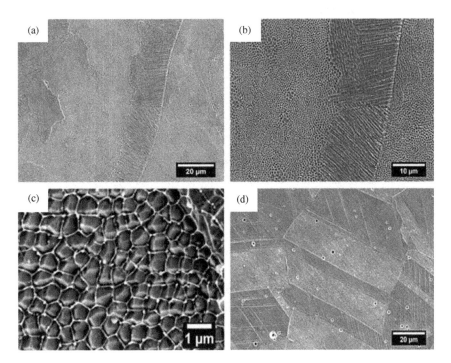

Fig. 8. SEM images of 316L SS: (a) SLM sample, (b) columnar and cellular subgrains in the coarser grain, (c) columnar and cellular subgrains at the center of the melt pool and (d) wrought sample.[34] Figures reprinted with permission from Elsevier^{TM}.

cellular subgrains was evident in the SLM sample as shown in Fig. 8(c). On the other hand, the wrought sample showed equiaxed faceted coarse austenitic grain structure as shown in Fig. 8(d). Twin boundaries can also be seen in these coarse grains (~40–70 µm) of wrought 316L SS. Various studies have explained the difference in the microstructural features of AM 316L SS in comparison to the wrought counterpart; however, the results reported in these studies displayed consensus on the microstructural variation owing to the operating conditions applied during the selective laser melting process of 316L SS.[32,34–36]

Kazemipour et al.[37] also compared the microstructural features of the SLM and wrought 316L SS samples. Side and top views of the samples were observed by using SEM as illustrated in Fig. 9. In both views, the selective laser melting-fabricated sample possessed a melt pool with a woven network but the shape of the melt pool was elongated. An oval-shaped melt pool was apparent in the top view and overlapped semicircular grain boundaries were evident in the side view. The microstructure of the wrought 316L SS sample reveals regular polygonal equiaxed austenite grains having intragranular twin boundaries highlighting the previous mechanical deformation.

Bartolomeu et al.[38] analyzed the microstructure of 316L SS manufactured by different processing routes, i.e. casting, hot pressing (HP) and selective laser melting. A comparison of the effects of processing parameters on the microstructures is shown in Fig. 10. Figure 10(a) shows that rectangular-shaped grains with an average size (length) of 91 ± 17 µm were produced during casting, while in Fig. 10(b) equiaxed grains of size 25 ± 4 µm were produced by the HP technique. Figure 10(c) shows the SLM sample microstructure with laser molten pool as reported in other studies.[39,40]

Comparing the cooling rates, 0.5, 3 and 10^3–10^8 K/s mainly corresponded to the casting, HP and selective laser melting processes and are considered responsible for such significant variations in the microstructure.[41] Stendal et al.[42] also compared the microstructures of both AM and wrought samples by electron backscattered diffraction (EBSD) technique as shown in Fig. 11. The wrought 316L SS was cast and rolled while the AM sample was produced by using

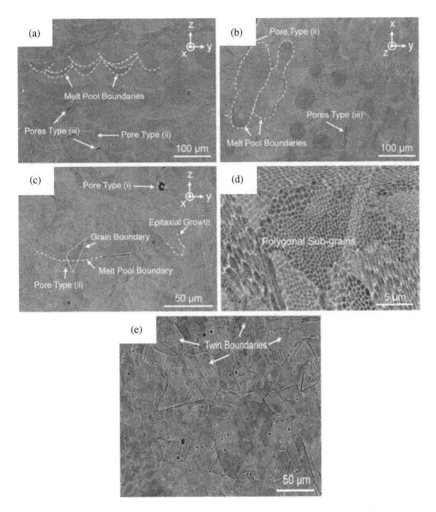

Fig. 9. SEM microstructures of 316L SS: (a), (c) side views, (b), (d) top views of the SLM sample and (e) the wrought manufactured sample.[37] Figures reprinted with permission from Springer™.

the selective laser melting technique. The EBSD scan of the AM sample shows grain elongation in the build direction with an average grain size of 10 μm, whereas the average grain size of the wrought sample was measured to be 20 μm. Compared to the wrought 316L SS, a significant number of grains of AM 316L SS presented (101) orientation in which the planar density of (111) grains was found to be high.

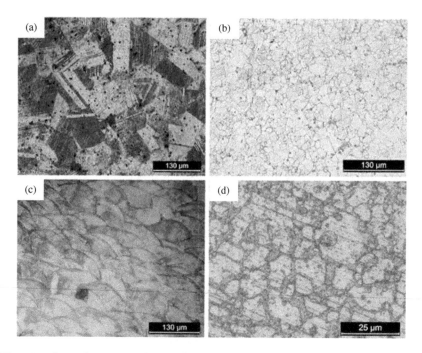

Fig. 10. Optical micrographs of samples etched with aqua regia reagent, manufactured by (a) casting, (b) HP, (c) selective laser melting and (d) selective laser melting (etched with the reagents picric acid and HCl).[38] Figures reprinted with permission from Elsevier[TM].

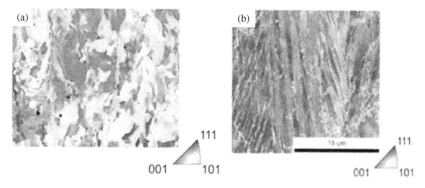

Fig. 11. EBSD scans of 316L SS: (a) AM sample and (b) wrought sample.[42] Figures reprinted with permission from Springer[TM].

Chao et al.[43] performed annealing of the SLM 316L SS and compared the microstructural features with the as-received and annealed wrought 316L SS sample.

Figure 12(a) shows the microstructure of the as-printed (SLM) 316L SS, indicating the formation of a polygonal granular structure containing intragranular fine-grained structure. In Figs. 12(c) and 12(h) are shown the band contrast maps of annealed SLM and wrought 316L SS samples that were obtained by EBSD in which the grain boundaries are represented by the green, black and red lines with the misorientation angle (θ) > 15°, 5° ≤ θ <15° and

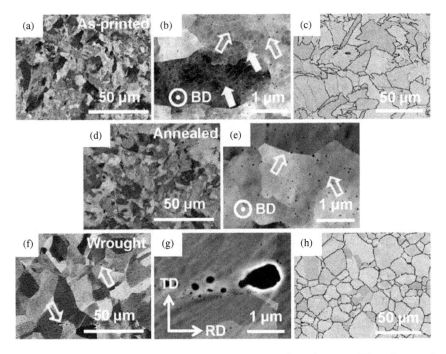

Fig. 12. (a), (b) SEM micrographs of as-printed (SLM) 316L SS at low and high magnifications, (c) the band structure of as-printed 316L SS, (d) the microstructure of SLM 313L SS obtained after annealing, (e) low- and high-magnification images of the annealed SLM 316L SS, (f), (g) microstructures of wrought annealed 316L SS and (h) band structure of annealed 316L SS (wrought) obtained from the EBSD analysis.[43] Figures reprinted with permission from Elsevier[TM].

60° <111> Σ3 twin boundaries, respectively. The EBSD images of as-printed and annealed SLM 316L SS samples are shown in Figs. 12(b) and 12(e), which depict the presence of nano-inclusions as highlighted by the solid and open arrows. On the other hand, the microstructure of the wrought 316L SS sample consists of the conventional recrystallized microstructure having equiaxed grains and free from δ-ferrite and σ-phase. Figure 12(f) shows the typical coarse austenitic grain structure of the wrought sample. The build-up, rolling and transverse directions in the wrought sample are shown by BD, RD and TD, respectively. The presence of nonmetallic inclusions and relatively high density of subgrain boundaries indicated the mechanically deformed microstructure [Fig. 12(g)]. In another study,[25] the microstructural features of the as-quenched 316L SS were compared with the SLM 316L SS produced at different laser beam powers. As a function of laser beam power, the grain sizes and morphologies of the grain boundaries of SLM 316L SS were compared with the grain structure of the quenched 316L SS as evaluated from the EBSD analysis (Fig. 13).

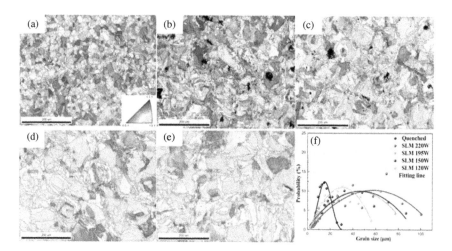

Fig. 13. EBSD micrographs of the SLM 316L SS: (a) quenched sample; samples manufactured at (b) 120 W, (c) 150 W, (d) 195 W and (e) 220 W; and (f) the trends showing the distribution of grain size.[25] Figures reprinted with permission from Elsevier[TM].

More regular polygon grains formed in the wrought 316L SS sample, whereas irregular-shaped granular structure was formed in 316L SS fabricated by the selective laser melting process. By increasing the laser beam power, the grain size was increased and the depth and diameter of the voids were also increased due to incomplete melting of the powder particles. In other words, the quenching of the wrought 316L SS sample produced more regular and polygonal-shaped grains compared to AM 316L SS sample grains, which were found to be coarse and irregular. This difference in the microstructures solely depends on the temperature gradient during AM 316L SS fabrication by laser scanning. In Figs. 13(b)–13(e) of AM 316L SS, the presence of black spots in the microstructure indicates the formation of porosity and a decrease in the pore size was observed with an increase in laser power > 150 W. However, at the laser powers of 195 W and 220 W, no porosity in the microstructure was observed as shown in Figs. 13(d) and 13(e). This effect is attributed to the complete melting of the powder particles at larger laser power resulting in the creation of fewer voids and porosities in the microstructure. Figure 13(f) presents the distribution of grain sizes and it can be seen that relatively fine grains were formed in the quenched 316L SS sample. On the other hand, the grain size of AM 316L SS exhibited a direct relation with the laser power indicating the formation of large-sized melt pool and solidification on the pre-solidified deposited layer that established a favorable thermal gradient for grain growth. By increasing the laser power, the size of the grains increased with a substantial decrease in porosity and void density. Shrestha et al.[44] investigated the effects of layer orientation and surface roughness on the mechanical properties and fatigue life of 316L stainless steel fabricated via the LB-PBF additive manufacturing process. In the as-built surface conditions, the vertical and horizontal specimens were manufactured. The microstructural features of LB-PBF 316L SS diagonally (D), vertically (V) and horizontally (H) built specimens relative to a plane perpendicular to the build plate were examined using a digital microscope and are shown in Figs. 14(a)–14(c), respectively.

194 M. Nabeel et al.

In additive manufacturing, a layer-by-layer building of the material results in the epitaxial growth of the grain structure. For instance, the pre-solidified layer provides an undercooling effect and promotes the formation of nuclei and the growth of the deposited metal. To verify these effects, the microstructural features of LB-PBF 316L SS D, V and H built specimens, on a plane perpendicular to the build plate, were observed using a digital microscope as shown in Figs. 14(a)–14(c), respectively. The red dashed lines in the figure indicated that the solidification of the melt pool in each specimen was parallel to the direction of the build plate. The intercept method was applied to approximately 50 grains and the aspect ratio was found to be the highest for the horizontally built specimen and remained the lowest for the vertically built specimen. The directional orientation of the grains associated with the solidification

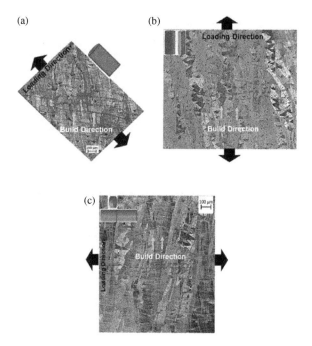

Fig. 14. Microstructures of LB-PBF 316L SS specimens built in the (a) diagonal, (b) vertical and (c) horizontal directions with respect to a plane perpendicular to the build plate.[44] Figures reprinted with permission from Elsevier[TM].

process could adversely affect the mechanical properties. These results indicated an increase in the anisotropic mechanical behavior of LB-PBF 316L SS.[45] For instance, it was reported that the specimen built vertically presented a larger elongation percentage before failure due to the orientation of the grains parallelly to the loading direction. On the other hand, the highest fatigue resistance was registered by a horizontally built specimen with the lowest fatigue life reported by the diagonally built specimen.

It has also been investigated the effects of layer orientation and surface roughness on the mechanical properties and fatigue life of 316L stainless steel fabricated via the LB-PBF process.[44] In the as-built surface conditions, the vertical and horizontal specimens were manufactured while in the machined surface conditions, specimens in the diagonal direction were fabricated. In additive manufacturing, a layer-by-layer building of the specimen results in the epitaxial growth of the specimen, i.e. when a layer is solidified over the previously solidified layer, the growth occurs in the perpendicular direction relative to the build direction. To observe the effects of layer orientation, the intercept method was used in which almost 50 grain boundaries are intercepted by a line at different locations. A clear variation in the grain size of each specimen was observed. The aspect ratio was also calculated, and the aspect ratio of the H built specimen was found to be higher than the V built specimen. The orientation of these elongated grains to the loading direction (i.e. longitudinal direction of specimens) varies among the H, V, and D built specimens, which resulted in the anisotropicity in the mechanical properties.[45,46] It was observed that the V built specimen showed the highest elongation to failure due to the parallel orientation of the grains to the loading direction. On the other hand, highest fatigue resistance was observed by the H built specimen, whereas the D built specimen exhibited the lowest fatigue strength.

2.2. *Comparison of microstructural differences*

The XRD patterns of both AM and wrought 316L SS samples presented almost identical peaks validating the presence of the

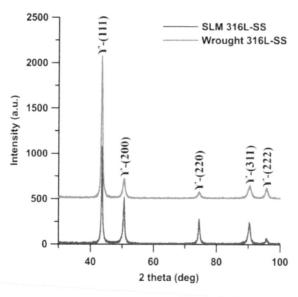

Fig. 15. The XRD patterns of the SLM and wrought 316L SS samples.[37] Figure reprinted with permission from Springer^{TM}.

austenitic (γ) phase as shown in Fig. 15. However, the intensity of the diffraction peak originated at approximately 43° corresponding to the γ-(111) plane was found to be higher for the wrought sample compared to the AM sample. On the other hand, the intensities of the peaks associated with the γ-(200), γ-(220) and γ-(311) planes were more intense for the SLM sample compared to the wrought sample. These results indicated that the γ-(111) plane orientation of the grain was suppressed in the case of the AM sample. These variations have been related to laser power and build direction. For instance, high power of the laser and 90° build direction preferred the (220) plane orientation, whereas low power of the laser and build direction of 45° resulted in the preferred plane orientation of (111).[34]

The absence of any peak associated with the impurity phase indicated the formation of a pure austenitic phase as evident from the XRD patterns of the AM and wrought 316L SS samples as shown in Fig. 16. The broader diffraction peaks in the XRD pattern of the SLM sample indicated the formation of fine crystallite size,

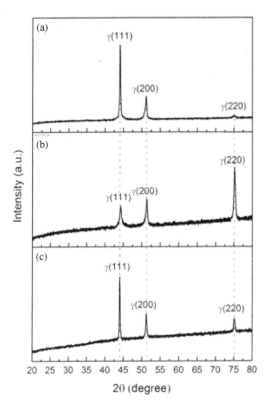

Fig. 16. Comparative X-ray diffraction patterns of (a) powder, (b) SLM and (c) wrought 316L SS samples.[34] Figure reprinted with permission from Elsevier™.

which is evident in the micrographs shown previously (Figs. 8 and 9).[34,46] Chao et al.[43] also explained the difference in the diffraction patterns of the wrought and SLM 316L SS samples. TEM–EDS analysis revealed the presence of nonmetallic irregularly shaped MnS inclusions in the vicinity of austenite grain boundaries in the wrought sample as shown in Fig. 17. The line EDS mapping indicated large concentrations of Mn, Al, Si and S species across the precipitate present in the wrought samples [Fig. 17(a)]. However, in the case of the SLM sample, very fine particle was mostly composed of Mn- and O-enriched phase as shown in Fig. 17(b).

The XRD patterns of the SLM 316L SS samples manufactured at various scanning speeds (800, 1083, 1200 and 1400 mm/s) were studied and the results are shown in Fig. 18. A typical crystal

198   M. Nabeel et al.

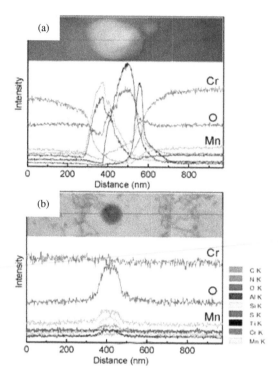

Fig. 17. TEM–EDS line profiles of the inclusions in (a) wrought and (b) as-printed 316L stainless steel samples.[43] Figures reprinted with permission from Elsevier^{TM}.

orientation without the formation of the martensitic phase was exhibited by these samples. No significant difference in the XRD patterns was observed. It was reported that the SLM samples produced at various sweep rates resulted in the formation of low-angle grain boundaries which restricted the nucleation of martensite in the microstructure. In another study, the XRD patterns of the SLM 316L SS manufactured at different laser powers were compared with the conventionally produced quenched 316L SS as shown in Fig. 19. The major diffraction peaks at approximately 43°, 51° and 74° were attributed to the γ-(111), γ-(200) and γ-(220) peaks, respectively, of the 316L stainless steel. In the quenched 316L SS, the intensities of the peaks associated with the γ-(111) and γ-(200) planes were significantly higher than in the SLM sample. The peak intensity of the

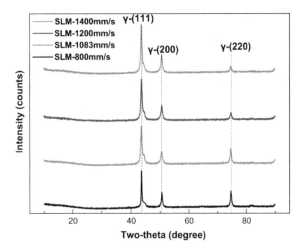

Fig. 18. XRD patterns of the SLM 316L SS formed under different laser scanning speeds.[24] Figure reprinted with permission from Springer[TM].

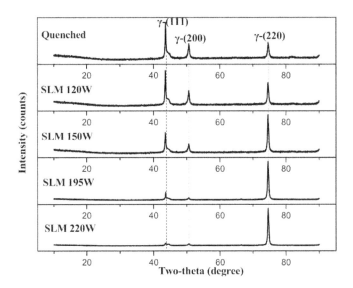

Fig. 19. XRD spectra of the as-quenched and SLM 316L SS samples produced at varying laser powers.[25] Figure reprinted with permission from Elsevier[TM].

γ-(220) plane was directly related to laser power. However, the intensity of the diffraction peak associated with the γ-(111) plane was greatly decreased with the increase in laser power.

2.3. Overview of the corrosion behavior of AM 316L SS in biological media

Recently, many studies explained the corrosion properties of AM 316L SS and compared its performance against the wrought stainless steel. For instance, the corrosion behaviors of additively manufactured and wrought manufactured 316L SS samples in the phosphate-buffered solution (PBS) with and without citrate ions were investigated. It is reported that the passivation behavior of 316L SS was strongly influenced by the presence of citrate ions in the PBS due to the formation of complex species. Citric acid does not provoke the active dissolution of 316L SS at low pH due to its capability to passivate the surface. To evaluate the susceptibility of wrought and AM stainless steel samples toward localized corrosion, the cyclic potentiodynamic polarization (CPP) tests were conducted (as shown in Fig. 20) and the quantitative information extracted from these tests is given in Table 8.

Compared to the wrought 316L SS, the improved passive film stability of the AM sample corresponded to its better corrosion resistance under oxidizing conditions. The E_{corr} potential of the SLM sample was more positive than that of the wrought sample indicating the SLM sample has lower susceptibility to corrosion.

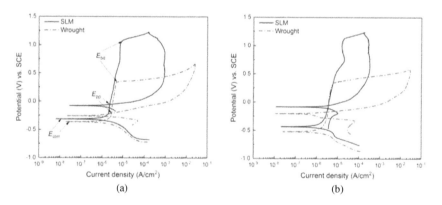

Fig. 20. CPP curves of the wrought and additively manufactured 316L SS samples obtained in (a) PBS solution and (b) citrate-containing PBS solution.[34] Figures reprinted with permission from Elsevier^{TM}.

Table 8. Quantitative data of the wrought and AM 316L SS samples evaluated from the CPP curves that compare the effects of pH, Cl^- and citrate ions on the corrosion behaviors of wrought and SLM 316L SS samples.[34]

Electrolyte	Sample	E_{corr} (mV$_{SCE}$)	I_{corr} (μA/cm^2)	I_{pass} (μA/cm^2)	E_{bd} (mV$_{SCE}$)	E_{pp} (mV$_{SCE}$)	$E_{bd} - E_{corr}$ (mV)	$E_{pp} - E_{corr}$ (mV)
PBS (pH = 7.2)	Wrought	−351.3 ± 11.4	0.79 ± 0.05	2.26 ± 0.06	347.3 ± 13.8	−204.3 ± 12.5	698.7 ± 11.8	147 ± 7.8
	SLM	−308.3 ± 15.1	0.69 ± 0.04	1.97 ± 0.06	1084 ± 9.3	−67 ± 10.6	1393 ± 22.9	241.3 ± 23
Citrate-containing PBS (pH = 7.2)	Wrought	−520.7 ± 23	1.16 ± 0.04	2.71 ± 0.06	328.4 ± 16.2	−177 ± 12.3	849 ± 8.7	343.7 ± 34.7
	SLM	−428.4 ± 12.9	1.07 ± 0.03	2.65 ± 0.04	1093.3 ± 8.3	−64.3 ± 8.7	1521.6 ± 11.6	364 ± 20.8
Pure citrate buffer (pH = 6)	Wrought	−473.3 ± 19.1	0.30 ± 0.04	1.62 ± 0.04	407.6 ± 14.2	54 ± 8.5	881 ± 32.2	419 ± 26.7
	SLM	−435 ± 10.5	0.21 ± 0.03	1.52 ± 0.05	1006.7 ± 13.2	—	1441.7 ± 6.8	—
Citrate + Cl^- (pH = 6)	Wrought	−503.7 ± 15	0.92 ± 0.02	2.89 ± 0.05	308.3 ± 15.9	−113 ± 14.2	812 ± 30.8	390.6 ± 4
	SLM	−445 ± ±12.1	0.91 ± 0.03	2.68 ± 0.06	1015 ± 8.2	—	1460 ± 19.1	—
Citrate + Cl^- (pH = 7.2)	Wrought	−504 ± 9.8	1.11 ± 0.03	2.87 ± 0.05	305.3 ± 13.8	−181.7 ± 7.8	809.3 ± 8.1	322.3 ± 5
	SLM	−423 ± 7.2	1.05 ± 0.04	2.47 ± 0.04	930.4 ± 22.5	−89 ± 8.9	1353.3 ± 24.2	334 ± 8.5

The wrought sample had a higher corrosion current density compared to the SLM sample. The appreciably high resistance to passive film breakdown and the comparatively large passive film protection capability exhibited by the SLM sample are a representation of its better performance in PBS solution containing citrate and chloride species.

The extended passive film stability and positive film breakdown potential (0.85 V versus SCE) of the AM 316L SS in highly acidic (pH = 1) and oxidizing environments were also evaluated. Compared to the wrought 316L SS, the improved passive film stability of the AM 316L SS was attributed to the fine cellular grain structure that facilitated the formation of a compact and uniform passive film.[47] The cyclic voltammetry (CV) tests of the wrought 316L SS and SLM 316L SS were performed to investigate the interaction of ionic species on the surface (i.e. citrate and Cl⁻ ions) as shown in Fig. 21. For the SLM samples, two anodic peaks and two cathodic peaks at almost similar potential were observed in both electrolytes. Anodic peaks indicated the active-to-passive state transition and peak A_1 corresponded to the active dissolution of Fe and formation of $Fe(OH)_2$. The origin of peak A_2 was claimed to be associated with the oxidation of $Fe(OH)_2$ to $Fe(OH)_3$ species. At potential > 0.3 V versus SCE, a slight increase in current highlighted the oxidation of

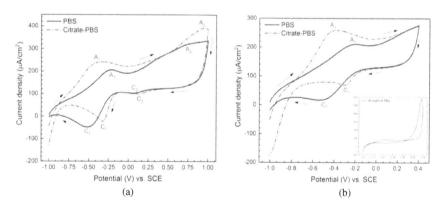

Fig. 21. Cyclic voltammograms of 316L SS at 50-mV/s scan rate in citrate-containing PBS and PBS solutions: (a) SLM sample and (b) wrought sample.[34] Figures reprinted with permission from Elsevier™.

Fe(OH)$_2$. However, the stable Cr(OH)$_3$ species contributed to the passive character of the SLM 316L SS. While in the case of the wrought sample, the appearance of A$_1$ and C$_1$ peaks represented the active dissolution of Fe from the surface of 316L SS into the PBS and citrate-containing PBS solutions. In other words, the relatively large anodic current and absence of the A$_2$ peak presented by the wrought sample compared to SLM 316L SS corresponded to its active dissolution. These trends were attributed to the presence of nonmetallic inclusions in the microstructure of wrought 316L SS that triggered its aggressive localized dissolution into the PBS solution containing citrate ions as shown in Fig. 22.

According to the above discussion, it was concluded that the citrate ions induce a considerable influence on the corrosion behavior of both wrought and SLM samples. However, the AM 316L SS demonstrated superior corrosion resistance in aggressive biological media compared to the wrought 316L SS. A large number of pits of variable size were formed on the surface of the wrought sample compared to the SLM sample. The large pitting density and formation of relatively large-size pits can be observed on the surface of wrought 316L SS [Fig. 22(b)].

The widths of the largest pits formed on the wrought and SLM samples were found to be approximately 105 μm and 74 μm, respectively. These results indicate the relatively larger pitting tendency

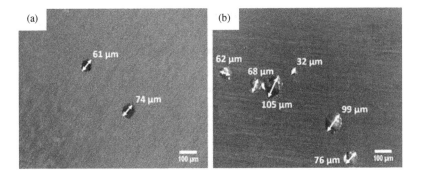

Fig. 22. Micrographs showing the pitting of (a) SLM and (b) wrought 316L SS samples after cyclic polarization tests in PBS solution.[34] Figures reprinted with permission from Elsevier^{TM}.

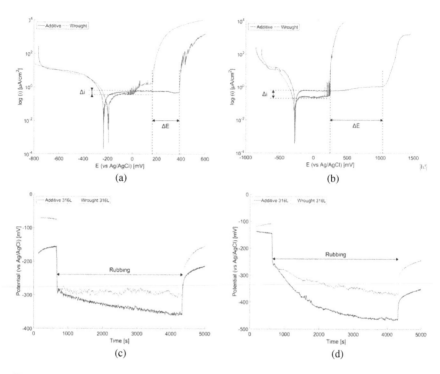

Fig. 23. Polarization curves and OCP trends of the AM 316L SS and wrought 316L SS in (a), (c) 0.9-wt.% NaCl solution and (b), (d) simulated body fluid.[42] Figures reprinted with permission from Springer^{TM}.

of the wrought sample compared to the AM 316L SS sample in PBS solution. Stendal et al.[42] compared the corrosion behaviors of the wrought and AM 316L stainless steel samples in 0.9-wt.% NaCl solution and a simulated body fluid containing protein, i.e. albumin. The wrought and AM samples were produced by the casting and selective laser melting processes, respectively. Figure 23 represents the polarization curves of the wrought and AM 316L SS samples in 0.9-wt.% NaCl solution and a simulated body fluid. The extended passive range and a larger breakdown potential of the AM 316L stainless steel sample in both 0.9-wt.% NaCl solution and simulated body fluid represented the greater stability of the passive film. These results also indicated that the passive film could withstand highly oxidizing conditions and could effectively reduce the dissolution of the AM 316L stainless steel. However, during mechanical scratching/rubbing of the passivated surface, the OCP was shifted to more negative potentials [Fig. 23(c)], which indicated the active localized dissolution tendency of the surface at the damaged area in the passive film. Compared to the wrought sample, the more negative shift in the OCP of the AM sample after mechanical rubbing and its slow repassivation tendency in both electrolytes are evident in Figs. 23(c) and 23(d). In other words, compared to the wrought 316L SS, the relatively more negative OCP of the AM sample after mechanical rubbing and sluggish tendency to regain the initial OCP represented the slow repassivation of the passive film. Maróti et al.[48] studied the corrosion behavior of the AM $X_2CrNiMo17$-12-2 steel (316L SS) in 0.9-wt.% NaCl solution as a function of exposure time. The polarization tests of the AM 316L SS were conducted in saline solution and the tests were repeated after 1, 2, 3, 4 and 5 h according to the ASTM G102 standard as shown in Fig. 24.

The corrosion rate of AM 316L SS was measured in 0.9-wt.% NaCl and seawater solutions. The corrosion rates of AM 316L SS

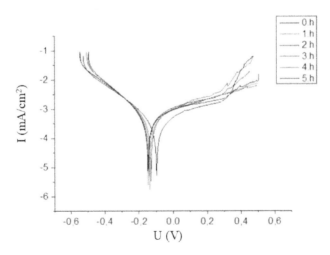

Fig. 24. Tafel polarization curves of AM 316L SS obtained in the 0.9-wt.% NaCl solution in 0–5 h at body temperature.[48] Figure reprinted with permission from SpringerLink^{TM}.

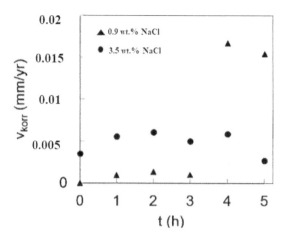

Fig. 25. The corrosion rates of AM 316L SS specimens as a function of time.[48] Figure reproduced with permission from Springer™.

reported were compared with the values (obtained in 3.5-wt.% NaCl solution) found in the literature. These results indicated that in physiological 0.9-wt.% saline solution, the corrosion rate did not exceed the acceptable corrosion rate limit of 0.13 mm/year for orthopedic implants at 37°C during the initial 3 h as shown in Fig. 25. However, the corrosion rate of 316L SS was measured to be ~0.005 mm/year and did not change during 5 h of immersion. The microstructural features of the AM 316L SS were also observed after 5 h of exposure to 0.9-wt.% saline solution as shown in Fig. 26. The presence of spherical-shaped particles and curved pattern dissolution marks was evident on the surface after exposure to physiological 0.9-wt.% saline solution at body temperature. These features were associated with the intrinsic microstructural features of the AM 316L SS that could form during the layer-by-layer deposition of the metal and directional solidification of the melt pool.

Chao et al.[43] also studied the effects of short-term annealing on the electrochemical properties of the wrought and SLM 316L SS samples. Cyclic potentiodynamic polarization tests were performed in 0.6-M NaCl at 25°C as shown in Fig. 27. It can be observed that the SLM–AM 316L SS sample without heat treatment presented a larger passive potential window compared to the wrought counterpart.

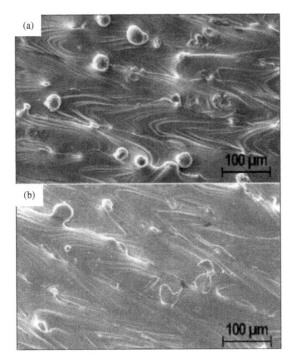

Fig. 26. The fractured surfaces of additively manufactured 316L SS (a) before and (b) after the corrosion test.[48] Figures reprinted with permission from SpringerLink.™

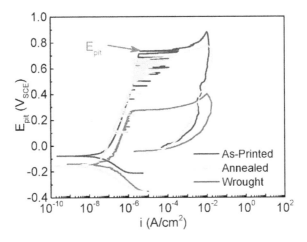

Fig. 27. CPP curves of 316L SS in 0.6-M NaCl for the wrought, SLM annealed and as-printed samples.[43] Figure reprinted with permission from Elsevier™.

The pitting potential (E_{pit}) of AM 316L SS sample was found to be 0.74 ± 0.02 V_{SCE}, whereas the wrought sample registered a low (0.23 ± 0.03 V_{SCE}) passive film breakdown potential. On releasing the residual stresses by short-term annealing of the as-printed 316L SS, the E_{pit} reduced to 0.56 ± 0.07 V_{SCE} but was still greater than that of wrought 316L SS.

Lodhi et al.[33] state that in stainless steel the diffusion of chromium on the surface is reduced due to refined grain structure and ultimately leads to the thickening of the oxide film that resulted in the improvement of the corrosion resistance of 316L SS.[49] Corrosion properties of 316L SS manufactured by the additive manufacturing process are influenced by the processing parameters and resultant microstructure that may develop during the transient cooling cycle.[50,51] The properties of additively manufactured 316L SS were compared with the wrought manufactured 316L SS. Charge transfer resistance (R_{ct}) and breakdown potential were also determined for the wrought and AM 316L SS samples in all three electrolytes (serum, 0.9-M NaCl and PBS).[33] The histograms shown in Fig. 28 represent the breakdown potentials and charge transfer resistances of both the wrought and AM 316L SS samples. It has been evaluated that compared to the wrought sample, the AM 316L SS exhibited larger passive film breakdown potential and limited dissolution

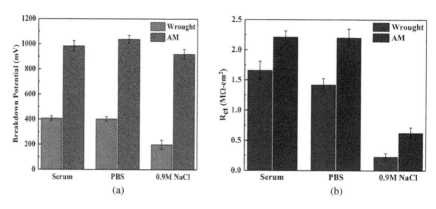

Fig. 28. Breakdown potentials and charge transfer resistances (R_{ct}) of 316L SS in serum, PBS and 0.9-M NaCl.[33] Figures reprinted with permission from Elsevier^{TM}.

tendency in all electrolytes. Overall, the 0.9-wt.% NaCl was found to be more aggressive compared to the serum and PBS solutions. However, the relatively better barrier characteristics of the AM sample compared to the wrought 316L SS were attributed to the fine microstructure and formation of a compact passive film on the former material. The highest R_{ct} shown by the AM sample in serum was due to the specific adsorption of proteins on the dense oxide film. The low R_{ct} of the wrought and AM 316L SS samples in 0.9-M NaCl and PBS indicated the localized dissolution of the passive oxide film due to the presence of a large amount of Cl⁻ species in the electrolyte.

In another study, the effect of different laser beam scanning speeds (800, 1083, 1200 and 1400 mm/s) on the corrosion tendency of the SLM–AM 316L SS was investigated as shown in Fig 29(a).[24] To determine the electrochemical properties of the AM samples, potentiodynamic polarization tests were performed and discussed rigorously in this study.

As shown in Fig. 29(a), the SLM sample manufactured at 800-mm/s laser beam speed showed the maximum corrosion resistance in 3.5-wt.% NaCl solution. On the other hand, the SLM sample manufactured at 1400-mm/s speed represented poor corrosion

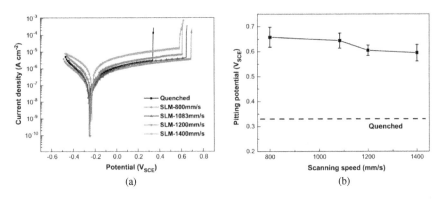

Fig. 29. (a) Potentiodynamic polarization curves of 316L SS at different laser beam speeds and (b) the pitting potentials of 316L SS in 3.5-wt.% NaCl solution evaluated from the potentiodynamic tests (potential scan rate was selected to be 0.1667 mV/s).[24] Figures reprinted with permission from Springer™.

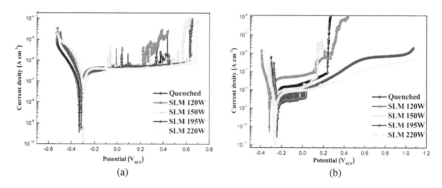

Fig. 30. Potentiodynamic polarization curves of the SLM and quenched 316L SS samples showing the trends just after immersion into SBF (a) and after 288 h of exposure (b) at 37°C.[25] Figures reprinted with permission from Elsevier^{TM}.

resistance compared to the quenched sample as shown in Fig. 29(b). The pitting potentials of all samples were higher than the quenched sample. Kong et al.[25] investigated the effects of laser power on the corrosion properties of the SLM–AM 316L SS and compared them with the traditionally quenched sample. Potentiodynamic polarization curves of the SLM and quenched 316L SS samples obtained just after exposure and after 288 h of immersion into SBF are shown in Fig. 30.

A relatively positive pitting potential of the as-quenched sample (435.9 mV$_{SCE}$) compared to the SLM 120-W sample (220.7 mV$_{SCE}$) after 1 h of immersion into SBF was observed, which decreased to 273.3 mV$_{SCE}$ and 125.8 mV$_{SCE}$, respectively, after 288 h of immersion. However, the pitting potential of the SLM 150-W sample was relatively more positive (573.8 mV$_{SCE}$) and decreased to 138.1 mV$_{SCE}$ in 288 h. The potentiodynamic polarization test performed on these samples also revealed a considerable difference in the pitting potentials. For instance, with the increase in laser power, an increase in the pitting potential was observed. This improvement in the pitting potential was attributed to the annihilation of MnS at high laser power during the selective laser melting process. On the other hand, at low laser power, the low pitting potential was related to the formation of a porous structure that acted as the pit

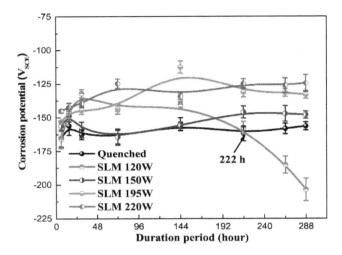

Fig. 31. Corrosion potentials of the SLM and quenched 316L SS samples immersed into SBF.[25] Figure reprinted with permission from Elsevier™.

initiation site. With the increase in laser power, the shift of corrosion potential to a more positive value highlighted the increase in corrosion resistance of SLM 316L SS in SBF as shown in Fig. 31. The as-quenched sample presented relatively negative potential compared to the SLM 316L SS samples produced at higher laser powers (> 120 W), e.g. SLM 220-W sample indicating an improvement in the corrosion resistance with an increase in laser power.

2.4. *Biocompatibility of the AM 316L SS: In-vitro and in-vivo analyses*

Jøraholmen *et al.*[52] examined the cell viability and biocompatibility of the AM 316L SS. The powder bed fusion method was used to manufacture 316L SS followed by magnetically assisted finishing (MAF) of the surface. During the MAF treatment, there was a uniform removal of material resulting in improved surface smoothness and the induction of less residual stresses. For instance, the sample without MAF treatment presented an arithmetic roughness of 0.25 µm compared to the high value (0.62 µm) registered by the as-produced sample. Mesenchymal stem cells (MSCs) were cultured on these finished surfaces and incubated for 48 h. After 48 h, no release of ions from the surface was observed and normal morphology of the cells and proliferation indicated the effective biocompatibility of the MAF-treated AM 316L SS as shown in Fig. 32. The morphology of the MSCs and adhesion on the surface of as-produced and MAF-treated 316L SS samples were estimated by examining the cell morphology and measuring the surface roughness. These results indicated the beneficial effect of the post-MAF treatment of AM 316L SS that could increase the cell proliferation and adhesion to the surface corresponding to its improved biocompatibility.

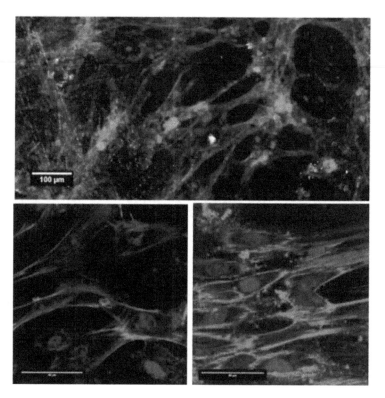

Fig. 32. Morphologies of the mesenchymal stem cells on the as-produced and MAF-treated SLM 316L SS samples.[52] Figures reprinted with permission from ECM[TM].

In another study, the biocompatibility of the SLM 316L SS with and without hydroxyapatite (HA) coating was measured and compared with the wrought 316L SS. The L929 cells were cultured and the proliferation tendency was determined after 24, 72 and 120 h of incubation.[53] Both disc-shaped SLM and wrought 316L SS samples were prepared and washed with PBS solution before immersion into the cell culture medium for cell proliferation analysis on the surface of these samples. The relative growth rate (RGR) of the cells was determined on each sample according to the GB/T16886.5-2003 standard. All samples presented good biocompatibility as per the standard test and may qualify for applications as implant materials. The cell proliferation tendencies of the 316L/HA, as-printed and as-cast 316L stainless steel samples were compared with the controlled Tissue Culture Polystyrenes (TCPS) medium after 24, 72 and 120 h of incubation. The AM 316L SS sample with and without coating showed the highest biocompatibility compared to the as-cast samples as evaluated from the cell viability study shown in Fig. 33. In the control medium, the improved cell viability of the

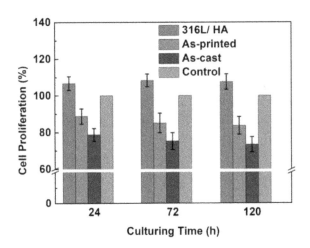

Fig. 33. Histograms showing the cell viability percentages of 316L/HA, AM and as-cast 316L stainless steel samples.[53] Figure reprinted with permission from MDPI™.

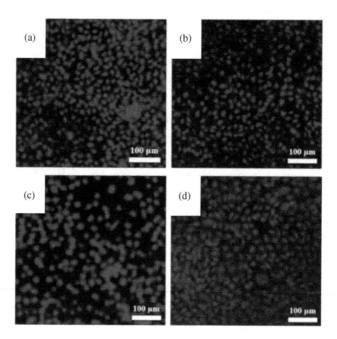

Fig. 34. Cell morphologies of L929 cells on (a) 316L SS with hydroxyapatite coating, (b) as-printed 316L SS, (c) as-cast 316L SS and (d) TCPS.[53] Figures reprinted with permission from MDPI™.

HA-coated 316L SS (> 100%) highlighted the beneficial effects of HA coatings on cell proliferation.

The cell morphologies on the 316L/HA, without HA coating, as-cast and TCPS samples after 120 h of incubation are shown in Fig. 34. It is evident from Fig. 34(a) that 316L/HA sample presented improved cell viability compared to the AM and as-cast 316L SS samples. The toxicity level of the as-printed 316L stainless steel was found to be increased under extended exposure and an appreciable decrease in the cell viability was observed after 120 h of incubation as shown in Fig. 35. However, significantly improved cell viability on the 316L/HA sample resulted in the highest cell density, which corresponded to its better performance compared to the as-printed 316L SS.

Comparatively low release of metallic ions, i.e. Fe, Cr, Mn, Ni and Mo, was observed in the case of 316L/HA sample than in

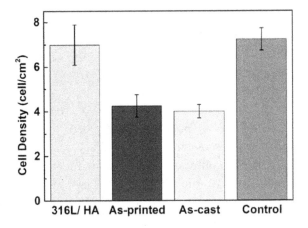

Fig. 35. Cell densities on the 316L SS and control samples after 120 h of incubation.[53] Figure reprinted with permission from MDPI[TM].

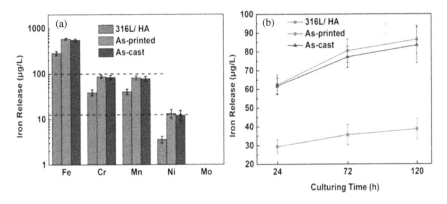

Fig. 36. (a) Quantitative values of metallic ions released from the 316L/HA, as-cast and as-printed 316L SS samples after 120 h of exposure and (b) the Fe release from 316L/HA, as-printed and as-cast 316L SS samples as a function of exposure time.[53] Figures reprinted with permission from MDPI[TM].

as-printed and as-cast samples as shown in Fig. 36(a). The low rate of ion release from the surface of 316L/HA sample corresponded to the barrier characteristics of the hydroxyapatite coating. Figure 36(b) shows the release of Fe from the sample surface after extended exposure to the culture medium. Significantly low release of Fe was observed with HA coating and it increased from 30 μg/L to 40 μg/L during 120 h of exposure. On the other hand, printed (without coating) and as-cast samples showed almost similar trends and exhibited approximately two times higher iron release under the same conditions. These results also validated the excellent bioactivity of 316L/HA sample.

The effect of laser beam scanning speed on the *in-vitro* biocompatibility of SLM 316L SS was investigated by Shang et al.[54] Biocompatibility was assessed by cytotoxicity assay on 15 samples of the SLM 316L SS manufactured in a cylindrical shape having 10-mm thickness and 10-mm diameter. These SLM samples were produced at a constant laser power of 195 W for each group and a hatch spacing of 0.09 mm with a layer thickness of 0.02 mm, and were further divided into five classes based on five different scanning speeds, i.e. 800, 900, 1050, 1100 and 1200 mm/s. The proliferation of the Human Embryonic Kidney (HEK) 293T cells was determined by performing a cytotoxicity CCK-8 assay and by using a cell counting kit. As a function of scan speed, the cell proliferation inhibition rate was determined after 24 h and 72 h of incubation as shown in Fig. 37.

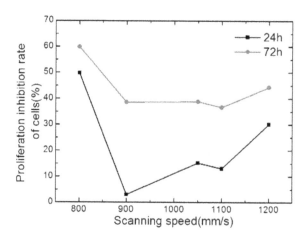

Fig. 37. Inhibition rates of HEK 293T cells as a function of various scanning speeds after 24 h and 72 h of incubation.[54] Figure reprinted with permission from SpringerLink.[TM]

After 24 h of incubation, the lowest inhibition rate was observed at 900-mm/s scan speed compared to the low and high scan speeds. However, under extended exposure (72 h), the inhibition rate was found to be independent of the scan speed within 900–1100 mm/s as shown in Fig. 37. Kong et al.[25] studied the effect of laser power on the biocompatibility of SLM 316L SS and compared it with the traditionally quenched 316L stainless steel. The cytotoxicity tests were conducted on the SLM samples produced at various laser powers and compared with the quenched 316L SS as shown in Fig. 38. The concentration of the ions released from these samples was measured after 288 h of immersion into the SBF. A large amount of Ni ions (~12 μg/L) were released from the SLM 316L SS sample prepared at low laser power (120 W). A decrease in Ni ion release was observed with an increase in laser power. On the other hand, the lowest Fe release (~2 μg/L) was measured at low laser power (120 W). There was found to be a negligible influence of laser power on the Cr and Mo ions released from the surface. The MC3T3-E1 cell morphology after exposure to these samples was also examined

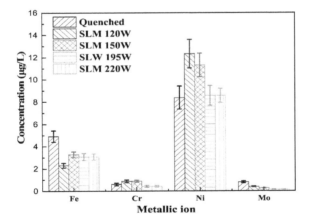

Fig. 38. Metal ion concentrations released from the SLM 316L SS samples produced at different laser powers measured after 288 h of immersion into the SBF solution. The solution volume used during the immersion test was 50 mL and the exposed sample surface area was 1 cm^2.[25] Figure reprinted with permission from Elsevier^{TM}.

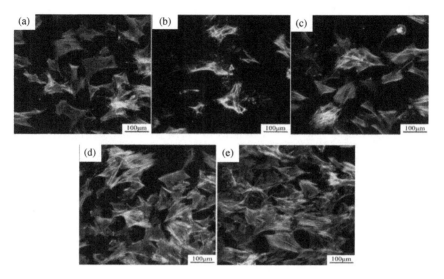

Fig. 39. (Color online) Morphologies of the cells proliferated on the surfaces of (a) quenched, (b) SLM 120-W, (c) SLM 150-W, (d) SLM 195-W and (e) SLM 220-W samples after 48 h of exposure, and the cells were stained to highlight the nuclei (blue) and F-actin (green).[25] Figures reprinted with permission from Elsevier^{TM}.

as shown in Fig. 39 and cells were stained to highlight the F-actin and nuclei in the green and blue colors, respectively.

The improved cell proliferation with an increase in laser power and under extended exposure to the culture medium was determined (as shown in Fig. 40) by using the CCK-8 assay kit. From these results, it can be predicted that on the surface of the SLM samples, the cell proliferation tendency was found to be improved with an increase in laser power compared to the quenched 316L SS. This improvement in cell proliferation was attributed to the larger hindrance to the ions released from SLM 316L SS (decrease in dissolution rate) produced at high laser power.

3. Brief Overview of the Wrought and Additive Manufacturing Processes

This review compares the microstructural, mechanical, electrochemical and biological properties of the wrought and AM 316L

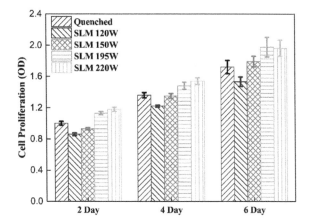

Fig. 40. Cell proliferations on the surfaces of the quenched and SLM 316L SS samples being manufactured at various laser powers after incubation for 2, 4 and 6 days before conducting the CCK-8 assay.[25] Figure reprinted with permission from Elsevier™.

stainless steel samples for orthopedic implant applications. Various additive manufacturing processes, i.e. direct metal laser printing, selective laser melting, electron beam melting, etc., can be used to manufacture intricate 316L stainless steel parts. There are many limitations to the production of the same metallic parts by conventional methods. For instance, conventionally, orthopedic implants are made in parts due to their complex design followed by joining to construct a single unit. Under the aggressive body environment and severe mechanical stresses, these implants could experience premature failure. On the other hand, via additive manufacturing route, by adjusting the operation parameters, the microstructural features can be modified, which could essentially tune the mechanical and electrochemical properties of the implant materials. As discussed above, the AM 316L SS has shown superior mechanical, electrochemical and biological properties compared to the wrought counterpart. This improvement in the properties was associated with the unique microstructural features of the AM 316L SS. Table 9 highlights the comparison of the corrosion properties of 316L SS samples produced via wrought and additive

Table 9. General comparison of the corrosion properties of wrought manufactured and AM 316L stainless steel samples.

Additive manufacturing	Wrought manufacturing
High breakdown potential	Low breakdown potential
High charge transfer resistance R_{ct}	Low charge transfer resistance R_{ct}
A more dense oxide film	A less dense oxide film
More stability of the passive film	Less stability of the passive film
High pitting resistance	Low pitting resistance
Corrosion potential (E_{corr}) is more positive	Corrosion potential (E_{corr}) is less positive

manufacturing processes. The significantly large breakdown potential and charge transfer resistance of the AM 316L SS were attributed to the formation of a relatively uniform and compact passive oxide film on its surface compared to wrought 316L SS.

The other benefit of additive manufacturing technology is the rapid production of customized implants of complex geometries for patients who need urgent bone replacement and fixation. This feature, namely the rapid production of the customized orthopedic implants with superior properties, makes the additive manufacturing process superior compared to the wrought or traditional manufacturing processes. Also, by changing the processing parameters during additive manufacturing process, complex shapes and low-density implants having micro- and macro-porous structures can be produced.

The microstructure of 316L SS is very important to understand the mechanical, electrochemical and biological performances of the materials. The intrinsic microstructure of the wrought 316L SS has a typical austenitic and equiaxed coarse grain structure having sharp grain boundaries. On the other hand, the microstructure of the AM 316L SS consists of fine, cellular and columnar-shaped subgrains. Grain size is also different in these two counterparts as it highly depends on the cooling rate. This can be explained in terms of the slow cooling rate associated with the wrought manufacturing processes, i.e. casting (0.5 K/s), compared to the accelerated cooling

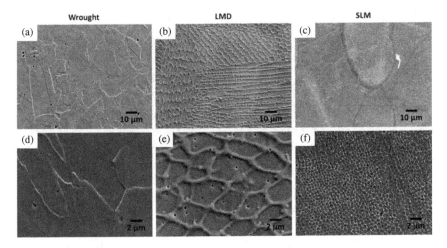

Fig. 41. SEM micrographs of 316L SS: (a), (d) wrought sample, (b), (e) laser metal deposition sample and (c), (f) SLM sample.[55] Figures reprinted with permission from Elsevier^{TM}.

rate of the 316L SS during the additive manufacturing process (103–108 K/s). This greatly influences the grain size as shown in Fig. 41. The wrought 316L SS sample shows polygonal equiaxed coarser grain structure compared to the finer grain structure observed in the laser metal deposition (LMD) and SLM samples.

In the additive manufacturing process, layer-by-layer deposition of the materials induces localized post-heat treatment of the previously solidified layers. Under these conditions, the grain structure in the AM sample grows perpendicular to the build plate, i.e. epitaxial growth, and consists of fine cellular dendrites due to the fast cooling of the molten metal. The XRD pattern of the AM 316L SS sample showed only three austenitic phase signature diffraction peaks. No additional diffraction peaks were observed validating the high metallurgical purity of the AM 316L SS. Compared to the wrought sample, a more intense XRD peak associated with the (220) plane was observed in the case of AM 316L SS. On the other hand, a more intense peak associated with the γ-(111) plane orientation was evident in the XRD pattern of the wrought 316L SS. The comparatively broader XRD peaks of SLM 316L SS also highlighted the formation of fine-grained structure.

Generally, AM 316L SS has superior mechanical properties compared to the wrought counterpart under optimized operation parameters of the additive manufacturing process. Overall, the tensile properties of AM 316L SS are better than the wrought sample due to the formation of fine-grained structure in the former. However, the low hardness of the laser-sintered AM 316L SS was associated with the porosity which can be eliminated by changing the processing parameters.

4. Main Challenges Associated with the Additive Manufacturing Process

Additive manufacturing process has certain advantages in the development of orthopedic implants of precise dimensions due to its versatility, the ease of design customization according to patients' needs, the production of intricate and complex geometries in a single unit and being highly adaptive and assistive technology for the surgeons and undertrained technical staff. However, there are still many challenges associated with additive manufacturing technology such as a lack in the standardization of the process. For instance, there exist wide variations in the manufacturing methods and processing parameters for different metallic materials. On the other hand, the main issue with additive manufacturing technology is the limitations in the manufacturing of multimaterial objects and it is envisioned that the mechanical, electrical and biological properties of complex biomedical components and implants can be fine-tuned by adopting an integrated additive manufacturing process. Similarly, the existing additive manufacturing technology is unable to produce the whole object in one step, e.g. total hip replacement implant, which is made up of dissimilar materials, i.e. metallic, ceramic and polymeric parts, owing to their multifunctional role in the body.

There are few types of structural steel, i.e. tool steel, 17-4 PH, AISI 304, 316L SS, etc., that have been produced via the additive manufacturing process but still, there is no standardized recipe available, and with a wide variation in the additive manufacturing process parameters one cannot expect the achievement of unified microstructure features and related properties. Also, via the additive manufacturing process, the formation of a fully dense structure is still a challenge and the presence of micropores could deleteriously affect the mechanical and corrosion properties, i.e. fatigue performance, pitting resistance, etc. However, under optimized manufacturing conditions for a specific alloy, the lack of powder particle fusion and development of porosity can be reduced or eliminated by selecting the proper laser power and scan speed. The rapid cooling of the molten pool and solidification pattern could also induce residual stresses in the AM implants that could lead to their premature failure under applied stresses and exposure to the aggressive biological environment. To avoid such failures, post-heat treatment, surface finishing and surface texturing procedures have been applied to improve the longevity of the implants but these post-treatment procedures could improve one property by compromising the others. For instance, the post-thermal treatments could adversely affect the mechanical strength and corrosion resistance of the 316L SS as learnt from the studies published recently.

The processing parameters of the selective laser melting process, i.e. the laser power, scan speed, hatch distance and post-treatments, can be optimized to adjust the total heat input and to tune the resultant microstructural features of a specific alloy. These optimized conditions may not be suitable for the manufacturing of other alloys and rigorous hit-and-trial experimentation may be required. Currently, for the production of specific alloys with a diverse range of compositions, the optimization of the processing parameters is a major challenge for the additive manufacturing industry, which requires a rigorous amount of work. It is proposed that by the adoption of machine learning and data science algorithms and modeling tools integrated with additive manufacturing, one could effectively address the challenges associated with the process parameter optimization and in the development of a wide range of metallic materials of varying compositions and properties. The important benefit of additive manufacturing process is the cost-effective and rapid production of complex designs and intricate shapes as a single unit thus eliminating the post-assembly and joining steps associated with the conventional manufacturing processes. Also, surface patterning and application of coatings, i.e. hydroxyapatite, poly-L-lactide and bone morphogenic coating (BMP), over specific structures and morphologies of 316L SS could be made to improve its biological properties.

5. Biological Response and Reactions Sequence of the Immune System

For more than a century, pure metal and metal alloys are being used in many medical specialties, where the only available option for treatment is the insertion of implants or medical devices. These are regulated by the U.S. Food & Drug Administration (FDA) as Class-II (moderate-risk) or Class-III (high-risk) devices and are monitored at pre/post-marketing stages for their safety, efficacy, toxicity and biocompatibility.[56] The metallic implants usually release trace elements, which can be either essential or nonessential depending upon the physiological requirement of the body. Toxicity is caused by the dissolution of metallic implants due to the release of metal ions that could exceed the dietary intake requirements. However, the toxicity caused by these ions is very different or maybe unknown.[57] Generally, the metal ions released from implants are usually well tolerated by the body due to its ability to maintain homeostasis. However, it is found that cytotoxic moieties, i.e. Ni, Cr and Co, may be released from the implant due to corrosion and could affect the surrounding tissues by chemically changing the tissue environment and by affecting cellular metabolism in the body. Among the various metallic implants, stainless steel, specifically 316L SS, is frequently used owing to its good strength and applied as a temporary implant for fracture fixation and bone support that is removed after rehabilitation.[58-60] The release of metallic ions from 316L SS caused by the biological environment, i.e. Ni (concentration > 11.7 ppm), could inhibit the cell growth and cause morphological changes leading to cells' death. Similarly, excessive Fe release from stainless steel attracts bacterial infections and therefore, is not suggested for long-term *in-vivo* use.[61-63] For instance, increased levels of Ni and Cr in the serum and urine samples have been measured after insertion of stainless steel spinal implants and the presence of macroscopic corrosion products was evident after extended exposure. Ni allergy is also well known and may cause several allergic reactions in humans including pruritus, scalp psoriasis, rashes, etc.[64]

The biocompatibility of any material depends on two main factors; the *host immune response* and the *degradation of biomaterial* in the natural body fluids. Stainless steel implants after insertion could cause hypersensitivity reactions which can be local or systemic and usually cause type-IV hypersensitivity reaction, a delayed reaction involving T-lymphocytes. The released metal ions of implants act as haptens which after making complexes with the internal body protein become antigens able to elicit an allergic response. Dermatitis is one of the clinical manifestations presenting erythema, vesicles, papules, pruritis, eczema, etc.[65,66]

5.1. *Host immune response*

When a medical device or biomaterial is implanted, the human body perceives it as a foreign body and a set of reactions take place sequentially to combat by activating the defense system of the body.[67] Upon direct physical contact between the implant and body tissues, the blood proteins get adsorbed and form a layer on the implant surface much before its interaction with the cells. For instance, adsorption of water molecules, protein, cells and bacterial adhesion, macrophage activation and tissue formation reactions are initiated at the implant–body interface with the temporal and spatial hierarchies as shown in Fig. 42. The adsorption of protein layer is responsible for the activation of the immune system and the coagulation system initiates the inflammatory response. Initially, the innate immunity of the human body is activated along with an acute inflammatory response and its severity depends on the structural composition and exposed sites of metals after implantation.[68,69]

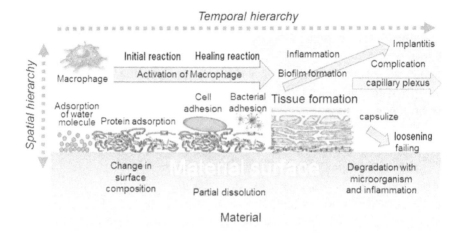

Fig. 42. Schematic of interfacial reactions and host body response after inserting an implant.[70]

The sequence of inflammatory reactions during wound healing or implantation is shown schematically in Fig. 43 and briefly explained in the following subsections.

5.1.1. *The acute inflammatory response*

An acute inflammatory response resembles a normal healing process involving molecular and cellular mechanisms. The host response initiates within seconds to minutes immediately after implantation and lasts from minutes to days based on the extent of injury.[72] The intrinsic coagulation pathway is adopted by the activation of Factor XII leading to the formation of thrombin, although in low quantity but sufficient enough to cause platelet activation to start coagulation forming a phospholipid layer.[73,74] The activation of integrin-binding sites at phagocytes is induced by fibrin/fibrinogen leading to more inflammatory and coagulatory responses.[75] Additionally, platelet activation at the implant site is characterized by extrinsic coagulation pathways along with host tissue factors. Also, the activation of the complement system causes the release of large amounts of C3a and C5a, the anaphylatoxins, from the implant

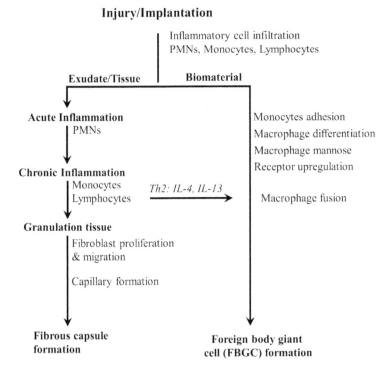

Fig. 43. Inflammatory reactions sequence and wound healing process resulting in the formation of Foreign Body Giant Cells (FBGCs). During transient chronic inflammation, T_h2-lymphocytes produce IL-4 and IL-3 resulting in the fusion of mono/macrophage to form FBGC.[71]

surface inducing inflammation.[76,77] Another inflammatory pathway being activated upon the uptake of metal-released debris by the cells is "Pattern Recognition Receptors" (PRRs), which is recognized by macrophages like immune cells once the extracellular matrix components comprising danger signals deposit a layer on the implant surface.[78] Initially, polymorphonuclear cells migrate at the implant site rapidly as the first line of defense in response to chemoattractants secreted by the damaged, endothelial cells and activated platelets. Histamine release by mast cell activation and degranulation through monocytes mobilization further employs more PMNs at the implant surface.[79,80] These PMNs are involved in phagocytosis of infectious cells along with the liberation of chemokines and cytokines, interleukin-8, monocytes and macrophages, lymphocytes and immature dendritic cells (DCs), further modulating the intensity and duration of inflammatory response via cell apoptosis in the absence of activation signals.[81,82] The polymorphonuclear cells are short-lived and undergo apoptosis during the inflammatory process, and no PMNs are found at the implant site after two days of implantation.[83]

5.1.2. Chronic inflammation

Chronic inflammation arises from the persistent stimuli at the site of implantation and depends on the physical and chemical properties of the implant and its motion at the implant site. Chronic inflammation is localized at the implant site and is characterized by the presence of mononuclear cells such as lymphocytes and monocytes along with angiogenesis and connective tissue proliferation at the implant site.[84,85] The monocytic cells differentiate into macrophages after migration from blood to the implant surface, get attached and release chemokines namely interleukin-8, MCP-1 and MIP-1β activating the inflammatory response further. These macrophages have a variety of functions including the maintenance of homeostasis and response to certain endogenous stimuli leading to the generation of various macrophages. The induction of macrophage activation is caused by Natural Killer (NK) cells, helper T-cells and/or Tumor Necrosis Factor (TNF-α); and these macrophages secrete cytokines, free nitrogen radicals, reactive oxygen species (ROS), neutral proteases, complement factors and coagulation factors as their antimicrobial response to the host defense mechanism.[86-88] The particles of implant debris greater than 10 μm get surrounded by macrophages and Foreign Body Giant Cells responding to various signals such as apoptosis, interleukin-10 and immune complex responses. Amongst the three types of macrophages, the regulatory macrophages cause immunosuppression by releasing interleukin-10, restricting inflammation and suppressing the immune response. The wound-healing macrophages participate in tissue regeneration via migration and proliferation of fibroblasts. In this way, all the three macrophages execute their functions in immune response with the classically activated macrophages in the initial phase and the other two in the later resolution phase.[89–92]

5.2. Granulation tissue

Granulation tissue is apparent after 3–5 days of post-implantation once both (acute and chronic) inflammatory responses settle, and is characterized by macrophages, infiltrating fibroblast and angiogenesis at the implant site. This granulation tissue has a soft granular appearance and its pink wound-healing surface depicts its name.[79,84] The small blood vessels generation via angiogenesis or neovascularization takes place along with fibroblasts proliferation, which is actively involved in the synthesis of proteoglycans initially and type-1 collagen predominately in the later stage, thus forming a fibrous capsule. However, some fibroblasts differentiate into myofibroblasts and cause wound contraction in the granulation tissue.[93]

5.3. FBGCs formation

Protein adsorption followed by cell adhesion and surface chemistry of the implant are two important parameters that play a pivotal role in foreign body response (FBR) and monocyte macrophages adhesion. This FBR comprises giant cells and the granulation tissue and this process resembles a natural wound-healing mechanism. As the FBR depends on the surface property of the implant, therefore, flat surfaces of implants result in the deposition of one–two-cell-thick layer of macrophages, while a rough surface results in the deposition of several layers of macrophages and FBGCs and FBR persist at the tissue–implant interface for a lifetime.[94] Isolation of the implant and the FBR from the external tissues is carried out by fibrous encapsulation of the bioimplant. The FBGC cells, originating from the tissue macrophages coalescence, are multinucleated, abundantly found at the implant surface and present till the lifetime of the implant; however, whether they are activated or not is unknown. These FBGCs cells are found to play an important role in the biodegradation of implants.[95] Generally, the acute and chronic inflammatory responses characterized by neutrophils and lymphocytes/monocytes, respectively, resolve within two weeks of implantation depending on the nature, type and location of the biomaterial.[84]

5.4. *Fibrosis*

Fibrosis is the final stage of the healing process also known as fibrous encapsulation by connective tissues except for tissue-engineered devices such as porous materials implanted in bones and the materials implanted in parenchymal tissues. Regeneration (substitution of injured tissue by parenchymal cells) and replacement (which forms a fibrous capsule by replacing connective tissues) are the two main processes involved in tissue repair at the implant site. The regeneration capacity of cells varies depending upon the nature of cells, e.g. the cells from the nerve, skeletal muscle and cardiac muscles are the permanent types of cells and follow the process of inflammation and ultimately lead to fibrosis, however, stable cells (parenchymal and mesenchymal) and labile cells (epithelial, lymphoid and hematopoietic) might follow the same process to fibrosis formation but additionally can follow the inflammatory exudates resolution, resulting in the restoration of normal tissue structure.[84]

The formation of the provisional matrix is an indicator of wound healing as its extent depicts the implant failure or success. There are two intentions involved in wound healing; the first type of healing process occurs when there is less or no space between the implant and body tissue; however, the second type is characterized by the presence of a large space between the tissue and implant, hence leading to the formation and accumulation of high level of the provisional matrix. This extensive provisional matrix formation ultimately causes the failure of the implant through various

Table 10. Components of the inflammatory (innate) and adaptive immune responses.[71]

Components
Complement cascade components
Immunoglobulins
Cellular components
Macrophages
NK cells
Dendritic cells
Cells with dual phagocytic and antigen-presenting capabilities

mechanisms involving the formation of fibrous capsules. There are two types of inflammatory responses, innate and adaptive, and both of them share some common features and the main components of these responses are given in Table 10. In the absence of an adaptive immune response, humoral immunity and cell-mediated immunity are activated sharing similar components leading to an innate inflammatory response.

5.5. *The adaptive immune response*

This is an acquired immune response of the human body to protect it from any foreign body or material, which is carried out by various recognition mechanisms that elicit humoral and cellular responses after recognizing the foreign agent to be "Non-self" and produce highly specific antigen response along with long-term memory development.[96,97] The humoral response is usually mediated by antibody molecules, complement proteins, various growth factors, cytokines, chemokines and cell mediators, which are usually formed by the cells from the immune system. This, in turn, coordinates with the same cells of cell-mediated response involved in adaptive immunity.[96,97] A strong immune response requires close coordination and controlled balance between the two systems, i.e. the soluble factors and cellular components.[97] The cells of immune response originate from bone marrow stem cells and are primarily B-lymphocytes and T-lymphocytes differing in morphology, function and membrane antibodies. However, their common function is to eliminate any foreign agent by the maintenance of cell surface receptors.

Once the implant is inserted, the intensity of host response is a determining factor for the outcome of the biological implant performance. Any degradation product, change in surface morphology or wear and tear in biomaterial can activate the human immune response.[98] The interaction between the immune system and the biomaterial is dependent on tissues surrounding the implant, driving the innate response followed by adaptive defence.[99] The complement proteins C3b and C3bi and antibodies bind to the foreign material and become opsonins, facilitating phagocytosis by neutrophils and macrophages, which present surface receptors for the complement component C3b. Similarly, C5a is a chemotactic complement component for the migration of neutrophils, macrophages and monocytes along with other inflammatory cells to the implant site. The classic and alternate pathways of the complement system produce a membrane attack complex, which can kill the microbes. The protein inhibitors of the cell membrane from the host cells closely control the complement cascade and protect the host cells from damage. The main component of the cell-mediated immunity is T-lymphocytes, which have a significant role in the migration and activation of lymphocytes and effector T-cells' function. The interaction of antigens with antigen-presenting cells (APCs) results in the formation of special types and specific functions of lymphocytes. B-lymphocytes carry cell surface immunoglobulins that recognize and bind with the soluble antigens and form a complex of major histocompatibility components with peptides. T-cells after activation produce their growth factor interleukin-2 thus promoting the T-cells' growth in the autocrine signaling avenue. The chemotactic agent, cytokines play the role of messenger molecule in the immune system and have a wide variety of functions and interact with different types of cells. However, notably, interleukin-1, TNF-α and IFN-γ have pleiotropic effects thus mediating, regulating and activating the responses by a variety of cells.

Immunotoxicity results in dysfunction of the immune system and arises when humoral or cell-mediated immunity is compromised. Due to this the immune system becomes unable to combat microbial infections, i.e. neoplastic disorders, any other tissue damage due to autoimmunity or hypersensitivity can cause either immunosuppression or immune-potentiation.[100] The hypersensitivity reaction is classified into four categories; type-I (anaphylactic), type-II (cytotoxic) and type-III (immune-complex) are the immediate hypersensitivity reactions as they occur within 24 h, whereas type-IV hypersensitivity response is a delayed-type cell-mediated reactions as listed in Table 11.[101] Hypersensitivity is also called allergy which happens when the reaction to an antigen is increased due to pre-exposure, leading to an adverse effect. The most commonly occurring hypersensitivity reactions are type-I and type-IV reactions with bioimplants and medical devices; however, type-II and type-III reactions are rare to happen in patients except in the case of tissue-engineered grafts, which can elicit type-II and type-III responses due to their surfaces presenting as antigens of any type (protein, extracellular matrix component or any other cell).

Table 11. Immunological effects and responses.[71]

Effect	Response
Hypersensitivity	Histopathological changes
Type-I: anaphylactic	Humoral responses
Type-II: cytotoxic	Host resistance
Type-III: immune-complex	Clinical symptoms
Type-IV: cell-mediated (delayed)	Cellular responses
Chronic inflammation	T-cells
Immunosuppression	NK cells
Immunostimulation	Macrophages
Autoimmunity	Granulocytes

Type-I hypersensitivity reaction is immunoglobulin E-mediated immune response comprising sensitization of the host with antigen, and upon re-exposure to the same antigen effect stage causes mast cells degranulation and histamine release with various inflammatory mediators.[101] Type-IV hypersensitivity reaction is a T-lymphocytes-mediated delayed response that can lead to cell and tissue injury due to the release of cytokines and various other mediators from the previously sensitized T-cells. CD4 T-cells and CD8 T-cells are the main mediators of classic delayed hypersensitivity and direct cell toxicity, respectively.

The biomaterials may present with strong antigens at their surface, capable of stimulating chronic inflammatory responses that can last from weeks to months. Therefore, the persistence of this chronic inflammatory response causes immunotoxicity leading to granuloma formation along with various immunological disorders not limited to autoimmune disease.[84]

5.6. *The bone healing process after implantation*

A variety of specialized cells with varying characteristics can be found at the implant site. Initially, mesenchymal tissues of the implant site recruit fibroblast cells deposition at the surface and secrete immature collagen.[72] This recruitment is carried out under the influence of signal transduction molecules derived from the bone and blood cells, which initiate a chain of reactions for bone formation and resorption, important factors for bone remodeling and healing.[102] Osteoprogenitor cells, found in the inner layer of the periosteum, can differentiate into osteoblasts leading to not only osteogenesis but also angiogenesis at the bone healing site. This process is carried out under the influence of bone morphogenetic protein (BMP-2). These osteoblasts result in the production of collagen, proteoglycan and glycoproteins, which are important bone-forming components upon differentiation resulting in the expression of Alkaline Phosphatase (ALP) known as a specific marker of bone formation. Osteocytes, the mature bone cells, communicate with each other through cytoplasmic processes and get trapped in lacunae. On the other hand, osteoclast cells are bone-resorbing cells derived from macrophages that become multinucleated by stimulating the signaling molecules and appear as woven bone upon injury.[103,104] It is very difficult to demarcate between the injury-related and Foreign Body Giant Cell response osteoclastic activities because both resemble morphologically, except that the former contains calcitonin receptors at the surface.

6. Summary

This review compares the microstructural, mechanical, corrosion and biological properties of the AM and conventionally produced wrought 316L SS samples used for orthopedic applications. It is deduced that the AM 316L SS exhibits better mechanical, corrosion resistance and biological responses owing to its unique and refined microstructural features compared to the wrought 316L SS. However, the formation of some internal defects inherited by the additive manufacturing process, such as the high density of dislocations accumulated across the subgrain boundaries, the partial fusion of the powder particles, generation of residual stresses and porosity in the resultant microstructure, could deteriorate the mechanical properties of 316L SS. Caused by the additive manufacturing process, the increased porosity in the microstructure could not only adversely affect the mechanical properties of the implant but also trigger pitting corrosion. There are many limitations related to the wrought manufacturing technology such as the hindrance in fabricating orthopedic implants with complex designs of different sizes in less time. On the other hand, additive manufacturing is an adaptive and assistive technology capable of producing customized implants of intricate shape/design according to the patient's requirement. Depending on the additive manufacturing process parameters, i.e. laser beam power and scanning speed, the exceptionally fast cooling rate (> 100 K/s) and directional solidification of the molten metal, a fine, cellular and columnar-shaped subgranular structure with negligible amount of nonmetallic inclusions presents better passive film stability and a large pitting potential (> 0.6 V versus SCE) compared to the wrought 316L SS in the same biological media. Similarly, the SLM 316L SS produced at high laser power presented larger corrosion resistance and improved biocompatibility compared to wrought 316L stainless steel as evident from the improvement in the *in-vitro* cell proliferation on its surface. Also, there is an urgent need to address various challenges associated with the additive manufacturing processes, i.e. a lack in the standardization of the additive manufacturing processes and a wide variation in the manufacturing methods and processing parameters. The mechanical or corrosion degradation of the implant materials could trigger an adverse biological response. The expanded overview of the possible inflammatory reactions in the wound-healing process, immunological host response and *in-vivo* reactions sequence are also presented to assess the issues associated with the biocompatibility of the implant material.

Acknowledgments

The valuable support of Ms. Hira Hafeez and Ms. Fatima Shaukat from the Institute of Metallurgy and Materials Engineering, University of the Punjab, Lahore during the compilation of this paper is appreciated and highly acknowledged.

References

1. R. Singh, S. Singh and M. S. J. Hashmi, *Reference Module in Materials Science and Materials Engineering*, Elsevier, Amsterdam, 2016, pp. 1–31.
2. J. Ni, H. Ling, S. Zhang, Z. Wang, Z. Peng, C. Benyshek, R. Zan, A. K. Miri, Z. Li, X. Zhang, J. Lee, K. J. Lee, H. J. Kim, P. Tebon, T. Hoffman, M. R. Dokmeci, N. Ashammakhi, X. Li and A. Khademhosseini, *Mater. Today Bio*, 2019, **3**, 100024.
3. L. Bai, C. Gong, X. Chen, Y. Sun, J. Zhang, L. Cai, S. Zhu and S. Q. Xie, *Metals*, 2019, **9**, 1004.

4. L. Yuan, S. Ding and C. Wen, *Bioact. Mater.*, 2019, **4**, 56–70.
5. W. S. W. Harun, M. S. I. N. Kamariah, N. Muhamad, S. A. C. Ghani, F. Ahmad and Z. Mohamed, *Powder Technol.*, 2018, **327**, 128–151.
6. E. A. Friis, D. Abel and T. Woods, *Mechanical Testing of Orthopaedic Implants*, ed. E. Friis, Woodhead Publishing, Duxford, 2017, pp. 49–60.
7. A. Sedrakyan, E. W. Paxton, C. Phillips, R. Namba, T. Funahashi, T. Barber, T. Sculco, D. Padgett, T. Wright and D. Marinac-Dabic, *J. Bone Joint Surg.*, 2011, **93**, 1–12.
8. ASTM, *Annual Book of ASTM Standards*, ASTM International, Philadelphia, 2004, vol. 3.01, p. 1–62.
9. W. A. Lane, *Br. Med. J.*, 1895, **1**, 861–863.
10. S. Nag and R. Banerjee, in *ASM Handbook*, ed. R. J. Narayan, ASM International, 2012, vol. 23, pp. 6–17.
11. S. Singh and S. Ramakrishna, *Curr. Opin. Biomed. Eng.*, 2017, **2**, 105–115.
12. M. J. Polinski and P. M. Berg, *Integrated Production of Patient-Specific Implants and Instrumentation*, Publication No. US 2010/0217270 A1, Google Patents, 2010.
13. W. Vancraen and M. Janssens, *Additive Manufacturing Flow for the Production of Patient-specific Devices Comprising Unique Patient-specific Identifiers*, Publication No. US 2012/0116203 A1, Google Patents, 2012.
14. J. Schroeder, S. Goodman and K.-J. Kim, *Personal Fit Medical Implants and Orthopedic Surgical Instruments and Methods for Making*, Publication No. WO 2010/120990 A1, Google Patents, 2007.
15. R. Bibb, D. Eggbeer, P. Evans, A. Bocca and A. Sugar, *Rapid Prototyp. J.*, 2009, **15**, 346–354.
16. Q. Chen and G. A. Thouas, *Mater. Sci. Eng. R, Rep.*, 2015, **87**, 1–57.
17. G. Manivasagam, D. Dhinasekaran and A. Rajamanickam, *Recent Pat. Corros. Sci.*, 2010, **2**, 40–54.
18. N. S. Manam, W. S. W. Harun, D. N. A. Shri, S. A. C. Ghani, T. Kurniawan, M. H. Ismail and M. H. I. Ibrahim, *J. Alloys Compd.*, 2017, **701**, 698–715.

19. K. B. Tayyab, A. Farooq, A. A. Alvi, A. B. Nadeem and K. Deen, *Int. J. Miner. Metall. Mater.*, 2021, **28**, 440–449.
20. V. P. Mantripragada, B. Lecka-Czernik, N. A. Ebraheim and A. C. Jayasuriya, *J. Biomed. Mater. Res. A*, 2013, **101**, 3349–3364.
21. A. Röttger, J. Boes, W. Theisen, M. Thiele, C. Esen, A. Edelmann and R. Hellmann, *Int. J. Adv. Manuf. Technol.*, 2020, **108**, 769–783.
22. H. M. Hamza, K. M. Deen, A. Khaliq, E. Asselin and W. Haider, *Crit. Rev. Solid State Mater. Sci.*, 2021, **47**, 46–98.
23. D. Kong, C. Dong, X. Ni, L. Zhang, J. Yao, C. Man, X. Cheng, K. Xiao and X. Li, *J. Mater. Sci. Technol.*, 2019, **35**, 1499–1507.
24. X. Ni, D. Kong, W. Wu, L. Zhang, C. Dong, B. He, L. Lu, K. Wu and D. Zhu, *J. Mater. Eng. Perform.*, 2018, **27**, 3667–3677.
25. D. Kong, X. Ni, C. Dong, X. Lei, L. Zhang, C. Man, J. Yao, X. Cheng and X. Li, *Mater. Des.*, 2018, **152**, 88–101.
26. S. Nahata and O. B. Ozdoganlar, *Procedia Manuf.*, 2019, **34**, 772–779.
27. L. Wang, X. Zhao, M. H. Ding, H. Zheng, H. S. Zhang, B. Zhang, X. Q. Li and G. Y. Wu, *Appl. Surf. Sci.*, 2015, **340**, 113–119.
28. K. Antony, N. Arivazhagan and K. Senthilkumaran, *J. Manuf. Process.*, 2014, **16**, 345–355.
29. A. Buford and T. Goswami, *Mater. Des.*, 2004, **25**, 385–393.
30. S. Mahathanabodee, T. Palathai, S. Raadnui, R. Tongsri and N. Sombatsompop, *Wear*, 2014, **316**, 37–48.
31. T. Vilaro, C. Colin, J. D. Bartout, L. Nazé and M. Sennour, *Mater. Sci. Eng. A*, 2012, **534**, 446–451.
32. K. Saeidi, X. Gao, Y. Zhong and Z. J. Shen, *Mater. Sci. Eng. A*, 2015, **625**, 221–229.
33. M. J. K. Lodhi, K. M. Deen, M. C. Greenlee-Wacker and W. Haider, *Addit. Manuf.*, 2019, **27**, 8–19.
34. N. S. Al-Mamun, K. M. Deen, W. Haider, E. Asselin and I. Shabib, *Addit. Manuf.*, 2020, **34**, 101237.
35. D. Wang, C. Song, Y. Yang and Y. Bai, *Mater. Des.*, 2016, **100**, 291–299.
36. Y. Zhong, L. Liu, S. Wikman, D. Cui and Z. Shen, *J. Nucl. Mater.*, 2016, **470**, 170–178.
37. M. Kazemipour, M. Mohammadi, E. Mfoumou and A. M. Nasiri, *JOM*, 2019, **71**, 3230–3240.

38. F. Bartolomeu, M. Buciumeanu, E. Pinto, N. Alves, O. Carvalho, F. S. Silva and G. Miranda, *Addit. Manuf.*, 2017, **16**, 81–89.
39. A. Mertens, S. Reginster, Q. Contrepois, T. Dormal, O. Lemaire and J. Lecomte-Beckers, *Mater. Sci. Forum*, 2014, **783–786**, 898–903.
40. C. Brecher *et al.*, *Integrative Production Technology for High-Wage Countries*, ed. C. Brecher, Springer, Heidelberg, 2012, pp. 17–76.
41. M. M. Jabbari Behnam, P. Davami and N. Varahram, *Mater. Sci. Eng. A*, 2010, **528**, 583–588.
42. J. Stendal, O. Fergani, H. Yamaguchi and N. Espallargas, *J. Bio-Tribo-Corros.*, 2018, **4**, 9.
43. Q. Chao, V. Cruz, S. Thomas, N. Birbilis, P. Collins, A. Taylor, P. D. Hodgson and D. Fabijanic, *Scr. Mater.*, 2017, **141**, 94–98.
44. R. Shrestha, J. Simsiriwong and N. Shamsaei, *Addit. Manuf.*, 2019, **28**, 23–38.
45. A. Yadollahi, N. Shamsaei, S. M. Thompson, A. Elwany and L. Bian, *Int. J. Fatigue*, 2017, **94**, 218–235.
46. A. J. Saldivar-Garcia and H. F. Lopez, *Metall. Mater. Trans. A*, 2004, **35**, 2517–2523.
47. K. M. Deen, M. J. K. Lodhi, E. Asselin and W. Haider, *J. Phys. Chem. C*, 2020, **124**, 21435–21445.
48. J. E. Maróti, D. M. Kemény and D. Károly, *Acta Mater. Transylvanica*, 2019, **2**, 55–60.
49. T. Balusamy, S. Kumar and T. S. N. Sankara Narayanan, *Corros. Sci.*, 2010, **52**, 3826–3834.
50. X. Chen, J. Li, X. Cheng, H. Wang and Z. Huang, *Mater. Sci. Eng. A*, 2018, **715**, 307–314.
51. M. J. K. Lodhi, K. M. Deen and W. Haider, *Materialia*, 2018, **2**, 111–121.
52. T. Jøraholmen, M. Westrin and O. Fergani, *Eur. Cells Mater.*, 2017, **33**, 17.
53. J. Luo, X. Jia, R. Gu, P. Zhou, Y. Huang, J. Sun and M. Yan, *Metals*, 2018, **8**, 548.
54. Y. Shang, Y. Yuan, Y. Zhang, D. Li and Y. Li, *Proc 2016 3rd Int Conf Mechatronics and Mechanical Engineering*, 2017, pp. 01009:1–01009:4.
55. R. I. Revilla, M. Van Calster, M. Raes, G. Arroud, F. Andreatta, L. Pyl, P. Guillaume and I. De Graeve, *Corros. Sci.*, 2020, **176**, 108914.

56. U.S. Food & Drug Administration, *Biological Response to Metal Implants*, Technical Report, 2019.
57. R. P. Brown, B. A. Fowler, S. Fustinoni and M. Nordberg, *Handbook on the Toxicology of Metals*, eds. G. F. Nordberg, B. A. Fowler and M. Nordberg, Academic Press, San Diego, 4th edn., 2015, pp. 113–122.
58. N. Eliaz and K. Hakshur, *Degradation of Implant Materials*, ed. N. Eliaz, Springer, New York, 2012, pp. 253–302.
59. S. V. Bhat, *Biomaterials*, Springer, Dordrecht, 2002, pp. 92–111.
60. H. Matusiewicz, *Acta Biomater.*, 2014, **10**, 2379–2403.
61. K. Yang and Y. Ren, *Sci. Technol. Adv. Mater.*, 2010, **11**, 014105.
62. G. Herting, I. O. Wallinder and C. Leygraf, *J. Environ. Monit.*, 2008, **10**, 1092–1098.
63. Y. T. Konttinen, I. Milošev, R. Trebše, P. Rantanen, R. Linden, V. M. Tiainen and S. Virtanen, *Joint Replacement Technology*, ed. P. A. Revell, Woodhead Publishing Series in Biomaterials, Woodhead Publishing, 2008, pp. 115–162.
64. J. Univers, C. Long, S. A. Tonks and M. B. Freeman, *J. Vasc. Surg.*, 2018, **67**, 615–617.
65. C. Svedman, H. Möller, B. Gruvberger, C. G. Gustavsson, J. Dahlin, L. Persson and M. Bruze, *Contact Dermatitis*, 2014, **71**, 92–97.
66. J. Wawrzynski, J. A. Gil, A. D. Goodman and G. R. Waryasz, *Rheumatol. Ther.*, 2017, **4**, 45–56.
67. J. M. Anderson, *ASAIO Trans.*, 1988, **34**, 101–107.
68. C. J. Wilson, R. E. Clegg, D. I. Leavesley and M. J. Pearcy, *Tissue Eng.*, 2005, **11**, 1–18.
69. S. Kargozar, S. Ramakrishna and M. Mozafari, *Curr. Opin. Biomed. Eng.*, 2019, **10**, 181–190.
70. T. Hanawa, *Front. Bioeng. Biotechnol.*, 2019, **7**, 170:1–170:13.
71. J. M. Anderson, *Principles of Regenerative Medicine*, eds. A. Atala, R. Lanza, A. G. Mikos and R. Nerem, Academic Press, Boston, 3rd edn, 2019, pp. 675–694.
72. K. M. R. Nuss and B. von Rechenberg, *Open Orthop. J.*, 2008, **2**, 66–78.
73. M. B. Gorbet and M. V. Sefton, *The Biomaterials: Silver Jubilee Compendium*, ed. D. F. Williams, Elsevier Science, Oxford, 2004, pp. 219–241.
74. C. Sperling, M. Fischer, M. F. Maitz and C. Werner, *Biomaterials*, 2009, **30**, 4447–4456.

75. W.-J. Hu, J. W. Eaton, T. P. Ugarova and L. Tang, *Blood*, 2001, **98**, 1231–1238.
76. M. Fischer, C. Sperling, P. Tengvall and C. Werner, *Biomaterials*, 2010, **31**, 2498–2507.
77. J. V. Sarma and P. A. Ward, *Cell Tissue Res.*, 2011, **343**, 227–235.
78. S. B. Goodman, Y. T. Konttinen and M. Takagi, *J. Long-Term Eff. Med. Implants*, 2014, **24**, 253–257.
79. L. Tang, T. A. Jennings and J. W. Eaton, *Proc. Natl. Acad. Sci. USA*, 1998, **95**, 8841–8846.
80. J. Zdolsek, J. W. Eaton and L. Tang, *J. Transl. Med.*, 2007, **5**, 31.
81. J. T. Kirk, A. K. McNally and J. M. Anderson, *J. Biomed. Mater. Res. A*, 2010, **94**, 683–687.
82. G. Salehi, A. Behnamghader and M. Mozafari, *Handbook of Biomaterials Biocompatibility*, ed. M. Mozafari, Woodhead Publishing, 2020, pp. 453–471.
83. D. W. Gilroy, *Drug Discov. Today, Ther. Strateg.*, 2004, **1**, 313–319.
84. J. Anderson, *Principles of Regenerative Medicine*, Academic Press, San Diego, 2008, pp. 704–723.
85. H. B. Fleit, *Pathobiology of Human Disease*, eds. L. M. McManus and R. N. Mitchell, Academic Press, San Diego, 2014, pp. 300–314.
86. J. A. Jones, D. T. Chang, H. Meyerson, E. Colton, I. K. Kwon, T. Matsuda and J. M. Anderson, *J. Biomed. Mater. Res. A*, 2007, **83**, 585–596.
87. F. O. Martinez, A. Sica, A. Mantovani and M. Locati, *Front. Biosci.*, 2008, **13**, 453–461.
88. D. M. Mosser and J. P. Edwards, *Nat. Rev. Immunol.*, 2008, **8**, 958–969.
89. D. M. Mosser, *J. Leukoc. Biol.*, 2003, **73**, 209–212.
90. J. S. Gerber and D. M. Mosser, *J. Immunol.*, 2001, **166**, 6861–6868.
91. G. Broughton II, J. E. Janis and C. E. Attinger, *Plast. Reconstr. Surg.*, 2006, **117**, 12S–34S.
92. G. J. Bellingan, P. Xu, H. Cooksley, H. Cauldwell, A. Shock, S. Bottoms, C. Haslett, S. E. Mutsaers and G. J. Laurent, *J. Exp. Med.*, 2002, **196**, 1515–1521.
93. J. M. Anderson, *Cardiovasc. Pathol.*, 1993, **2**, 33–41.
94. C.-C. Chu, J. A. von Fraunhofer and H. P. Greisler (eds.), *Wound Closure Biomaterials and Devices*, CRC Press, Boca Raton, 2018.

95. M. A. Niekrasz and C. L. Wardrip, *Nonhuman Primates in Biomedical Research*, eds. C. R. Abee, K. Mansfield, S. Tardif and T. Morris, Academic Press, Boston, 2nd edn., 2012, pp. 339–358.
96. C. A. Janeway, Jr., P. Travers, M. Walport and M. J. Shlomchik, *Immunobiology: The Immune System in Health and Disease*, Garland Science, New York, 5th edn., 2001.
97. E. Mariani, G. Lisignoli, R. M. Borzì and L. Pulsatelli, *Int. J. Mol. Sci.*, 2019, **20**, 636.
98. C. Esche, C. Stellato and L. A. Beck, *J. Invest. Dermatol.*, 2005, **125**, 615–628.
99. J. M. Anderson, A. Rodriguez and D. T. Chang, *Semin. Immunol.*, 2008, **20**, 86–100.
100. B. Reid, M. Gibson, A. Singh, J. Taube, C. Furlong, M. Murcia and J. Elisseeff, *J. Tissue Eng. Regen. Med.*, 2015, **9**, 315–318.
101. D. Trizio *et al.*, *Food Chem. Toxicol.*, 1988, **26**, 527–539.
102. K. N. Marwa K, *StatPearls* [Internet], StatPearls Publishing, Treasure Island, 2021, pp. 1–14.
103. B. D. Ratner, A. S. Hoffman, F. J. Schoen and J. E. Lemons, *Biomaterials Science: An Introduction to Materials in Medicine*, Academic Press, San Diego, 2004.
104. S. C. Marks, Jr. and S. N. Popoff, *Am. J. Anat.*, 1988, **183**, 1–44.
105. H. Meyer, *JAMA*, 2001, **286**, 95.

© 2025 World Scientific Publishing Company
https://doi.org/10.1142/9789819806409_0005

Cortisol Biosensors: From Sensing Principles to Applications*

Yuki Tanaka,[†] Nur Asinah binte Mohamed Salleh,[†] Khin Moh Moh Aung,[†] Xiaodi Su[†,‡,§,∥] and Laura Sutarlie[†,¶,∥]

[†]*Institute of Materials Research and Engineering (IMRE), Agency for Science Technology and Research (A*STAR), 2 Fusionopolis Way Innovis, #08-03, Singapore 138634, Republic of Singapore*
[‡]*Department of Chemistry, National University of Singapore Block S8, Level 3, 3 Science Drive 3, Singapore 117543, Republic of Singapore*
[§]*xd-su@imre.a-star.edu.sg*
[¶]*laura-sutarlie@imre.a-star.edu.sg*

Stress detection and monitoring have attracted substantial research interests due to stress being a risk factor for health disorders and economic burdens. In particular, the steroid hormone cortisol plays an important role both as an indicator of stress and a coordinator of downstream physiological responses. Recent years have witnessed a flourishing of cortisol biosensors and bioassays based on various physical principles. In this review, we first provide an overview of cortisol function and its presence in different biological matrices. Next, we discuss the existing range of cortisol biosensors, from their sensing principles (i.e. chromogenic, nanoparticle-based colorimetric and fluorometric, surface-enhanced Raman spectroscopy, surface plasma resonance spectroscopy, and electrochemical sensors), performances (sensitivity, selectivity, portability, etc.), and applications. We

---

[∥]Corresponding authors.
*To cite this article, please refer to its earlier version published in the *World Scientific Annual Review of Functional Materials*, Volume 1, 2330001 (2023), DOI: 10.1142/S2810922823300015.

particularly correlate the sensing performances and their suitability for point-of-care diagnostics with sensor principles and the use of different affinity ligands, such as antibodies, aptamers, molecular imprint, and even 2D materials such as MXenes. Finally, we discuss the challenges and perspectives of future high-performing cortisol sensors for a wider range of applications in human and animal stress monitoring.

Keywords: Stress monitoring; cortisol; hormone; biosensors; nanoparticle.

1. Introduction

Stress is a recognized risk factor for health disorders and economic burden, hence stress detection and monitoring have attracted significant research interests over the years. Stress was first coined as "the non-specific response of the body to any demand" by Hans Seyle in 1936.[1,2] Stress is a biological response in humans and other living organisms towards stressors,[3] which threaten the homeostatic equilibriums of organisms.[4-6] These stressors maybe biological (e.g. starvation), physical (e.g. injury, temperature fluctuations), or social (e.g. isolation or exposure to novelties) in nature.[4] The symptoms of stress are variable depending on the duration of stress. In general, short-term stressors induce acute stress, while long-term stressors induce chronic stress.[4] When the stressor does not dissipate and continues over time, the allostatic load and deleterious effects[7] on the organism's biological welfare are pronounced. Stress is the leading cause of up to 75–90% of diseases which encompasses cardiovascular and metabolic, along with mental health disorders.[8] For example, psychologically-related stress disorders, such as Acute Stress Disorder (ASD) and Post Traumatic Stress Disorder (PTSD), are evident following experiences of traumatic events.[9-11] Affective disorders such as depression are also triggered by chronic stress.[12] This problem is worsened in immunocompromised individuals, rendering them susceptible to other pathogenic attacks.[13] As such, analytical methods that can detect and monitor stress levels are useful for assessing the well-being of organisms.

Common methods to measure stress levels experienced by an individual include clinical behavioral tests, e.g. the Trier Social

Stress Test (TSST) and the Perceived Stress Scale (PSS).[1] Alternatively detection of biomarkers can provide a more quantitative method of monitoring stress levels.[1,13,14] Stress biomarkers are molecules by which stress can be identified and provide insights into an organism's well-being.[1,15] These stress biomarkers are present in biofluids such as sweat,[16] saliva,[17] and tears,[18] which can be easily collected via non-invasive procedures. Common stress biomarkers are hormones produced from the sympathetic-adrenal-medullary (SAM) and hypothalamic-pituitary-adrenal (HPA), which are the two main systems controlling the biological response to stress.[1,4,19] The SAM system responds to stress by mediating a flight or fight response, via the release of catecholamines (e.g. dopamine, epinephrine).[4] Catecholamines constrict blood vessels and increase blood pressure.[4] Meanwhile, activation of the HPA system releases glucocorticoids into the bloodstream, exerting widespread effects on metabolism, immune, and brain function.[4,20] Cortisol is the major glucocorticoid released in fishes,[21] pigs and mammals, whereas corticosterone is the primary glucocorticoid found in rodents and rabbits.[4]

Amongst the stress hormones, cortisol is regarded as the gold standard for assessing the activity of the HPA.[1] Cortisol levels fluctuate in a circadian rhythm throughout the day, with levels peaking in the morning before decreasing towards the evening.[22] Cortisol release during acute stress helps to mobilize energy stores, promote cognitive function and regulate the immune system.[5] However, sustained cortisol elevation due to chronic stress can impair not only metabolic health but also mental health, causing depression, emotional instability, and irritability.[23] Anxiety disorders such as ASD and PTSD are similarly associated with high levels of cortisol and disordered HPA axis function.[1,9,24]

While glucocorticoid receptor activation tends to have anti-inflammatory effects, high levels of cortisol are known to inhibit the HPA and provoke the release of proinflammatory cytokines.[4,25] Elevated cortisol is also associated with hippocampal neuropathy[26] and poorer memory, which are phenotypes also observed in many neuropsychiatric disorders.[23,24,27] Further, Cushing's Syndrome, caused by sustained cortisol elevation, is characterized by excess weight gain, memory loss, and depression.[28] Hence, cortisol monitoring may be a useful indicator of chronic stress and associated diseases. Similarly in animals, cortisol (or corticosterone) levels may be used to assess their health and welfare.[29]

Conventional methods for cortisol detection include immunoassays[30-38] and high-performance liquid chromatography (HPLC).[34,39-41] Commercial radioimmunoassay (RAI) and enzyme-linked immunosorbent assay (ELISA) are widely available (Abcam Cortisol ELISA Kit,[42] ThermoFisher Scientific Cortisol Competitive Human ELISA kit,[43] Cusabio Cortisol ELISA kit,[44] and IBL Cortisol RIA kit,[45] etc.). Immunoassays (e.g. RIA and ELISA) rely on antibody–antigen formation for detection. RIA relies on radioisotopes while ELISA utilizes enzymes, such as horseradish peroxidase (HRP) to generate detection signals. Radioactive (RIA) is commonly found in a competitive assay format, while ELISA can be in competitive or sandwiched assay formats. In both assay formats, a captured antibody is bound to the surface of a substrate. In competitive assays, the sample containing the antigen is first mixed with an antibody. A secondary antibody labeled with either a RIA or ELISA component competes with the antibody–antigen complex to bind to the capture antibody. The signal obtained (either in the form of radioactivity for RIA or color change for HRP-ELISA) is thus inversely proportionate to the quantity of analyte. In sandwiched assays, the sample containing the antigen and a labeled antibody are sequentially added to the capture antibody. Successful binding of the analyte to the antibodies is observed through the addition of HRP which registers an optical signal. For a small molecule such as cortisol, competition assays are preferred over sandwiched assays as they attain higher sensitivities.[46]

HPLC is an analytical technique that relies on the separation of compounds based on relative affinities to the stationary and mobile phases in the column.[47] Sample pre-treatment is typically performed first via solid-phase extraction or liquid–liquid extraction to remove interfering components and concentrate the sample.[48–50] The polarity of the mobile phase is optimized through the choice of aqueous/organic solvents for improving analyte retention in the column. Various studies have reported the use of HPLC for detecting cortisol in human biofluids[41] and animal samples.[39,40]

In recent years, biosensors have emerged as promising tools for cortisol detection. Several publications have specifically reviewed electrochemical-based cortisol sensors and their related wearable forms.[51–53] In this review, we aim to provide a wider perspective by presenting a broader range of cortisol biosensors, including chromogenic sensors, nanomaterial-based colorimetric and/or fluorometric sensors, surface plasmon resonance (SPR) sensors, surface-enhanced Raman spectroscopic (SERS) and electrochemical sensors. Firstly, we provide some background information about cortisol found in different biological matrices. Next, we discuss a range of cortisol sensors that can be classified as either affinity ligand-based biosensors or direct molecular identity-based sensors, e.g. Raman sensors. The sensing performances of the assays, applications in different mediums, and potential uses for point-of-care diagnostics will be discussed. Finally, we provide insights into the challenges of existing cortisol biosensing assays and the further work needed to leverage their current successes.

2. Cortisol in Biological Fluids or Body Parts of Organisms

Cortisol can be detected in humans, mammals, cattle, fish, etc. Typically, cortisol is quantified from blood where it exists at relatively high concentrations. However, cortisol can also be found and measured from various biological fluids/body parts such as saliva, hair in humans/mammals, and scales, fins, or mucus in fish.[21] Depending on the biological matrices, cortisol exists in different forms: Free form and bound form (to proteins). Total cortisol refers to the sum of bound and free forms. More than 90% of circulating cortisol are bound to proteins. Amongst the bound cortisol, more than 80% is bound to cortisol-binding globulin with high affinity and low capacity, while 10–15% is bound to albumin with low affinity and high capacity.[54] The free form of cortisol is responsible for mediating biological functions via binding and activation of the glucocorticoid receptor.[53] These include the coordination of body functions in response to stressors and maintenance of healthy immune systems.[23] Only the free form of cortisol is physiologically active, and its quantification is regarded to be of superior biological relevance.[53]

In response to stress, cortisol is directly released from the HPA system into blood. In blood samples, cortisol presents in bound and free forms.[23] The concentration of cortisol in blood varies by time of the day, species, age, and sex. For example, free and total serum cortisol levels were ~33% and ~18%, respectively, and higher in males than in females.[55] Up to 30–55% of bound cortisol was found in the plasma samples of salmonids, with the concentration being higher in females (45%) than males (37%).[56] However, blood sampling is invasive and stress-inducing. For instance, the plasma cortisol levels of infants increased significantly ($P < 0.0005$) in 30–60 min after venepuncture for blood sampling.[57] Similar observations were reported in dogs in 20 min following venepuncture.[58] Moreover, measurement of cortisol in blood samples requires complex sample treatment such as dialysis and ultrafiltration[53] to isolate free cortisol prior to quantification via HPLC or RIA.[59]

Non-invasive cortisol detection through alternative biological fluids or body parts (e.g. sweat, saliva, hair, etc.) has also been introduced. The liposolubility and low-molecular weight of cortisol promote its clearance from plasma and uptake into cells.[60] Salivary and sweat cortisol are regarded as reliable indicators of stress in organisms[13] due to their good correlation with blood cortisol.[53] Saliva sampling from animals is regarded as a well-tolerated procedure involving the saturation of absorbent material with saliva in the presence of stimulants such as food treats.[4] Furthermore, the samples maybe collected at specific time periods for behavioral studies.[4] One limitation is that salivary cortisol levels are much lower than those present in blood. The salivary cortisol content in dogs represents up to 7–12% of those found in plasma,[4,61] while salivary cortisol levels in human are 100 times lower than in blood.[62] Thus, sensitive biosensing assays with low LODs are required for measurement of salivary cortisol samples. Different approaches such as HPLC[63] and immunoassays[64,65] have been reported for efficient detection of cortisol in saliva.

Hair sampling confers the advantages of monitoring systemic cortisol levels over prolonged periods of time.[7] The first centimeter of hair closest to the scalp corresponds to the cortisol produced in the past month and this growth occurs proportionally.[53] Hair cortisol is commonly employed for stress monitoring and clinical diagnostics of diseases such as Prader–Willi syndrome,[66] Cushing syndrome,[67] and acute myocardial infarction.[68] Shukur et al. reported that hair cortisol levels were higher in patients with Prader–Willi syndrome (12.8 pg/mg of hair) as compared to the control group (3.8 pg/mg of hair).[66] In a study of 198 Rhesus macaques, Novak et al. reported that alopecia (>30% hair loss) was associated with elevated cortisol concentrations.[69] Cortisol in hair samples is typically quantified through immunoassays.[70,71]

Another method to quantify cortisol levels is through urine samples, as cortisol is metabolized in the liver and excreted into urine in its free form.[53] Urinary cortisol samples are less influenced by circadian rhythm and are employed for monitoring peripheral circulation.[72] Cortisol levels in urine have traditionally been quantified via immunoassays[73] and HPLC.[74] However, urine contains various steroid metabolites with similar molecular structures to cortisol (e.g. 5α-tetrahydrocortisol, cortisone), leading to cross-reactions in immunoassays.[72,73] Discrepancies in sensing performances were reported when using different brands of immunoassay kits to analyze urinary cortisol samples.[73] In comparison, measurement of urinary cortisol via chromatographic-based detection yields small bias (<10% between batches) with low interferences.[75] It is noted that urinary cortisol levels cannot be correlated with blood cortisol due to renal tubular reabsorption and secretion.[13]

Cortisol may also be released into breast milk where it is used as a marker of perinatal psychosocial stress.[76] Some factors affecting breast milk cortisol levels include gestation period and delivery time.[77] Cortisol concentrations are higher in the gestation period and decrease by more than 50% after delivery.[77] Breast milk cortisol levels are also associated with various offspring behavioral phenotypes such as infant growth and temperament.[78] Higher cortisol levels in breast milk are associated with increased infant weight gain of rhesus macaque.[78] Steroids found in breast milk are typically protein-bound, although the protein may degrade after parturition.[79,80] In general, chromatographic techniques are preferred over immunoassays for detecting cortisol in breast milk[81–83] as immunoassay-based detection underestimates the cortisol level when high concentrations of corticosteroid binding globulin are present.[84] The organic solvents employed in chromatographic detection promote the release of cortisol from the proteins.[80] However, the solvent in chromatographic detection may lead to the formation of emulsions and loss of analyte.[85]

In fish, cortisol diffuses from the bloodstream and are released into the scales, fins, mucus, and tank water where their concentration maybe quantified.[21] It is noted that detection of cortisol in the above mediums measures the free cortisol.[21] The cortisol levels in scales have been used to assess chronic stress levels faced by fishes, and are influenced by water quality management (water supply, stocking density, feeding frequency) and external injuries.[86] Cortisol levels found in fish mucus and fins are well correlated with plasma content.[21] Mucus and fin samplings of cortisol were demonstrated to be effective tools for stress transport measurements in European sea bass, common coup and rainbow trout.[87] Various reported studies have quantified cortisol levels in fish tank water, fins, mucus, and scales through HPLC[21] and immunoassays.[33,88]

We summarize the type of cortisol found in different biological fluids and their measurement challenges in Table 1. Furthermore, Fig. 1 illustrates the typical cortisol levels found in various human biofluids and fish tissues (an animal example). Each type of sample requires different cortisol measurement sensitivity. In the following

Table 1. A summary of cortisol found in different biological fluids and their measurement challenge.

Biological fluids/ body parts	Cortisol presence	Cortisol measurement challenge
Blood	Free and bound cortisol	Invasive and stress-inducing sampling; requires complex sample pre-treatment prior to quantification.[53]
Saliva	Free cortisol	Requires ultrasensitive detection as salivary cortisol concentration is 100 times lower than in blood.[62]
Sweat	Free cortisol	Variations in sweat rate changes the concentration of biomarker.[89]
Hair	Free cortisol	Variations in hair growth rate (e.g. due to ethnicity) and sampling regions render it difficult to accurately estimate cortisol levels from specific time periods.[90]
Urine	Free cortisol	Complex sample matrix; cannot be correlated with blood plasma content due to renal tubular reabsorption and secretion.[13]
Breast milk	Free and bound cortisol	Difficulty in isolating biologically available cortisol from bound cortisol.

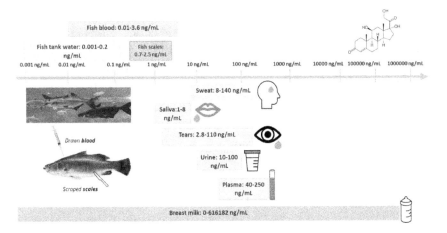

Fig. 1. Normal cortisol levels in various biological fluids from human[18,91–96] and in fish.[21]

sections, we will discuss a range of cortisol sensors, their applications and suitability for various types of cortisol samples.

3. Chromogenic Sensors

We have previously introduced conventional methods (HPLC, RIA, and ELISA) for cortisol detection. These methods are time-consuming and typically require trained personnel to operate the equipment, thereby limiting deployment for layman users. For HPLC, the use of organic solvents in large volumes is non-environmental friendly. In recent years, various biosensors have emerged as promising tools for cortisol detection. These include chromogenic sensors, nanoparticle-based sensors, surface plasmonic resonance (SPR) sensor, surface-enhanced Raman spectroscopy (SERS), and electrochemical detection. In this section, we will first discuss chromogenic sensors, and the rest will be discussed in Secs. 4–6.

Chromogenic sensors or assays rely on chemical reagents and their interactions with the analytes to produce visible color changes or fluorescent signals.[97] Usually, the interactions involve specific functional groups of the analytes. Four general chromogenic assays have been developed, relying on sulfuric acid, Porter–Silber reagent, Prussian blue, or blue tetrazolium as reagents for identification of steroid classes,[98] (e.g. corticosteroids that include the cortisol hormone) and progressively improved for detection of specific hormones.[95] However, their selectivity is still limited due to similar backbone structure/functional groups across steroids.

The first chromogenic assay involves corticosteroids interaction with concentrated or anhydrous acids to produce colored solutions with fluorescence under illumination. The formation of the colored products is due to the oxidation of the ketol functional groups in steroids by the acids.[99] For instance, for cortisol, the reaction product is a yellow–orange solution with green fluorescence.[99] Such fluorescence occurs due to the formation of charged steroid-anhydrous acids complexes involving the oxygen functional groups of the corticosteroid compounds at the C3, C11, and C17 positions.[100] The use of sulfuric acid promotes the formation of carbonium ions at these positions.[100] The green fluorescence for cortisol is mainly dictated by the formation of carbonium ion at the C11 position.[100] Additives, such as 2-methyl-4-pentanone, have been employed for sample extraction and signal amplification in the fluorometric-based chromogenic cortisol detection, yielding a LOD of 2.7 μM within a linear range of 2.7–27.6 μM.[100] However, this assay cannot overcome interference by 11-deoxycortisol.

In addition to the fluorometric detection, the colored products generated can be quantified by absorbance measurements.[101] For example, cortisol interaction with concentrated sulfuric acid yields a product with absorption peaks at 280 nm, 395 nm, and 475 nm.[102] This assay was used to identify the presence of cortisol in plasma of dairy cattle.[103] However, this technique is limited by cross-selectivity towards different steroids, such as cortisone and corticosterone.[99,104,105] To circumvent this limitation, the reaction conditions, such as the type of acids, reaction time, and temperature, have been varied.[104,105] Additives can also be added to improve the selectivity. For example, Schulz et al. added 2,6-di-tert-butyl-p-cresol in alkaline solutions and leveraged the formation of colored steroid-phenol chromogens as a basis for improving the selectivity to cortisol.[105] Particularly, the developed product is a blue color with peak absorption maxima at 625 nm, while interference cortisone developed a yellow-brown colored product, with an absorption peak at 471 nm. This method is applicable for pharmaceutical analysis to identify steroids in medicinal formulations.

The second chromogenic assay is based on Porter–Silber reaction that uses sulfuric acid and phenylhydrazine to identify steroids with a dihydroxyacetone group, such as cortisol.[101] The solution developed a yellow color with peak absorption at 410 nm upon the addition of steroids into the reagents in the presence of alcohol.[101,106] The reaction is postulated to occur by the Mattox rearrangement of dihydroxyacetone chains to yield an intermediate 20,21-ketal compound.[106,107] This intermediate then undergoes a reaction with phenylhydrazine to form the yellow-colored 21-phenylhydrazone compound.[106,108] This method is applicable for general identification of hydroxycorticosteroids in samples,[98] but limited by poor selectivity.

Thirdly, tetrazolium salt is employed for steroid detection, through reduction in alkaline mediums to form formazans.[109] For example, 3,3'-dianisole-bis-4,4'-(3,5-diphenyl)tetrazolium chloride (i.e. blue tetrazolium) is used in steroid identification due to the formation of a stable blue-colored product.[110,111] Identification of cortisol using blue tetrazolium is usually done by monitoring the absorption peak at either 525 nm[112] or 510 nm,[95] depending on different solvent systems employed.[95] Tetrazolium-based detection is hampered by poor selectivity as corticosteroids having an α-ketol side chain[113] (e.g. cortisol, hydrocortisone acetate) can participate in the reaction.

Lastly, Singh et al. leveraged the reduction properties of corticosteroids to form Prussian blue after reaction with iron (III) and potassium hexacynoferrate.[114] This mechanism is based on the oxidation of cortisol by iron (III) in acidic mediums to form an iron (II) product. Subsequent complexation of iron (II) with potassium hexacynoferrate (III) yields a blue-green iron (II) ferricynide complex, with a peak absorption at 780 nm. This method is applicable for the identification of corticosteroids in pharmaceutical formulations but limited by selectivity across different steroids, such as cortisol, prednisole, dexamethasone, and prednisolone, since they all exhibit an absorbance peak at 780 nm.[114]

In a more recent study, Tu et al. evaluated the four techniques: Sulfuric acid, Porter–Silber reagent, Prussian blue, or blue tetrazolium for detecting cortisol in sweat (Fig. 2).[95] The Porter–Silber method yielded a LOD of 0.145 μg/mL within a dynamic range of 0–70 μg/mL. The Prussian blue method was limited by the high standard deviation of the calibration curve, resulting from the wide distribution of particle sizes. For the above two detection methods, the time taken for color development was longer at 45–60 min. Although the concentrated sulfuric acid-based detection offered immediate color development, this method was hampered by its unstable color development and high LOD (3 μg/mL). Amongst the four methods explored, reaction with blue tetrazolium gave the lowest LOD (97 ng/mL) and a broad detection range (0.05–2 μg/mL) within a quick color development time of 10 min. Its detection

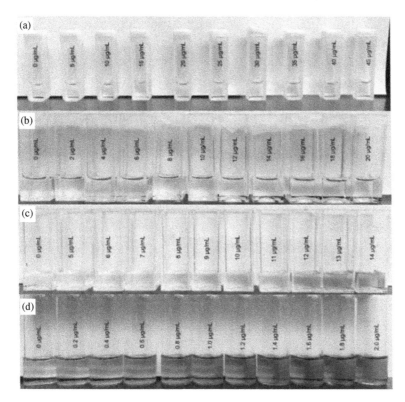

Fig. 2. Optical images of varying cortisol concentrations by using (a) Sulfuric acid (b) Porter–Silber, (c) Prussian blue, and (d) blue tetrazolium.

Source: Copyright 2020, *ACS Omega*.[95]

performance was further confirmed in artificial sweat, along with its ability to discriminate between apocrine and eccrine cortisol glands, and across genders.

4. Nanoparticle-Based Cortisol Sensors

Noble metals, such as gold nanoparticles (AuNP) and silver nanoparticles (AgNP), exhibit unique optical properties known as localized surface plasmon resonance (LSPR), which refers to the collective oscillation of electrons in resonance with the incident electromagnetic radiation.[115–118] The LSPR spectrum (or solution color) is determined by their morphologies (size and shape), interparticle distances, and surface coating. These properties can change due to chemical interactions occurring nearby.[119,120] These LSPR properties can support optical sensors of various principles, including particle aggregation-based colorimetric sensing, fluorescence quenching-based fluorometric and surface-enhanced Raman spectroscopy (SERS).

Semiconductor nanoparticles, such as quantum dots (QDs), have tunable fluorescence emission, achieved through control of morphologies,[121] chemical compositions,[121] and surface engineering.[122] In comparison to traditional fluorophores, QDs exhibit stable fluorescence emissions with higher intensity and narrower emission bands for multiplex detection.[123,124] The nanometer particulate nature of QDs promotes three-dimensional diffusion of analytes for faster detection, while the availability of functional groups on QDs confers multiple opportunities for probe attachments.[125] Leveraging their superior physicochemical properties, QDs have been widely employed in fluorometric-based assays for hormone detection.[126–128]

To achieve selective detection, nanoparticles are often coupled with bioreceptors, such as antibodies and aptamers, for selective binding to the target analyte. In general, antibodies display high binding affinity to the target analyte but are limited by their *in vivo* production.[129] Antibodies are commonly employed in sandwiched or competitive immunoassays alongside absorbance, fluorescence or chemiluminescent readouts.[130] On the other hand, aptamers are single-stranded DNA selected by an *in-vitro* process known as Systematic Evolution of Ligands by EXponential enrichment (SELEX).[129,131] They bind to target analytes with high selectivity by folding into tertiary structures.[132–134]

4.1. *Metal nanoparticle and aptamer-based colorimetric sensors*

Gold and silver nanoparticles will change their dispersion states in the presence of high salt concentrations, leading to a visual color change.[135] This principle has been coupled with aptamers for designing colorimetric sensors.[136–138] Typically, when aptamers are adsorbed to the surface of AuNP, the AuNPs are protected by the aptamer against salt-induced aggregation. However, when aptamer-analyte complex is formed, the conformation change of the aptamer reduces the protection effect, and thus the AuNPs undergo aggregation under the same salt conditions, leading to a color change from red to blue. The analyte concentration can be quantitatively correlated with the degree of AuNPs aggregation (i.e. degree of color change). In 2014, Martin et al. reported the first-known cortisol binding aptamer (Apt 15-1, 85 mer) (K_d = 6.9–16.1 µM) and demonstrated a colorimetric assay for cortisol as described above.[139] The aptasensor was responsive to 0–5 µM cortisol in a buffer system (50 mM Tris, 137 mM NaCl and 5 mM $MgCl_2$) with selectivity over other biomarkers such as norepinephrine, epinephrine and cholic acid.[139]

Dalirirad et al. subsequently reported the use of a truncated version of the 85-mer aptamer (denoted as Apt 15-1a, 40 mer) with AuNPs for detecting cortisol in sweat.[92] The authors reported a LOD of 1 ng/mL (2.7 nM) in artificial sweat with selectivity over other stress biomarkers, such as neuropeptide Y and serotonin. By tapping on the sensitivity and selectivity conferred by solution phase assays, the authors further developed a lateral-flow assay (LFA). In this biosensor strip, cysteamine was immobilized on the test zones of the nitrocellulose membrane (Fig. 3). An optimum amount of aptamer-coated AuNP was then used for loading onto the strip. Addition of cortisol desorbed the aptamers from the surface of AuNP and the free AuNP could be captured by the cysteamine. A quantitative relationship between the cortisol concentrations and the intensity of red line on the test zones was established. The sensor exhibited a fast detection time (5 min) with a LOD of 1 ng/mL within a detection range of 1–300 ng/mL. The obtained sensing performance is sufficient to cover the cortisol levels found in human sweat (Fig. 1). Coupled with the good stability of AuNP (~30 days) and strips (~10 days), the developed portable sensor is promising for POC diagnostics of stress. In 2020, the authors reported a duplex aptamer-conjugated AuNP for salivary

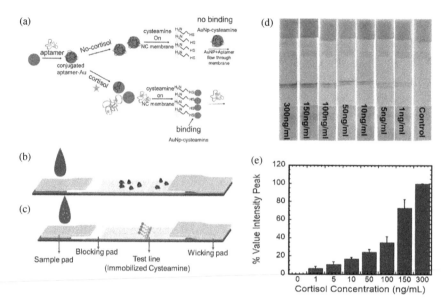

Fig. 3. Illustration of cortisol detection using AuNP-based lateral flow assay. (a) Basic concept. (b) In the absence of cortisol where no color change is observed. (c) In the presence of cortisol where the color change is observed. (d) Effect of cortisol concentrations on the visibility of test line. (e) Percentage value intensity peak as a function of cortisol concentration (0–300 ng/mL).
Source: Copyright 2019, *Sensors and Actuators B: Chemical*.[92]

cortisol detection in an LFA device.[140] The sensor exhibited a dynamic range of ~0.5–15 ng/mL and a LOD of 0.37 ng/mL which was validated by ELISA. High selectivity of detecting cortisol over other stress biomarkers in saliva was observed.

In comparison to the free aptamer form employed above, Tian *et al.* conjugated Apt 15-1 to AuNP and reported a cortisol detection range of 1–1000 nM, in a typical colorimetric solution-based assay.[141] They reported that the folded aptamers (with cortisol addition) had decreased stability in comparison to the unfolded aptamers in the presence of NaCl. Thus, a higher degree of aggregated AuNP reflected increased cortisol concentrations. The selectivity of the cortisol biosensor was confirmed in blood over other similar structured steroids such as cortisone, progesterone, and triamcinolone.

Most recently, Niu et al. evaluated the binding affinity and sensing responses of two commonly used cortisol aptamers: Apt 15-1a (40-mer) and Apt CSS.1 (42-mer) using an isothermal titration calorimetric (ITC) and a typical free aptamer -AuNP colorimetric assay.[142] Apt CSS.1 is a recently reported high affinity aptamer (Kd = 30–140 nM).[143] Through an ITC assay, the authors reported high binding affinity of cortisol to Apt CSS.1 (Kd = 245 nM) but failed to observe cortisol binding to Apt 15-1a.[142] Using an AuNP and free Apt CSS.1 colorimetric-based assay, a linear detection range of 1–12.5 μM cortisol was achieved.[142] However, the specificity of the sensor was challenged by other biomarkers such as 17β-estradiol, thymidine, and dopamine.[142] There are many other high binding affinity aptamers (Kd = 30–140 nM) for steroids (including cortisol)[143] via theoretical simulation but they have been less employed in biosensors.[142] Future experimental works to test these aptamers and development of highly selective and sensitive aptamers for cortisol are highly recommended.

4.2. Antibody-AuNP-based immunosensor

In 2019, Apilux et al. reported a paper-based competitive immunoassay for cortisol detection in serum.[46] In this assay, a cortisol-BSA conjugate was immobilized on the test zone. An anti-cortisol antibody was conjugated to AuNP and further mixed with test samples containing cortisol before dripping onto the detection zone. The free cortisol and the immobilized cortisol-BSA competed to bind with the anti-cortisol antibody-conjugated AuNP. The intensity of red color developed on the test zone was determined by the amount of AuNP-antibody conjugate. Thus, the cortisol levels in test samples were quantitatively and inversely related to the intensity of red color developed on the paper strip. The sensor was able to differentiate cortisol levels into three different ranges: <25 μg/dL (normal cortisol levels), 25–50 μg/dL ("warning" cortisol levels), and 50 μg/dL (high cortisol levels). A LOD of 21.5 and 25 μg/dL was obtained through image processing software and visual detection, respectively. Coupled with the low-cost and

4.3. Metal/metal-bearing nanoparticle sensor based on target-induced fluorescence quenching

Various steroid hormones induce fluorescence quenching of QDs when in close proximity.[125] This mechanism is associated with the charge transfer between the excited QDs and the steroid molecules.[125] Steroid molecules may exhibit electron-accepting capabilities due to their carbonyl and unsaturated ketone groups or electron-donating properties from their aromatic structure.[125,144] Cortisol is known to function as electron donors and its charge transfer to organic molecules such as 2,3-dichloro-5,6-dicyano-p-benzoquinone have been reported.[145] Likewise, cortisol may participate in charge transfer reactions with QDs.

To enhance the selectivity of such fluorescence-based assays, Liu et al. conjugated cortisol-selective aptamers or antibodies onto CdSe/ZnS QD carried by magnetic nanoparticles (MNPs).[125] The cortisol-induced fluorescence quenching was measured through PDMS microfluidic wells sealed above the QD-coated glass substrates. The aptamer-based and antibody-based nanosensors achieved a LOD of 1 nM and 100 pM, respectively, within a detection time of 20 min. The higher LOD of aptamer-based sensor was attributed to the larger dissociation constant of the cortisol-aptamer complex. Cortisol detection in spiked saliva samples was successfully demonstrated with a mean recovery rate of 85–107%. This facile, label-free sensor fabricated on solid substrates is promising for POC stress hormone diagnostics.

5. Surface Plasmon Resonance (SPR) Sensor

SPR is a phenomenon describing the oscillation of surface electrons upon exposure to incident light.[146] This technique measures changes in the refractive index of the surface of a planar metal film when molecular interactions occur.[147-149] Planar gold-based SPR relies on a prism to polarize light and excite electrons on the metal surface under total internal reflection conditions. Bioreceptors such as antibodies are commonly immobilized on the surface of the SPR sensor for selective binding with the analytes.[150–152] For small molecule analytes, such as cortisol, competitive assays have been used for its detection since direct binding of the analyte to the receptor-bound sensor surface does not lead to measurable changes in refractive index.[46]

Stevens et al. developed a portable SPR sensor for detecting cortisol in saliva by a competitive immunoassay.[93] This portable SPR sensor is a 6-channel system equipped with a flow filter to exclude larger-sized particulates from reaching the sensor's surface, thereby increasing its sensitivity. Meanwhile, an external flow promoted the diffusion of smaller-sized molecules into the sensor stream. A competitive immunoassay was exploited by flowing anti-cortisol antibody, followed by the cortisol samples over the SPR sensor channels coated with either cortisol-BSA or BSA. A LOD of 1.0 ng/mL was achieved in a detection range of 1.5–10 ng/mL, within less than 10 min.

Mitchell et al. reported microfluidic SPR detection of cortisol exploiting a competitive immunoassay.[153] A carboxymethylated dextran-modified SPR chip surface was coated with a derivatized cortisol. Cortisol-containing sample was then incubated with the antibody and injected into the SPR. The sample cortisol and the immobilized cortisol derivative compete the binding site on the antibody. The SPR signal induced by the antibody binding is inversely proportional to the cortisol concentration in the solution. A LOD as low as 49 pg/mL and a dynamic range of 91–934 pg/mL were reported for human saliva samples. The fast detection time (15 min) renders it promising for POC diagnostics and monitoring of stress-related disorders.

Optical fibers have been employed in SPR sensor construction.[154] Due to their compact structure, these optical fiber SPR sensors are used in narrow spaces for real-time detection and *in-situ* monitoring, for example, within fish tanks.[149,154] Leitao et al. reported the use of AuPd coated plastic optical fiber (POF) SPR sensor for detecting cortisol in PBS (Fig. 4).[155] By leveraging the strong interactions between gold and sulphur, AuPd was functionalized with a

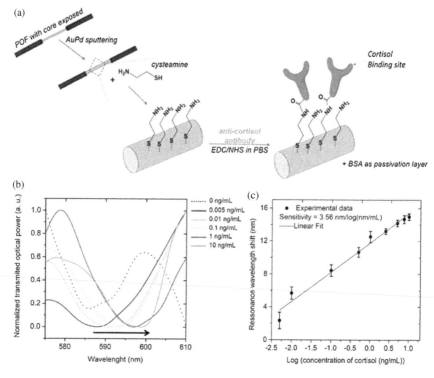

Fig. 4. (a) Schematic illustrating chemical functionalization of optical fiber. (b) Optical spectra of AuPd acquired for cortisol concentrations of 0–10 ng/mL in PBS. (c) SPR peak shift as a function of logarithm of cortisol concentration. *Source*: Copyright 2021, *Biotechnology Reports*.[155]

cysteamine linker. An anti-cortisol antibody was then covalently immobilized to the linker via EDC/NHS coupling. The developed SPR sensor demonstrated a 15 nm wavelength shift for cortisol concentration range of 0.005–10 ng/mL, with a LOD of 1 pg/mL. Selectivity of the sensor was confirmed through the small SPR shift (1 nm) when functionalized with anti-human chorionic gonadotropin antibody and tested against 10 ng/mL cortisol. Other reported optical-fiber-based SPR cortisol sensors include Au-coated tilted fiber Bragg grating[156] and SiO$_2$/SiC grating on Ag.[157]

Theoretical simulations indicate that the use of 2D materials such as MXenes as fiber optic probes in SPR may be promising for elevating the sensing performances.[158] MXenes are reportedly known for their high stability, biocompatibility and hydrophilicity.[158] In comparison to traditional 2D materials, the hydrophilic sites on MXenes are readily accessible for measurements of aqueous samples (e.g. fish tank water).[159] Sharma et al. simulated that the use of $Ti_3C_2O_2$ as probes can detect cortisol concentrations as low as 15.7 fg/mL, rendering it useful for stress monitoring in marine life.[159] Further research into synthesis of $Ti_3C_2O_2$ through top-down or bottom-up approaches and their integration into SPR assays for cortisol detection is recommended.

6. Surface-Enhanced Raman Spectroscopy Detection

Surface-enhanced Raman Spectroscopy (SERS) is an ultrasensitive detection method used for fingerprinting chemical compounds of interest.[160-163] This method capitalizes on the LSPR properties of metal nanostructures to increase their electric fields, which amplifies the Raman scattered light.[164,165] The characteristic Raman peak intensities of the analyte are greatly enhanced when in proximity with metal nanoparticles.[165,166] To date, SERS-based detection of cortisol has been reported using either direct strategies, where the Raman signatures of cortisol are directly measured,[167] or indirect strategies, where the signals are obtained from a Raman-active reporter molecule instead.[94,168]

In 2020, Moore et al. reported the first direct SERS detection of cortisol without employing Raman reporters.[167] The authors incubated AgNP with cortisol samples for 2 h prior to SERS measurement in the solution phase. The signature Raman peaks of cortisol were generated by density-functional theory (DFT) calculations and further confirmed experimentally. The SERS sensor demonstrated a response towards cortisol in the range of 125–250 nM, sufficient to cover physiologically relevant ranges. A LOD of 177 nM was achieved for cortisol prepared in ethanol solutions. Further testing of the SERS sensing performance in biological mediums (bovine serum albumin and PBS) was performed. For such complex sample matrices, principal component analysis (PCA) was employed to discern the specific Raman fingerprints of cortisol. A detection range of 125–2000 nM for cortisol in serum was reported.

Villa *et al.* demonstrated two indirect SERS competition immunoassays for detecting cortisol by using SERS nanotags either in the solution phase involving magnetic beads or on solid surface of gold-coated glass substrate (Fig. 5).[94] A cortisol antibody was used to functionalize the magnetic beads and the gold-coated glass substrate. The SERS nanotags were gold nanostars (AuNS) co-functionalized with a Raman reporter and cortisol-BSA conjugate (AuNS/Raman tag/BSA-cortisol). In the solution phase assay, cortisol

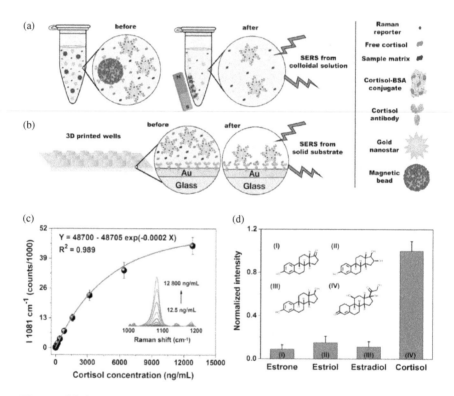

Fig. 5. (a) Solution-based SERS assay with cortisol antibody-conjugated magnetic beads as capture substrates. (b) Solid surface SERS assay using Au-coated glass functionalized with cortisol antibodies as capture substrates. (c-I) SERS signal as a function of varying cortisol concentrations in solution-based SERS assay, and (c-II) Selectivity for cortisol in solution-based SERS assay.

Source: Copyright 2020, *Biosensors and Bioelectronics*.[94]

antibody-conjugated magnetic beads bind to free cortisol in samples. Upon magnetic separation, the Raman signal of the unbounded SERS nanotags remaining in the solution increased proportionally to cortisol concentrations. In the absence of cortisol, the SERS nanotags were captured by the antibody-magnetic beads, leading to weaker Raman signals obtained from the solution. For the solid-phase immunoassay, a similar competition reaction is exploited. The Raman signal of AuNS/Raman tag/BSA-cortisol bound on the substrate was inversely related to cortisol levels. The LOD of the solution and solid-based immunoassays was 7 ng/mL and 3 ng/mL, respectively, sufficient to detect cortisol found in urine and serum (Fig. 1). The SERS sensing performance was further demonstrated using a portable Raman spectrophotometer, showing promising applications for point-of-care diagnostics.

Gao et al. reported an indirect SERS immunoassay for detecting cortisol in hair samples.[168] The sensor comprised a double-layered paper microfluidic device for sample collection and reaction, and a portable Raman spectrophotometer for detection. The first layer of the paper served as a sample pretreatment to remove hair residues, while the second layer comprised immobilized cortisol for immunoassay reactions. A SERS tag was designed by conjugating Au nanostars to 4-mercaptobenzoic acid and cortisol antibody. In a competitive immunoassay, the immobilized cortisol and free cortisol in the samples competed to bind with the antibody-bound Au nanostars. The nanosensor detected cortisol up to 1 pg/mL and its sensing performance was further validated by liquid chromatography-mass spectrometry (LCMS). Furthermore, the portable paper device was capable of collecting up to 48 samples at one time, rendering it useful for large-scale, on-site detection.

7. Electrochemical Detection

Electrochemical sensors have emerged as a promising analytical method for ultrasensitive, rapid and continuous detection.[169–171] In general, electrochemical sensors comprise a conductive transducer material (or electrode) coated with biomolecular recognition ligands that target the analyte of interest.[53] The sensor measures changes in the electrical signals of conductive materials upon adsorption of the analyte onto its surface via bioreceptors.[53] Electrochemical sensors have been employed for real-time monitoring due to their ability to measure small concentrations of analytes within a short time span. To date, a number of reviews have been published based on electrochemical and related real-time detection of cortisol.[1,51-53] Thus, this section will provide a short/brief overview of their recent developments.

In 2020, Ku et al. reported a smart contact lens, i.e. a graphene-based field-effect transistor (FET), for real-time detection of cortisol in tears[18] using the binding of cortisol to the antibody functionalized FET chip to generate electronic signals. The biosensor was further interfaced with a smartphone for programmable control. A LOD of 10 pg/mL was achieved, which is sufficient to cover the cortisol levels found in tears (Fig. 1). The biocompatibility of this immunoassay was confirmed with the use of live rabbits and human pilot experiments.

An et al. designed a wearable cortisol aptasensor (using Apt 15-1a) for rapid, real-time monitoring of cortisol in sweat.[172] The sensor patch comprised conductive polyacrylonitrile (PAN) nanofibers and poly(3,4-ethylenedioxythiophene) (PEDOT) channels on a polyester film in a liquid-ion site-gated FET system (Fig. 6). The cortisol-specific aptamers were further conjugated to PAN-PEDOT for achieving target-specific binding. The biosensor displayed a rapid analysis time (<5 s) with a low LOD of 10 pM and a linear response range of 1 pM – 10 μM. High selectivity of this aptasensor is reported in the presence of other hormones at 100× higher concentrations. Real-time monitoring of human sweat was further demonstrated with a detectable cortisol concentration of 1 nM.

Besides aptamers and antibodies, artificial receptors, such as molecular imprinted polymers (MIP), have been reported for cortisol detection.[173-177] MIP possess superior properties, such as

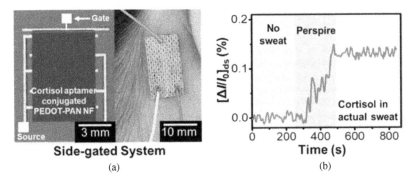

Fig. 6. (a) Illustration of side-gated cortisol aptasensor. (b) Real-time monitoring of cortisol eluted from actual sweat sample.
Source: Copyright 2022, ACS Sensors.[172]

high-recognition ability, low-cost, good durability, and stability. They have been employed for biosensing applications.[178] MIP are formed in the presence of the analyte, which after removal, leave cavities within the polymer template structure.[179] The formed cavity is similar in morphology and chemical functionality to the analyte for selective recognition.[179] Reversible interactions between the analyte and MIP are desirable for biosensing applications through mediation of non-covalent bonds in the polymerization process. Factors such as the content of cross-linking molecule, monomer and solvent are crucial parameters to control during MIP synthesis. Tang *et al.* reported an electrochemical sensor involving imprinted polypyrrole membrane for binding to cortisol selectively, upon which the electrical current response to embedded Prussian Blue redox probes decreased.[173] A porous and sweat absorbing polyvinyl alcohol (PVA) hydrogel was further designed to collect and wick up sweat samples eluted from fingertips. Overall, the MIP-based biosensor displayed rapid analysis time (3.5 min) and a LOD of 0.9×10^{-9} M and 0.2×10^{-9} M in PBS and artificial sweat, respectively. The sensor was further adapted to a wearable sensor patch for real-time monitoring of cortisol levels during exercise. Future opportunities include expanding the scope of detection to other biomarkers or stress hormones.

8. Comparison of Cortisol Biosensors

Table 2 summarizes the cortisol-sensing performances reported by the various biosensors. In general, electrochemical sensors exhibit the lowest LOD with the widest dynamic sensing range. Their reported sensing performances in terms of sensitivity and detection range suit the cortisol concentration range in biomedical and environmental samples, such as human/animal biofluids and aquaculture fish tanks. For nanoparticle-based colorimetric, SERS and SPR sensors, their LOD and sensing ranges are generally comparable in nanomolar range and cover the clinical concentration ranges of human biofluids. Among these three sensor types, SPR detection exhibits a performance edge through employment of novel sensing materials/sensor designs. For instance, the reported plasmonic unclad AuPd SPR sensor yields a LOD of 1 pg/mL within a sensing range of 0.005–10 ng/mL,[155] comparable to the performance of electrochemical sensors. Chromogenic sensors exhibit the highest LOD in micromolar range. Their sensing performances are applicable for qualitative (i.e. Yes/No test) rather than quantitative analysis of human biofluids, since they only cover the higher cortisol concentration ranges (Fig. 1).

Portable and user-friendly sensors are highly desirable for POC diagnostics/monitoring of stress-related diseases. The discussed cortisol biosensors have been tested in laboratory settings, with a handful reporting their portable form (Table 2). Among the different sensor forms, electrochemical sensors are most easily adaptable into portable devices including lab-on-a chip systems through nanofabrication techniques. Colorimetric sensors employing AuNP and aptamers are easily incorporated into LFA devices for portable, fast, and easy detection. The stability of such nanomaterials promotes good shelf-life of the portable device (60 days)[140] without compromising sensing performances. Furthermore, colorimetric sensors assay maybe programmed to be smartphone controlled, with quantifiable color readouts using mobile applications such as Color Picker.[180] Even so, color readouts may vary due to different types of mobile devices[181] and environmental lighting.[182] A higher

Table 2. A summary of reported cortisol biosensors.

Biosensor format	Reagents	Tested sample matrix	LOD*	Dynamic range*	Portability	Ref
Colorimetric	AuNP and free aptamer	Sweat	~1 ng.mL (2.7 nM)	1–300 ng/mL (2.7–827.7 nM)	Yes	92
Colorimetric	AuNP conjugated duplex aptamer	Saliva	0.37 ng/mL (1.0 nM)	~0.5–15 ng/mL (1.4–41.4 nM)	Yes	140
Colorimetric	AuNP conjugated aptamer	NA	—	1–1000 nM	No	141
Colorimetric	AuNP, cortisol antibody and cortisol-BSA	Serum	21.5 μg/dL (593.2 nM)	Differentiate cortisol levels into three ranges: <25 μg/dL (<689.7 nM), 25–50 μg/dL (689.7–1379.5 nM), >50 μg/dL (> 1379.5 nM)	Yes	46
Colorimetric	Aptamer/antibody conjugated QDs carried by magnetic NPs	PBS, Saliva	Aptamer-based sensor: ~1 nM Antibody-based sensor: ~100 pM	Aptamer-based sensor: 0.4–400 nM	Yes	125

(Continued)

Table 2. (Continued)

Biosensor format	Reagents	Tested sample matrix	LOD*	Dynamic range*	Portability	Ref
Chromogenic	Blue tetrazolium	Sweat	0.097 μg/mL (267.6 nM)	0–1.5 μg/mL (0–4138.4 nM)	No	95
Chromogenic	Porter–Silber with AuNP	Sweat	0.145 μg/mL (400 nM)	0–70 μg/mL (0–1.9 × 10^5 nM)	No	95
Chromogenic	Potassium hexacyanoferrate (III) and iron (III)	Sweat	0.731 μg/mL (2016.8 nM)	Undetermined	No	95
Chromogenic	Pure concentrated sulfuric acid	Sweat	3.00 μg/mL (8276.8 nM)	0–100 μg/mL (0–2.8 × 10^5 nM)	No	95
Chromogenic	Sulfuric acid and acetic acid	NA	2.7 μM	2.7–27.6 μM	No	100
SPR	Cortisol antibody, cortisol-BSA	Saliva	1.0 ng/mL (3.6 nM)	1.5–10 ng/mL (4.1–27.6 nM)	Yes	93
SPR	Carboxymethylated dextran surface, cortisol antibody	Buffer, saliva	Buffer: 13 pg/mL (0.04 nM) Saliva: 49 pg/mL (0.1 nM)	In saliva: 91–934 pg/mL (0.3–2.6 nM)	No	153
SPR	AuPd, cysteamine, anti-cortisol antibody, BSA	PBS	1 pg/mL (0.003 nM)	0.005–10 ng/mL (0.01–27.6 nM)	No	155

Method	Recognition element	Medium	Sensitivity/LOD	Range	Wearable	Ref.
SPR	Au-coated tilted fiber Bragg grating	PBS	0.275 ± 0.028 nm/ngmL^{-1}	0.1–10 ng/mL (0.3–27.6 nM)	No	156
SPR	SiO$_2$ or SiC on Ag grating	Saliva	NA	SiO2: 22.6 fg/mL (6.2×10^{-5} nM) SiC: 68.17fg/mL (1.9×10^{-4} nM)	No	157
SPR	Denaturalized BSA chip, cortisol antibody	PBS	5–100 ng/mL (13.8–275.9 nM)	1 ng/mL (2.7 nM)	No	187
SERS	AgNP	PBS, ethanol	In ethanol: 177 nM	In ethanol: 125–250 nM In PBS: 125–2000 nM	No	167
SERS	Au nanostars, cortisol antibody, cortisol-BSA	Urine, serum	Solution-phase: 7 ng/mL (19.3 nM) Solid-phase: 3 ng/mL (8.3 nM)	Solution-phase: 12.5–400 ng/mL (34.5–1103.6 nM)	Yes	94
Electrochemical	Graphene FET, cortisol antibody	Tears	10 pg/mL (0.03 nM)	1–40 ng/mL (2.7–110.4 nM)	Yes	18
Electrochemical	PAN-PEDOT with conjugated aptamer	PBS, Sweat, Saliva	In PBS: 10 pM In sweat: 1 nM	In PBS: 1 pM–10 μM	Yes	172
Electrochemical	MIP: poly (glycidylmethacrylate-co ethylene glycol dimethacrylate)	Sweat	2.0 ± 0.4 ng/mL (5.5 ± 1.1 nM)	10–66 ng/mL (27.6–182 nM)	Yes	175

(Continued)

Table 2. (*Continued*)

Biosensor format	Reagents	Tested sample matrix	LOD*	Dynamic range*	Portability	Ref
Electrochemical	MIP: Au-doped poly o-phenylenediamine	PBS, artificial saliva	200 fM	1 pM–500 nM	Potentially yes	176
Electrochemical	AuNPs@MIP@PANI@CNT/CNC@textile	Sweat	8.00 ng/mL (22.1 nM)	9.8–49.5 ng/mL (27–136.6 nM)	Yes	188
Electrochemical	ZnO coated carbon yarns, anti-cortisol antibodies	Sweat	Cyclic voltammetry: 0.45 fg/mL (1.2×10^{-6} nM) Differential pulse voltammetry: 0.098 fg/mL (2.7×10^{-7} nM)	1 fg/mL–1 μg/mL (2.8×10^{-6} – 2.8×10^{3} nM)	Potentially yes	189
Electrochemical	Glucose oxidase, anti-cortisol antibodies	Fish plasma	1.01 ng/mL (2.8 nM)	1.25–200 ng/mL (3.4–551.8 nM)	No	190

*The units have been converted to respective values in nM to facilitate data comparison. Molecular weight of cortisol = 362.46 g/mol.

accuracy of image analysis maybe achieved through controlled illumination of images via controlled chambers[182] or trained machine-learning algorithms in specifically designed mobile applications.[183] In addition, recent developments in portable detection systems such as Raman spectrophotometers and SPR biosystems, which are designed to be compact and transportable, allow for on-site detection.

In terms of sensing materials, the unique optical properties of nanomaterials have pivoted the advancement of colorimetric, SERS and SPR-based cortisol sensors. The LSPR properties of metal nanoparticles are utilized to generate color changes in relation to their aggregation/dispersion state based on aptamer-analyte interactions. Metal nanoparticles such as gold nanospheres can also increase the Raman scattering through electromagnetic enhancement by 10^2–10^3.[184] Further Raman scattering enhancement is achieved through the use of nanoparticles with edges such as nanocubes or nanostars. Newly emerging 2D nanomaterials such as MXenes could be explored in conjunction with high binding affinity aptamers, antibodies and/or SERS tag to develop high-performing sensors for cortisol detection. Research has demonstrated that MXene aptasensors can exhibit high sensitivity with a LOD in nanomolar to femtomolar range,[185] with potentials for scale-up in industrial processes.[186]

9. Conclusions and Perspectives

Research over the last decade has demonstrated that stress hormones play a critical role in regulating our bodily functions and that of other organisms. Some existing challenges include the complex sample matrices where cortisol is found, the low-cortisol level in some sample types, and the ability to differentiate cortisol from other similarly structured hormones. It is thus essential to develop selective, sensitive, rapid, and accurate sensors for timely detection of cortisol. In this review, we have presented an overview of cortisol-based sensing assays, their sensing performances in different mediums as well as their suitability for POC diagnostics. In general, colorimetric sensors, especially the colorimetric AuNP-aptamer-based assay offer a fast, simple, and low-cost detection method. However, they are hampered by their inadequate sensitivities for detection in environmental samples and animal biofluids. On the other hand, electrochemical-based sensors confer higher sensitivities and opportunities for *in-situ* monitoring but are higher in cost. They are further limited by stability issues such as electrode fouling.[191] In aqueous environments, the exposed hydrophobic coating on electrodes may undergo chemical reactions with aromatic/aliphatic molecules and proteins.[191] Electrode fouling that occurs by this hydrophobic reaction is known to be irreversible, limiting the sensor's performance.[191] The stability of electrochemical sensors is an issue worth investigating should we consider the aqueous sample matrix that cortisol is typically found in. Thus far, it remains a challenge to combine the simplicity and stability of colorimetric sensors with the sensitivities of electrochemical-based detection methods.

Another challenge of cortisol detection is the complex sample matrices, in which they are found, especially for POC diagnostics. The matrix of real-world samples is much more complicated than the artificial/simulated/spiked samples used in laboratory settings. Blood samples contain multiple chemical components such as proteins, glucose, mineral ions, while urine samples contain interfering steroid metabolites with similar structure to cortisol.[72,73] Fish tank water is known to contain several compounds such as faeces, feed, ammonia, and nitrates. Such chemical compounds may interfere with the binding of the cortisol analyte to the receptor and mask the sensing signal. For POC diagnostics, the calibration and quantification of sensing assays generated from laboratory should be validated on-site using real samples. Variations in temperature, Ph, and sample collection can all lead to differences in the sensing performances obtained in laboratory versus on-site. Careful validation and reproducibility characterization of the biosensing performances are thus recommended.

In addition, cortisol production differs based on multiple factors such as circadian rhythms and individual variability.[4] These factors should be carefully considered during data acquisition, processing, and interpretation.[4] Detecting cortisol alone may be insufficient for providing a holistic assessment of physiological responses, since it does not capture the body's ability to clear cortisol.[89] For instance, during exercise, variations in sweat rate may lead to fluctuating cortisol levels which changes the concentrations of the biomarker.[89] Improvements maybe made by concurrent measurements of other parameters such as flow-rate, pH, and temperature to account for these variables.[173] For comprehensive stress monitoring, quantitative measurements of cortisol should be complemented with analysis of other biomarkers released from the SAM and HPA systems.[4] An integrated platform for simultaneous monitoring of stress and biomarkers for pain, reward or pleasure (e.g. monoamines, endogenous opioids, oxytocin) is recommended for holistic assessment of one's well-being. Furthermore, the influences of environmental and social factors towards production of these biomarkers may be studied with these sensors.

Aside from cortisol detection in human biological matrices, the assessment of cortisol levels in animals is also an essential aspect of monitoring their welfare. Thus far, ELISA, RIA, and HPLC have been adopted for detecting cortisol in animal biofluids. However, these analytical techniques require sophisticated lab equipment with trained personnel and are higher in cost. In addition, cortisol levels in animals such as fish are generally lower than in humans, especially for non-invasive samples. As such, future development of cortisol sensors for animals' welfare should be made user-friendly and sensitive enough to reach the cortisol level in animals' samples.

Furthermore, the design of materials with advanced physicochemical properties alongside high binding bioreceptors is needed. The selection of sensitive chromogenic or colorimetric sensing reagents is essential towards achieving higher sensitivity and selectivity. One plausible consideration is the use of nanoparticles with catalytic activity (nanozymes) coupled with aptamers for selective detection and 3,3′,5,5′-tetramethylbenzidine (TMB) for signal amplification. With the use of suitably chosen materials, the developed product may exhibit improved signal intensity and homogeneity, especially desirable for portable sensors. Thus far, various bioreceptors such as aptamers and antibodies have been used to improve the selectivity of the sensing assays. Despite their successful employment in biosensing research, some challenges still need to be addressed. For example, the sensing performance using Apt 15-1/Apt 15-1a has been reported with discrepancies in binding affinities to cortisol by different research groups.[92,139,142] Further, theoretical research on more selective and sensitive bioreceptors is needed. MIP was also deployed as another artificial antibody,[173–177] with future opportunities for recycling of sensors. Moving forward, we anticipate the further advancement of cortisol biosensors for broadened application beyond human stress monitoring and their deployment as enabling tools for social science studies, in addition to their use as animal health and welfare indicators.

Acknowledgments

This work is supported by the Agriculture and Aquaculture Horizontal Technology Coordinating Offices (A2HTCO) Seed Fund (C211018003) under the Agency of Science, Technology and Research (A*STAR). LS and SX thank Dr. Caroline Wee from the Institute of Molecular and Cell Biology, A*STAR for reviewing this paper.

References

1. C. Samson and A. Koh, *Front. Bioeng. Biotechnol.* 2020, **8**, 1037.
2. L. Rochette and C. Vergely, in *Annales de Cardiologie et d'Angéiologie*, 2017, vol. 66, pp. 181–183.
3. T. Grandin, *Animals*, 2020, **10**, 363.
4. E. Chmelíková, P. Bolechová, H. Chaloupková, I. Svobodová, M. Jovičić and M. Sedmíková, *Domest. Anim. Endocrinol.* 2020, **72**, 106428.

5. G. Russell and S. Lightman, *Nat. Rev. Endocrinol.* 2019, **15**, 525–534.
6. C. D. Schetter and C. Dolbier, *Soc. Personal. Psychol. Compass*, 2011, **5**, 634–652.
7. S. M. Staufenbiel, B. W. J. H. Penninx, A. T. Spijker, B. M. Elzinga and E. F. C. van Rossum, *Psychoneuroendocrinology*, 2013, **38**, 1220–1235.
8. Y.-Z. Liu, Y.-X. Wang and C.-L. Jiang, *Front. Hum. Neurosci.* 2017, **11**, 1–11.
9. F. Bakhshian, A. Abolghasemi and M. Narimani, *Procedia-Social Behav. Sci.* 2013, **84**, 929–933.
10. N. Restauri and A. D. Sheridan, *J. Am. Coll. Radiol.* 2020, **17**, 921–926.
11. M. Nagasawa, K. Mogi and T. Kikusui, *Sci. Rep.* 2012, **2**, 724.
12. G. E. Tafet and C. B. Nemeroff, *J. Neuropsychiatry Clin. Neurosci.* 2015, **28**, 77–88.
13. K. Dhama, S. K. Latheef, M. Dadar, H. A. Samad, A. Munjal, R. Khandia, K. Karthik, R. Tiwari, M. I. Yatoo and P. Bhatt, *Front. Mol. Biosci.* 2019, **6**, 91.
14. A. J. Steckl and P. Ray, *ACS Sensors*, 2018, **3**, 2025–2044.
15. K. Strimbu and J. A. Tavel, *Curr. Opin. HIV AIDS*, 2010, **5**, 463.
16. B. A. Katchman, M. Zhu, J. Blain Christen and K. S. Anderson, *Proteomics — Clin. Appl.* 2018, **12**, 1800010.
17. L. Zhang, H. Xiao and D. T. Wong, *Mol. Diagn. Ther.* 2009, **13**, 245–259.
18. M. Ku, J. Kim, J.-E. Won, W. Kang, Y.-G. Park, J. Park, J.-H. Lee, J. Cheon, H. H. Lee and J.-U. Park, *Sci. Adv.* 2020, **6**, eabb2891.
19. R. Murison, *Neuroscience of Pain, Stress, and Emotion*, Elsevier, 2016, pp. 29–49.
20. S. Heimbürge, E. Kanitz and W. Otten, *Gen. Comp. Endocrinol.* 2019, **270**, 10–17.
21. B. Sadoul and B. Geffroy, *J. Fish Biol.* 2019, **94**, 540–555.
22. R. Rao and I. P. Androulakis, *Horm. Behav.* 2019, **110**, 77–89.
23. E. Dziurkowska and M. Wesolowski, *J. Clin. Med.* 2021, **10**, 5204.
24. B. S. McEwen, *Physiol. Rev.* 2007, **87**, 873–904.
25. M. N. Silverman and E. M. Sternberg, *Ann. N. Y. Acad. Sci.* 2012, **1261**, 55–63.
26. S. J. Lupien, M. De Leon, S. De Santi, A. Convit, C. Tarshish, N. P. V Nair, M. Thakur, B. S. McEwen, R. L. Hauger and M. J. Meaney, *Nat. Neurosci.* 1998, **1**, 69–73.

27. E. J. Kim, B. Pellman and J. J. Kim, *Learn. Mem.* 2015, **22**, 411–416.
28. R. R. Lonser, L. Nieman and E. H. Oldfield, *J. Neurosurg. JNS*, 2017, **126**, 404–417.
29. C. R. Ralph and A. J. Tilbrook, *J. Anim. Sci.* 2016, **94**, 457–470.
30. N. J. Cook, A. L. Schaefer, P. Lepage and S. D. Morgan Jones, *J. Agric. Food Chem.* 1997, **45**, 395–399.
31. M. Shimada, K. Takahashi, T. Ohkawa, M. Segawa and M. Higurashi, *Horm. Res. Paediatr.* 1995, **44**, 213–217.
32. T. P. Barry, A. F. Lapp, T. B. Kayes and J. A. Malison, *Aquaculture*, 1993, **117**, 351–363.
33. T. Ellis, J. D. James, C. Stewart and A. P. Scott, *J. Fish Biol.* 2004, **65**, 1233–1252.
34. J. Blahova, R. Dobšíková, Z. Svobodova and P. Kalab, *Acta Vet. Brno*, 2007, **76**, 59–64.
35. D. Proverbio, R. Perego, E. Spada, G. Bagnagatti de Giorgi, A. Belloli and D. Pravettoni, *Sci. World J.* 2013, **2013**, 216569.
36. H. W. Palm, E. Berchtold, B. Gille, U. Knaus, L. C. Wenzel and B. Baßmann, *Aquac. J.* 2022, **2**, 227–245.
37. R. He, Y. Su, A. Wang, B. Lei and K. Cui, *Aquac. Res.* 2020, **51**, 3495–3505.
38. R. Şemsi, U. Kökbaş, B. Arslan, E. Ergünol, L. Kayrın and A. Sepici Dinçel, *Appl. Biochem. Biotechnol.* 2022, **194**, 1166–1177.
39. S. Gong, Y.-L. Miao, G.-Z. Jiao, M.-J. Sun, H. Li, J. Lin, M.-J. Luo and J.-H. Tan, *PLoS One*, 2015, **10**, e0117503.
40. J. W. Turner Jr, R. Nemeth and C. Rogers, *Gen. Comp. Endocrinol.* 2003, **133**, 341–352.
41. J. Piwowarska, A. Chimiak, H. Matsumoto, A. Dziklińska, M. Radziwoń-Zaleska, W. Szelenberger and J. Pachecka, *Psychiatry Res.* 2012, **198**, 407–411.
42. Abcam, Cortisol ELISA Kit (ab108665), https://www.abcam.com/cortisol-elisa-kit-ab108665.html.
43. ThermoFisher, Cortisol Competitive Human ELISA Kit, https://www.thermofisher.com/elisa/product/Cortisol-Competitive-Human-ELISA-Kit/EIAHCOR#:~:text=TheCortisol Competitive ELISA research, cortisol in saliva and urine.
44. Cusabio, Cortisol (COR) ELISA kit, https://www.cusabio.com/ELISA-Kit/Cortisol-COR-ELISA-kit-1026974.html.
45. I. America, Cortisol RIA, https://www.ibl-america.com/cortisol-ria/.

46. A. Apilux, S. Rengpipat, W. Suwanjang and O. Chailapakul, *J. Pharm. Biomed. Anal.* 2020, **178**, 112925.
47. M. S. H. Akash and K. Rehman, *Essentials of Pharmaceutical Analysis*, Springer, 2020, pp. 175–184.
48. U. Turpeinen, H. Markkanen, M. Välimäki and U.-H. Stenman, *Clin. Chem.* 1997, **43**, 1386–1391.
49. R. Dawson, P. Kontur and A. Monjan, *Hormones*, 1984, **20**, 89–94.
50. L. J. Ney, K. L. Felmingham, R. Bruno, A. Matthews and D. S. Nichols, *J. Pharm. Biomed. Anal.* 2021, **201**, 114103.
51. A. Singh, A. Kaushik, R. Kumar, M. Nair and S. Bhansali, *Appl. Biochem. Biotechnol.* 2014, **174**, 1115–1126.
52. M. Sekar, R. Sriramprabha, P. K. Sekhar, S. Bhansali, N. Ponpandian, M. Pandiaraj and C. Viswanathan, *J. Electrochem. Soc.* 2020, **167**, 67508.
53. M. Zea, F. G. Bellagambi, H. Ben Halima, N. Zine, N. Jaffrezic-Renault, R. Villa, G. Gabriel and A. Errachid, *TrAC Trends Anal. Chem.* 2020, **132**, 116058.
54. I. Perogamvros, D. W. Ray and P. J. Trainer, *Nat. Rev. Endocrinol.* 2012, **8**, 717–727.
55. Y. Sofer, E. Osher, R. Limor, G. Shefer, Y. Marcus, I. Shapira, K. Tordjman, Y. Greenman, S. Berliner and N. Stern, *Endocr. Pract.* 2016, **22**, 1415–1421.
56. D. R. Idler and H. C. Freeman, *Gen. Comp. Endocrinol.* 1968, **11**, 366–372.
57. S. Mantagos, A. Koulouris and A. Vagenakis, *J. Clin. Endocrinol. Metab.* 1991, **72**, 214–216.
58. M. B. Hennessy, M. T. Williams, D. D. Miller, C. W. Douglas and V. L. Voith, *Appl. Anim. Behav. Sci.* 1998, **61**, 63–77.
59. P. Volin, *J. Chromatogr. B Biomed. Sci. Appl.* 1992, **584**, 147–155.
60. N. Ivković, Đ. Božović, M. Račić, D. Popović-Grubač and B. Davidović, *Acta Fac. Medicae Naissensis*, 2015, **32**, 91–99.
61. I. C. Vincent and A. R. Michell, *Res. Vet. Sci.* 1992, **53**, 342–345.
62. G. Morineau, A. Boudi, A. Barka, M. Gourmelen, F. Degeilh, N. Hardy, A. Al-Halnak, H. Soliman, J. P. Gosling and R. Julien, *Clin. Chem.* 1997, **43**, 1397–1407.
63. F. G. Bellagambi, I. Degano, S. Ghimenti, T. Lomonaco, V. Dini, M. Romanelli, F. Mastorci, A. Gemignani, P. Salvo and R. Fuoco, *Microchem. J.* 2018, **136**, 177–184.
64. M. F. Haussmann, C. M. Vleck and E. S. Farrar, *Adv. Physiol. Educ.* 2007, **31**, 110–115.

65. O. Thomsson, B. Ström-Holst, Y. Sjunnesson and A.-S. Bergqvist, *Acta Vet. Scand.* 2014, **56**, 55.
66. H. H. Shukur, Y. B. de Rijke, E. F. C. van Rossum, L. Hussain-Alkhateeb and C. Höybye, *BMC Endocr. Disord.* 2020, **20**, 1–7.
67. A. Hodes, M. B. Lodish, A. Tirosh, J. Meyer, E. Belyavskaya, C. Lyssikatos, K. Rosenberg, A. Demidowich, J. Swan and N. Jonas, *Endocrine*, 2017, **56**, 164–174.
68. T. Faresjö, S. Strömberg, M. Jones, A. Stomby, J.-E. Karlsson, C. J. Östgren, Å. Faresjö and E. Theodorsson, *Sci. Rep.* 2020, **10**, 22456.
69. M. A. Novak, M. T. Menard, S. N. El-Mallah, K. Rosenberg, C. K. Lutz, J. Worlein, K. Coleman and J. S. Meyer, *Am. J. Primatol.* 2017, **79**, e22547.
70. W. F. Albar, E. W. Russell, G. Koren, M. J. Rieder and S. H. van Umm, *Clin. Invest. Med.* 2013, **36**, E312–E316.
71. L. Manenschijn, R. G. P. M. van Kruysbergen, F. H. de Jong, J. W. Koper and E. F. C. van Rossum, *J. Clin. Endocrinol. Metab.* 2011, **96**, E1862–E1865.
72. R. Gatti, G. Antonelli, M. Prearo, P. Spinella, E. Cappellin and F. Elio, *Clin. Biochem.* 2009, **42**, 1205–1217.
73. H. Horie, T. Kidowaki, Y. Koyama, T. Endo, K. Homma, A. Kambegawa and N. Aoki, *Clin. Chim. Acta*, 2007, **378**, 66–70.
74. O. Al Sharef, J. Feely, P. V. Kavanagh, K. R. Scott and S. C. Sharma, *Biomed. Chromatogr.* 2007, **21**, 1201–1206.
75. J. E. Wear, L. J. Owen, K. Duxbury and B. G. Keevil, *J. Chromatogr. B*, 2007, **858**, 27–31.
76. A. Ziomkiewicz, M. Babiszewska, A. Apanasewicz, M. Piosek, P. Wychowaniec, A. Cierniak, O. Barbarska, M. Szołtysik, D. Danel and S. Wichary, *Sci. Rep.* 2021, **11**, 11576.
77. J. K. Kulski and P. E. Hartmann, *Aust. J. Exp. Biol. Med. Sci.* 1981, **59**, 769–778.
78. K. Hinde, A. L. Skibiel, A. B. Foster, L. Del Rosso, S. P. Mendoza and J. P. Capitanio, *Behav. Ecol.* 2015, **26**, 269–281.
79. D. W. Payne, L.-H. Peng and W. H. Pearlman, *J. Biol. Chem.* 1976, **251**, 5272–5279.
80. J. J. Hollanders, PhD Thesis, Research and graduation internal, Vrije Universiteit Amsterdam, 2020.
81. C. Hechler, R. Beijers, J. M. Riksen-Walraven and C. de Weerth, *Dev. Psychobiol.* 2018, **60**, 639–650.

82. M. Aparicio, P. D. Browne, C. Hechler, R. Beijers, J. M. Rodríguez, C. de Weerth and L. Fernández, *PLoS One*, 2020, **15**, e0233554.
83. B. van der Voorn, M. de Waard, J. B. van Goudoever, J. Rotteveel, A. C. Heijboer and M. J. J. Finken, *J. Nutr.* 2016, **146**, 2174–2179.
84. M. J. Vos, P. H. Bisschop, M. M. L. Deckers and E. Endert, *Clin. Chem. Lab. Med.* 2017, **55**, e262–e264.
85. B.-L. Sahlberg and M. Axelson, *J. Steroid Biochem.* 1986, **25**, 379–391.
86. L. Weirup, C. Schulz, H. Seibel and J. Aerts, *Aquaculture*, 2021, **543**, 736924.
87. D. Bertotto, C. Poltronieri, E. Negrato, D. Majolini, G. Radaelli and C. Simontacchi, *Aquac. Res.* 2010, **41**, 1261–1267.
88. X. Cai, J. Zhang, L. Lin, Y. Li, X. Liu and Z. Wang, *Aquac. Rep.* 2020, **18**, 100514.
89. J. R. Runyon, M. Jia, M. R. Goldstein, P. Skeath, L. Abrell, J. Chorover and E. M. Sternberg, *Int. J. Prog. Heal. Manag.* 2019, **10**, 1–11.
90. M. J. E. Greff, J. M. Levine, A. M. Abuzgaia, A. A. Elzagallaai, M. J. Rieder and S. H. M. van Uum, *Clin. Biochem.* 2019, **63**, 1–9.
91. E. Russell, G. Koren, M. Rieder and S. H. M. Van Uum, *Ther. Drug Monit.* 2014, **36**, 30–34.
92. S. Dalirirad and A. J. Steckl, *Sens. Actuat. B Chem.* 2019, **283**, 79–86.
93. R. C. Stevens, S. D. Soelberg, S. Near and C. E. Furlong, *Anal. Chem.* 2008, **80**, 6747–6751.
94. J. E. L. Villa, I. Garcia, D. Jimenez de Aberasturi, V. Pavlov, M. D. P. T. Sotomayor and L. M. Liz-Marzán, *Biosens. Bioelectron.* 2020, **165**, 112418.
95. E. Tu, P. Pearlmutter, M. Tiangco, G. Derose, L. Begdache and A. Koh, *ACS Omega*, 2020, **5**, 8211–8218.
96. B. van der Voorn, F. Martens, N. S. Peppelman, J. Rotteveel, M. A. Blankenstein, M. J. J. Finken and A. C. Heijboer, *Clin. Chim. Acta*, 2015, **444**, 154–155.
97. S. Krishnan and Z. ul Q. Syed, *Sens. Actuat. Rep.* 2022, **4**, 100078.
98. E. Furuya, V. Graef and O. Nishikaze, *Anal. Biochem.* 1978, **90**, 644–650.
99. H. Kalant, *Biochem. J.* 1958, **69**, 79–93.
100. D. Appel, R. D. Schmid, C.-A. Dragan, M. Bureik and V. B. Urlacher, *Anal. Bioanal. Chem.* 2005, **383**, 182–186.

101. C. C. Porter and R. H. Silber, *J. Biol. Chem.* 1950, **185**, 201–207.
102. A. Zaffaroni, *J. Am. Chem. Soc.* 1950, **72**, 3828.
103. V. L. Estergreen and G. K. Venkataseshu, *Steroids*, 1967, **10**, 83–92.
104. C. R. Szalkowski, M. G. O'Brien and W. J. Mader, *Anal. Chem.* 1955, **27**, 944–946.
105. E. P. Schulz and J. D. Neuss, *Anal. Chem.* 1957, **29**, 1662–1665.
106. M. L. Lewbart and V. R. Mattox, *J. Org. Chem.* 1964, **29**, 513–521.
107. V. R. Mattox, *J. Am. Chem. Soc.* 1952, **74**, 4340–4347.
108. R. Ashbel and A. M. Seligman, *Endocrinology*, 1949, **44**, 565–583.
109. I. E. Bush and M. M. Gale, *Analyst*, 1958, **83**, 532–536.
110. A. M. Rutenburg, R. Gofstein and A. M. Seligman, *Cancer Res.* 1950, **10**, 113–121.
111. W. J. Mader and R. R. Buck, *Anal. Chem.* 1952, **24**, 666–667.
112. D. E. Guttman, *J. Pharm. Sci.* 1966, **55**, 919–922.
113. S. Görög and P. Horváth, *Analyst*, 1978, **103**, 346–353.
114. D. K. Singh and R. Verma, *Iran. J. Pharmacol. Ther*, 2008, **7**, 61–65.
115. D. Vilela, M. C. González and A. Escarpa, *Anal. Chim. Acta*, 2012, **751**, 24–43.
116. X. Huang and M. A. El-Sayed, *J. Adv. Res.* 2010, **1**, 13–28.
117. K. M. M. Aung, Y. N. Tan, K. V. Desai and X. Su, *Aust. J. Chem.* 2011, **64**, 1288–1294.
118. Z. Jiang, G. Wen, Y. Luo, X. Zhang, Q. Liu and A. Liang, *Sci. Rep.* 2014, **4**, 5323.
119. E. Priyadarshini and N. Pradhan, *Sens. Actuat. B Chem.* 2017, **238**, 888–902.
120. A. Mizuno and A. Ono, *ACS Appl. Nano Mater.* 2021, **4**, 9721–9728.
121. A. M. Smith and S. Nie, *Acc. Chem. Res.* 2010, **43**, 190–200.
122. K. Zhang, Q. Mei, G. Guan, B. Liu, S. Wang and Z. Zhang, *Anal. Chem.* 2010, **82**, 9579–9586.
123. U. Resch-Genger, M. Grabolle, S. Cavaliere-Jaricot, R. Nitschke and T. Nann, *Nat. Meth.*, 2008, **5**, 763–775.
124. L. Sutarlie, S. Y. Ow and X. Su, *Biotechnol. J.* 2017, **12**, 1500459.
125. Y. Liu, B. Wu, E. K. Tanyi, S. Yeasmin and L.-J. Cheng, *Langmuir*, 2020, **36**, 7781–7788.
126. L. Sun, S. Li, W. Ding, Y. Yao, X. Yang and C. Yao, *J. Mater. Chem. B*, 2017, **5**, 9006–9014.

127. M. Zan, S. An, L. Cao, Y. Liu, L. Li, M. Ge, P. Liu, Z. Wu, W.-F. Dong and Q. Mei, *Appl. Surf. Sci.* 2021, **566**, 150686.
128. M. Chen, C. Grazon, P. Sensharma, T. T. Nguyen, Y. Feng, M. Chern, R. C. Baer, N. Varongchayakul, K. Cook, S. Lecommandoux, C. M. Klapperich, J. E. Galagan, A. M. Dennis and M. W. Grinstaff, *ACS Appl. Mater. Interfaces*, 2020, **12**, 43513–43521.
129. H. Y. Kong and J. Byun, *Biomol. Ther. (Seoul).*, 2013, **21**, 423.
130. P. Datta, *Accurate Results in the Clinical Laboratory*, Elsevier, 2nd edn, 2019, pp. 69–73.
131. K. Sefah, D. Shangguan, X. Xiong, M. B. O'Donoghue and W. Tan, *Nat. Protoc.* 2010, **5**, 1169–1185.
132. L. Wu, Y. Wang, X. Xu, Y. Liu, B. Lin, M. Zhang, J. Zhang, S. Wan, C. Yang and W. Tan, *Chem. Rev.* 2021, **121**, 12035–12105.
133. T. Bing, W. Zheng, X. Zhang, L. Shen, X. Liu, F. Wang, J. Cui, Z. Cao and D. Shangguan, *Sci. Rep.* 2017, **7**, 15467.
134. M. McKeague, R. Velu, A. De Girolamo, S. Valenzano, M. Pascale, M. Smith and M. C. DeRosa, *Toxins (Basel)*. 2016, **8**, 336.
135. L. Tan, K. G. Neoh, E.-T. Kang, W.-S. Choe and X. Su, *Anal. Biochem.* 2012, **421**, 725–731.
136. Y. Jiang, M. Shi, Y. Liu, S. Wan, C. Cui, L. Zhang and W. Tan, *Angew. Chemie Int. Ed.* 2017, **56**, 11916–11920.
137. O. A. Alsager, K. M. Alotaibi, A. M. Alswieleh and B. J. Alyamani, *Sci. Rep.* 2018, **8**, 12947.
138. G. Liu, M. Lu, X. Huang, T. Li and D. Xu, *Sensors*, 2018, **18(12)**, 4166.
139. J. A. Martin, J. L. Chávez, Y. Chushak, R. R. Chapleau, J. Hagen and N. Kelley-Loughnane, *Anal. Bioanal. Chem.* 2014, **406**, 4637–4647.
140. S. Dalirirad, D. Han and A. J. Steckl, *ACS Omega*, 2020, **5**, 32890–32898.
141. T. Wu, L. Ding, Y. Zhang and W. Fang, *IEEE Sens. J.* 2022, **22**, 12485–12492.
142. C. Niu, Y. Ding, C. Zhang and J. Liu, *Sens. Diagn.* 2022, **1**, 541–549.
143. K.-A. Yang, H. Chun, Y. Zhang, S. Pecic, N. Nakatsuka, A. M. Andrews, T. S. Worgall and M. N. Stojanovic, *ACS Chem. Biol.* 2017, **12**, 3103–3112.
144. A. C. Allison, M. E. Peover and T. A. Gough, *Life Sci.* 1962, **1**, 729–737.

145. A. M. A. Adam, H. A. Saad, M. S. Refat and M. S. Hegab, *J. Mol. Liq.* 2022, **357**, 119092.
146. J. Jeon, S. Uthaman, J. Lee, H. Hwang, G. Kim, P. J. Yoo, B. D. Hammock, C. S. Kim, Y.-S. Park and I.-K. Park, *Sens. Actuat. B. Chem.* 2018, **266**, 710–716.
147. X. Guo, *J. Biophotonics*, 2012, **5**, 483–501.
148. X. Su, C.-Y. Lin, S. J. O'Shea, H. F. Teh, W. Y. X. Peh and J. S. Thomsen, *Anal. Chem.* 2006, **78**, 5552–5558.
149. M. Piliarik, H. Vaisocherová and J. Homola, eds. A. Rasooly and K. E. Herold, Methods in Molecular Biology, Humana Press, Totowa, NJ, 2009, pp. 65–88.
150. D. Kotlarek, F. Curti, M. Vorobii, R. Corradini, M. Careri, W. Knoll, C. Rodriguez-Emmenegger and J. Dostálek, *Sens. Actuat. B. Chem.* 2020, **320**, 128380.
151. D. Sun, Y. Wu, S.-J. Chang, C.-J. Chen and J.-T. Liu, *Talanta*, 2021, **222**, 121466.
152. I. Mihai, A. Vezeanu, C. Polonschii, C. Albu, G.-L. Radu and A. Vasilescu, *Sens. Actuat. B. Chem.* 2015, **206**, 198–204.
153. J. S. Mitchell, T. E. Lowe and J. R. Ingram, *Analyst*, 2009, **134**, 380–386.
154. Y. Zhao, R. Tong, F. Xia and Y. Peng, *Biosens. Bioelectron.* 2019, **142**, 111505.
155. C. Leitão, A. Leal-Junior, A. R. Almeida, S. O. Pereira, F. M. Costa, J. L. Pinto and C. Marques, *Biotechnol. Rep.*, 2021, **29**, e00587.
156. C. Leitão, S. O. Pereira, N. Alberto, M. Lobry, M. Loyez, F. M. Costa, J. L. Pinto, C. Caucheteur and C. Marques, *IEEE Sens. J.* 2021, **21**, 3028–3034.
157. A. K. Pandey, A. K. Sharma and C. Marques, *Materials (Basel).* 2020, **13**, 1623.
158. L. Wu, Q. You, Y. Shan, S. Gan, Y. Zhao, X. Dai and Y. Xiang, *Sensors Actuat. B. Chem.* 2018, **277**, 210–215.
159. A. K. Sharma, B. Kaur and C. Marques, *Optik (Stuttg).* 2020, **218**, 164891.
160. Y. Tanaka, E. H. Khoo, N. A. binte Mohamed Salleh, S. L. Teo, S. Y. Ow, L. Sutarlie and X. Su, *Analyst*, 2021, **146**, 6924–6934.
161. K. Xu, R. Zhou, K. Takei and M. Hong, *Adv. Sci.* 2019, **6**, 1900925.
162. C. Zong, M. Xu, L.-J. Xu, T. Wei, X. Ma, X.-S. Zheng, R. Hu and B. Ren, *Chem. Rev.* 2018, **118**, 4946–4980.

163. N. A. binte Mohamed Salleh, Y. Tanaka, L. Sutarlie and X. Su, *Analyst*, 2022, **147**, 1756–1776.
164. S. Nie and S. R. Emory, *Science*, 1997, **275**, 1102–1106.
165. J.-H. Ryu, H. Y. Lee, J.-Y. Lee, H.-S. Kim, S.-H. Kim, H. S. Ahn, D. H. Ha and S. N. Yi, *Appl. Sci.* 2021, **11**, 11855.
166. J. Langer, D. Jimenez de Aberasturi, J. Aizpurua, R. A. Alvarez-Puebla, B. Auguié, J. J. Baumberg, G. C. Bazan, S. E. J. Bell, A. Boisen and A. G. Brolo, *ACS Nano*, 2019, **14**, 28–117.
167. T. J. Moore and B. Sharma, *Anal. Chem.* 2020, **92**, 2052–2057.
168. G. Zhigang, Z. Tingting, D. Jiu, L. Xiaorui, Q. Yueyang, L. Yao, L. Tingjiao, L. Yong, Z. Weijie and L. Bingcheng, *Acta Chim. Sin.* 2017, **75**, 355–359.
169. M. H. Hassan, R. Khan and S. Andreescu, *Electrochem. Sci. Adv.* 2021, **n/a**, e2100184.
170. S. Takamatsu, J. Lee, R. Asano, W. Tsugawa, K. Ikebukuro and K. Sode, *Sens. Actuat. B Chem.* 2021, **346**, 130554.
171. C. Jiang, Y. He and Y. Liu, *Analyst*, 2020, **145**, 5400–5413.
172. J. E. An, K. H. Kim, S. J. Park, S. E. Seo, J. Kim, S. Ha, J. Bae and O. S. Kwon, *ACS Sensors*, 2022, **7**, 99–108.
173. W. Tang, L. Yin, J. R. Sempionatto, J. Moon, H. Teymourian and J. Wang, *Adv. Mater.* 2021, **33**, 2008465.
174. P. Manickam, S. K. Pasha, S. A. Snipes and S. Bhansali, *J. Electrochem. Soc.* 2016, **164**, B54.
175. S. M. Mugo and J. Alberkant, *Anal. Bioanal. Chem.* 2020, **412**, 1825–1833.
176. S. Yeasmin, B. Wu, Y. Liu, A. Ullah and L.-J. Cheng, *Biosens. Bioelectron.* 2022, **206**, 114142.
177. O. Parlak, S. T. Keene, A. Marais, V. F. Curto and A. Salleo, *Sci. Adv.* 2018, **4**, eaar2904.
178. N. Nawaz, N. K. Abu Bakar, H. N. Muhammad Ekramul Mahmud and N. S. Jamaludin, *Anal. Biochem.* 2021, **630**, 114328.
179. E. Daniels, Y. L. Mustafa, C. Herdes and H. S. Leese, *ACS Appl. Bio Mater.* 2021, **4**, 7243–7253.
180. J. Yue, Q. Lv, W. Wang and Q. Zhang, *Talanta Open*, 2022, **5**, 100099.
181. D. de Fez, M. J. Luque, M. C. García-Domene, V. Camps and D. Piñero, *Optom. Vis. Sci.* 2016, **93**, 85–93.
182. Y. Fan, J. Li, Y. Guo, L. Xie and G. Zhang, *Measurement*, 2021, **171**, 108829.

183. B. Khanal, P. Pokhrel, B. Khanal and B. Giri, *ACS Omega*, 2021, **6**, 33837–33845.
184. V. Joseph, A. Matschulat, J. Polte, S. Rolf, F. Emmerling and J. Kneipp, *J. Raman Spectrosc.* 2011, **42**, 1736–1742.
185. Q. ul ain Zahra, S. Ullah, F. Shahzad, B. Qiu, X. Fang, A. Ammar, Z. Luo and S. Abbas Zaidi, *Prog. Mater. Sci.* 2022, **129**, 100967.
186. C. E. Shuck, A. Sarycheva, M. Anayee, A. Levitt, Y. Zhu, S. Uzun, V. Balitskiy, V. Zahorodna, O. Gogotsi and Y. Gogotsi, *Adv. Eng. Mater.* 2020, **22**, 1901241.
187. X. Chen, L. Zhang and D. Cui, *Micro Nano Lett.* 2016, **11**, 20–23.
188. S. M. Mugo, W. Lu and S. Robertson, *Biosensors*, 2022, **12**, 854.
189. S. Madhu, A. J. Anthuuvan, S. Ramasamy, P. Manickam, S. Bhansali, P. Nagamony and V. Chinnuswamy, *ACS Appl. Electron. Mater.* 2020, **2**, 499–509.
190. H. Wu, H. Ohnuki, S. Ota, M. Murata, Y. Yoshiura and H. Endo, *Biosens. Bioelectron.* 2017, **93**, 57–64.
191. B. L. Hanssen, S. Siraj and D. K. Y. Wong, *Rev. Anal. Chem.* 2016, **35**, 1–28.

… 2025 World Scientific Publishing Company
https://doi.org/10.1142/9789819806409_0006

Recent Advances in Ocular Therapy by Hydrogel Biomaterials*

Lan Zheng[†,‡], Yi Han[†,‡], Enyi Ye[§,¶], Qiang Zhu[§], Xian Jun Loh[§], Zibiao Li[§,¶,∥,**,‡‡] and Cheng Li[†,‡,††,‡‡]

[†]*Eye Institute & Affiliated Xiamen Eye Center, School of Medicine, Xiamen University Xiamen, Fujian 361102, P. R. China*
[‡]*Fujian Provincial Key Laboratory of Ophthalmology and Visual Science & Ocular Surface and Corneal Diseases, Xiamen, Fujian 361102, P. R. China*
[§]*Institute of Materials Research and Engineering (IMRE), Agency for Science, Technology and Research (A*STAR), Singapore 138634, Singapore*
[¶]*Institute of Sustainability for Chemicals, Energy and Environment (ISCE²) Agency for Science, Technology and Research (A*STAR) 2 Fusionopolis Way, Singapore 138634, Singapore*
[∥]*Department of Materials Science and Engineering, National University of Singapore 9 Engineering Drive 1, Singapore 117576, Singapore*
[**]*lizb@imre.a-star.edu.sg*
[††]*cheng-li@xmu.edu.cn*

Current clinical practice in ocular disease treatment dosage forms primarily relies on eye drops or eye ointments, which face significant challenges in terms of low bioavailability profiles, rapid removal from

---

[‡‡]Corresponding authors.
*To cite this article, please refer to its earlier version published in the *World Scientific Annual Review of Functional Materials*, Volume 1, 2230002 (2023), DOI: 10.1142/S2810922822300021.

the administration site, and thus ineffective therapeutic efficiency. Hydrogel has several distinct properties in semi-solid thermodynamics and viscoelasticity, as well as diverse functions and performance in biocompatibility and degradation, making it extremely promising for overcoming the challenges in current ocular treatment. In this review, the most recent developments in the use of hydrogel biomaterials in ocular therapy are presented. These sophisticated hydrogel biomaterials with diverse functions, aimed at therapeutic administration for ocular treatment, are further classified into several active domains, including drug delivery system, surface repair patch, tissue-engineered cornea, intraocular lens, and vitreous substitute. Finally, the possible strategies for future design of multifunctional hydrogels by combining materials science with biological interface are proposed.

Keywords: Hydrogels; drug delivery system; repair patch; tissue-engineered cornea; intraocular lens; vitreous substitute.

1. Introduction

Hydrogel has the ability to absorb a large quantity of water without disintegrating. Because of the cross-linked network, hydrogel has high mechanical and stability characteristics.[1,2] In addition, it has good physical and chemical qualities in terms of transparency, water content, flexibility, and drug loading capacity.[3] The high water content in hydrogel also makes it comparable to the composition of the extracellular matrix, which is important for maintaining the activity of the small molecules, proteins, peptides, etc.[4-7] Since their discovery in the 1960s, hydrogels have been extensively researched in a variety of fields, including medicine and biology, catalyst carriers, water environmental protection, agricultural drought resistance, and food preservation.[8] In biomedical area, a range of particular hydrogels have been designed and employed in cell transplantation systems, improving wound healing, cartilage/bone regeneration, and drug delivery systems.[7] Recently, hydrogels are emerging as a biological alternative for contact lenses, hygiene products, tissue engineering scaffolds, wound dressings, and biosensors, indicating a promising future in the development of ophthalmic devices.[7] As a result, it is timely for us to outline the most recent advancements in the use of hydrogel in ophthalmic applications.[9-12]

Eyeball is identified as a delicate and complex organ, and the structure of the eyeball mainly includes cornea, sclera, retina, aqueous humor, lens, and vitreous.[13] Any pathological events occurring to the eyeball can cause a threat to human health and the quality of life.[14] Ocular surface diseases commonly include dry eye, infectious and immune keratitis or conjunctivitis, which if not promptly treated may even cause blindness.[15] The most popular method to handle the ocular surface diseases is in the form of application of eye drops, which accounts for more than 90% of all ophthalmic formulas. In the ophthalmic drug delivery system, only about 5% of eye drops are bioavailable, resulting from blinking, tears, and ocular surface clearance mechanism, which leads to difficulty in delivering drugs to the cornea or to the posterior part of the eye tissue. If topical administration is done frequently, drug-induced ocular surface injury is a hurdle that is hard to avoid.[16] Hydrogel has now matured as a sustained-release system, and the delivery of drugs in the form of contact lenses has been successfully developed.[17] In addition, there are vast reports describing the application of patch to promote repair, or in tissue engineering of the cornea,[3,18] or lens, vitreous body, and retina that form the intraocular organs. Due to the unique structure of the eyeball, it is difficult to transport the drug to the intraocular organs. Intravitreal injection is the most commonly used way, which significantly increases the drug concentration in the retina and reduces side effects. Periocular route is considered as the most effective route, and the pathway includes the retrobulbar, peribulbar, subtenon, and subconjunctival routes. Implantation is an emerging method of drug delivery to target the posterior segment of the eye, which meets various treatment needs. In addition, corneal stroma injection is only used in certain special treatments.[19-22] There are still many problems encountered in the existing ocular administration routes. Most of the topical administration routes flush out the drug through various mechanisms, resulting in low bioavailability of the drug. Besides, the special structure of the cornea also restricts the entry of drugs.[23] On the other hand, the administration method of intraocular injection and implantation has the risk of causing side effects such as endophthalmitis.[24] Therefore, it is necessary to find new ways of drug delivery.[25] Currently, there are many studies on the replacement of intraocular tissues with hydrogels. Replacement materials for the vitreous and lenses need to meet the requirements of good biocompatibility, high transparency, and water content, and hydrogels very suitably meet these criteria.[26]

This review aims to introduce the application of hydrogel materials in ophthalmology, from the ocular surface to the inside of the eye. We summarized the role of hydrogel as a drug delivery system, for the promotion of repair and tissue engineering of cornea in ocular surface. Furthermore, the role of hydrogel as a biological substitute of the lens and vitreous is also discussed (Fig. 1).

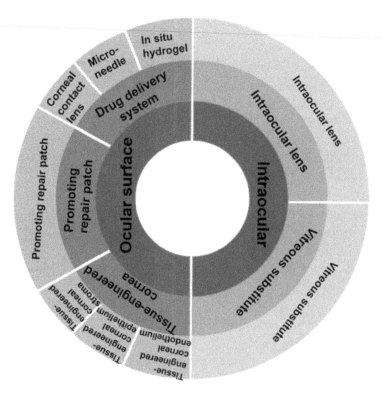

Fig. 1. An overview of applications of hydrogels in ophthalmology.

2. Therapeutic Administration for the Ocular Treatment

The eyeball can be divided into two parts, the eyeball wall and the eye contents.[27] The eyeball wall is mainly composed of three layers of membranes; the outer membrane includes the cornea and the sclera, the middle membrane is the uvea that is composed of the iris, ciliary body, and choroid, and the innermost layer is the retina. The contents of the eyeball are generally composed of lens and vitreous.[28] Among them, the outer membrane is thin, and it mainly plays a role in maintaining the shape of the eyeball and protecting the eyeball. The cornea is smooth and transparent, and participates in the refraction of light.[29] The sclera consists of irregular sheets of collagen fibers, which are connected to the cornea. The outer two-third parts of the sclera are connected to the sheath, while the inner one-third part forms the lamina cribrosa.[30]

The iris is located in the front of the lens, and lies behind the cornea and the aqueous humor.[31] The iris is attached to the ciliary body at its edge, which is called the iris root.[13,30,32] The ciliary body is a complex, highly specialized tissue consisting of different cells, and the surface is refined into ciliary processes.[33] The ciliary body can be divided into two parts, the anterior part and the posterior part, and the posterior part is connected to the vitreous body.[34] The choroid, attached to the sclera, is anterior to the ciliary body.[2] The retina belongs to the central nervous system, and its development is similar to that of the cerebral cortex.[35,36] The central retina, which consists of the macula, results in the highest vision for humans.[37,38]

The vitreous and lens form an important proportion of the eye. The lens is located at the back of the iris and the front of the vitreous, whose major function is to focus on objects to form a clear image on the retina.[39–41] The vitreous body is the largest structure of the eyeball, located in the space between the retina and the lens.[42] The production of vision is inseparable from the coordinated operation of various parts of the eye, and each part plays an irreplaceable role. However, an ocular disease may bring very serious consequences, therefore timely and effectively treatment is of great significance.

The most common way to handle ocular diseases is by administration of eye drops; however, their utilization rates seem inefficient. Furthermore, other methods for ophthalmic drug delivery include subconjunctival injection, corneal stroma injection, intravitreal injection, and implantation. Next, we focus on ocular administration combined with the hydrogel delivery system.

2.1. *Eye drops*

Eye drops own the major advantages as a local drug delivery method, including convenience and simplicity. Eye drops using hydrogel not only possess the advantages of conventional eye drops, but also enhance the retention time of the drug on the ocular surface.[43,44] Lai *et al.* designed multifunctional glutathione-dependent hydrogel eye drops for the treatment of glaucoma, which showed ocular biocompatibility, efficient antioxidant activity, strong adhesion, and high cellular permeability. When compared with other eye drops in animal experiments, the hydrogel eye drops offered a better therapeutic effect.[45] Chouhan *et al.* designed self-healing hydrogel eye drops to achieve a continuous delivery of decorin for preventing corneal scarring. The drops improved the microstructure of the hydrogel for admission into the ocular surface. Additionally, they observed the solid–liquid–solid transitions of hydrogel on cornea after blinking, and found that it could resist the ocular clearance.[46] Although eye drops are one of the oldest methods of administration of drugs into the eye, there is no doubt that they will remain the principal method for ophthalmology in the future.

2.2. *Subconjunctival injection*

Subconjunctival injection is safe and easy for achieving a sustained drug release on the ocular surface.[47] However, a study reported that RNA nanoparticles were quickly eliminated from the cornea after subconjunctival injection, indicating that there is a need to develop more stable drug delivery systems.[48] Rong et al. prepared a dual-controlled release system by adding insulin-loaded chitosan nanoparticles (ICNs) to the polylacticcoglycolic acid–polyethylene glycol–polylacticcoglycolic acid (PLGA–PEG–PLGA) hydrogel. After injecting the nanoparticles under the conjunctival, they found no obvious damage to the function, structure, and neurons of the retina. This proved that hydrogel is safe for application through subconjunctival injection.[49] Zhong et al. developed a bio-printing based on digital light processing to produce hydrogel microstructures of the conjunctival stem cells. Such bioprinting water coagulation was suitable for the scalable dynamic suspension culture of conjunctival stem cells, and could be successfully delivered to the bulbar conjunctival epithelium using minimally invasive subconjunctival injection.[50] Subconjunctival injection of hydrogels will soon become the focus of the ocular administration research in near future.

2.3. Vitreous injection

It is difficult for ocular surface administration to deliver drugs into the intraocular surface; as a result, vitreous injection is the most common method to deliver drugs in the clinic.[51] However, vitreous injection may cause retinal detachment, bleeding, endophthalmitis, and cataract. In addition, the drugs injected into the vitreous quickly get diffused, requiring repeated injections.[52,53] Materials with good biocompatibility, such as hydrogels, have been reported as a vehicle for vitreous injection. Yu et al. evaluated an injectable chemically cross-linked hydrogel for controlling the release of bevacizumab in the vitreous. They found that a transparent gel could be observed after the vitreous injection, and there was no eye bleeding, retinal detachment, inflammation, or other pathological changes.[54]

2.4. Intraocular implants

Intraocular implantation is a controlled release drug delivery system via surgical implantation. This mode of administration can release the drug into the vitreous body and retina directly and continuously, thereby maintaining the drug concentration in the vitreous body in a stable range. Intraocular implants are also highly targeted delivery systems avoiding the blood–eye barrier. However, if the implant is unbefitting, it easily causes adverse reactions in the body. Therefore, it is important to choose low-toxicity, biodegradable implants.[55,56] In order to improve the retinal pigment epithelium transplantation technology, Gandhi et al. implanted a fibrin hydrogel under the retina as a scaffold. They found that the hydrogel stent was degraded by the eighth week, and there were no toxic side effects to the retina.[57]

3. Hydrogel as a Multifunctional Material for Different Ocular Surface Usage Scenarios

3.1. Drug delivery system

The topical administration of eye drops is a common method in the treatment of ocular diseases.[58] However, the flushing effects of tear fluid or the blink reflex greatly reduce the retention time of the drug on the ocular surface, resulting in a reduction of the efficacy of the drug.[44] Besides, the tight junction of the corneal epithelium or the conjunctival epithelium brings in a limited permeability of the ocular drugs; additionally, the blood–eye barrier layout also acts as an obstacle for bioavailability.[59,60] The epithelial cell layer and the endothelial cell layer contain plenty of lipids, which restricts the absorption of hydrophilic drugs. On the contrary, the corneal stroma is rich in water, which poses an adverse impact on the hydrophobic drug. As a result, the bioavailability of the ocular drug is terrible, below 5%.[61] In such cases, frequent administration of ocular drugs is required, which in turn may generate drug toxicity. Researches have been devoted to enhance the retention time of the drug on the ocular surface, meanwhile overcoming various barriers of the cornea to reach the effective site.[62] As hypotoxic and biocompatible polymers, hydrogels may be helpful in the retention of drugs on the ocular surface. For drugs with poor hydrophilicity, hydrogels can play a role of drug vehicle, in order to obtain better effects.[63] In this subsection, the hydrogels for drug delivery are summed up as the corneal contact lens or *in-situ* gel formation.

3.1.1. *Corneal contact lenses*

Corneal contact lens is a kind of repair equipment for ocular surface, and currently its main application is to correct vision or be used as a type of therapeutic method with drug loading in ocular surface diseases.[64] As of now, the most common polymer hydrogel ophthalmic formulation is the vision-correction soft disposable contact lens. Contact lenses derived from hydrogels exhibit abundant advantages, such as good biocompatibility, long-lasting abrasion resistance, durability, and reusability. This shows that they have great potential for further development.[16]

Currently, there are two methods for making therapeutic contact lenses. The first method is to pre-soak the contact lens in the medicine, and release it on the ocular surface after it is fully absorbed. The second is to put the contact lens on the ocular surface and then topically administer the eye drops, which can also achieve a slow-release effect.[65] The disadvantage of the second method is that the amount of drugs loaded is not enough, and it rarely reaches the part that needs to be effective.[66] In the production of hydrogel-based contact lenses, the most commonly used method is to pour the medicated hydrogel into the prepared mold.[67] Various materials can be used as molds. A hydrogel contact lens was customized from a Petri dish and a steel ball as an eyeball mold by Childs *et al.*[68] Figure 2(a) depicts a schematic for the creation of the hydrogel contact lens. Eyeball molds were placed in the bottom of a Petri dish, into which the hydrogel was poured. Following polymerization of the hydrogel, the mold was taken out and the

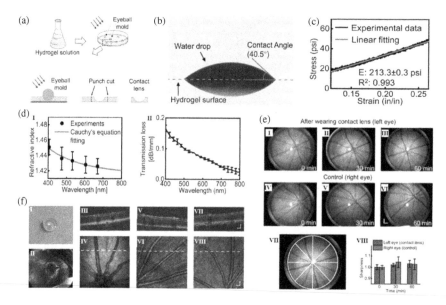

Fig. 2. (a) Schematic for the creation of hydrogel contact lenses using a Petri dish and a steel ball mold. (b) Hydrophobicity tests. (c) The Young's modulus of the hydrogel being measured by the cylinder compression test. (d) Optical properties of hydrogels: (i) refractive indices and (ii) transmission losses. (e) Fundus images of rat eyes with and without the hydrogel contact lens: (i)–(iii) fundus images prior to insertion and 30-min and 60-min post-insertion of the hydrogel contact lens; (iv)–(vi) fundus images of normal rat eyes; and (vii), (viii) sharpness analyses of the retinal images. (f) OCT images of the rat retina with and without the hydrogel contact lens: (i) the hydrogel contact lens; (ii) rat eyes with the hydrogel contact lens; (iii), (iv) B-scan and retinal image of rat eyes with the hydrogel contact lens; (v), (vi) B-scan and retinal image of rat eyes with the commercial contact lens; and (vii), (viii) B-scan and retinal image of a normal rat eye.

Source: Copyright 2016, *Scientific Reports*.[68]

dented part of the hydrogel was retrieved. The results of performance tests show that in comparison with commercial contact lenses, the hydrogel contact lens exhibited comparable hydrophobicity [Fig. 2(b)], consistent Young's modulus [Fig. 2(c)], similar refractive index, and better transparency [Fig. 2(d)]. *In-vivo* OCT and fundus images of rat eyes with and without contact lenses are shown in Fig. 2(f), revealing that the contact lens made from this mold can be fitted to the rat eye. Moreover, Fig. 2(e) shows no corneal opacity after wearing the lens for an extended amount of time and no significant effect on the rat cornea. Therefore, the preparation method and hydrogel are suitable for contact lenses.

Huang et al. designed a drug delivery system using hydrogel contact lenses with comprehensive antifungal functions in the treatment of fungal keratitis. This contact lens consisted of four parts, quaternized chitosan (HTCC), silver nanoparticles, and graphene oxide (GO). The hydrogel was cross-linked by the electrostatic interaction between GO and HTCC, and the antifungal drug voriconazole was loaded on GO. Figure 3(a) describes the detailed preparation of the hybrid hydrogel contact lens, which was applied to an animal model of fungal keratitis. Figures 3(b) and 3(c) show that the hybrid hydrogel contact lens exhibited a better therapeutic effect on fungal keratitis, suggesting that this drug delivery system can offer great potential for rapid and effective treatment. The ingredients gave the contact lenses a strong antifungal ability and a sustained drug-release capacity. The hydrogel possessed effective mechanical properties, allowing maintenance of the shape of the contact lens [Fig. 3(d)]. Moreover, as can be seen from Figs. 3(e) and 3(f), GO exhibited efficient loading capacity and retention of hydrophobic drugs, followed by promotion of their sustained release from the contact lens. Co-culture of different materials with corneal epithelial cells for 24 h demonstrated that the materials had no toxic effects on the cells [Fig. 3(g)], which is crucial for contact lenses.[69]

Jung et al. dispersed the glaucoma drug timolol into propoxylated glyceryl triacylate (PGT) nanoparticles to create a contact lens that could treat glaucoma in the long term. The drug-loading method used was immersion of the contact lens in a drug solution. It was demonstrated that at room temperature, this drug-loaded hydrogel released timolol into PBS for an extended period of time [Fig. 4(a)], with an average release rate of 4.8 g/day. Even with a certain amount of drug loss during packaging and storage, the

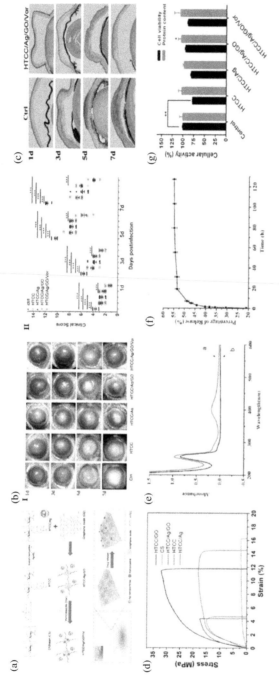

Fig. 3. (a) Synthesis and drug release of drug-loaded hydrogels. (b) A fungal keratitis mouse model that is established and the therapeutic effects being observed in each group for 7 days: (i) representative photos and (ii) clinical scores. (c) H&E staining. (d) Stress–strain curves. (e) UV–vis spectra of HTCC/Ag/GO/Vor and voriconazole in PBS: HTCC/Ag/GO/Vor (curve a) and voriconazole (curve b). (f) Release profiles of voriconazole in PBS. (g) Cell cytotoxicities of different materials.

Source: Copyright 2016, ACS Nano.[69]

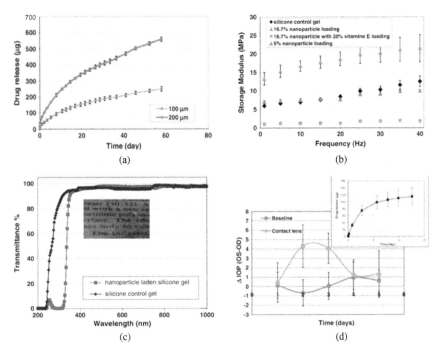

Fig. 4. (a) Release of 5% particles prepared at a 1:1 ratio of drug to PGT. (b) Frequency-dependent storage moduli of gels containing PGT nanoparticles (with or without vitamin E). Data are expressed as mean ± standard deviation (SD) (n = 3). (c) Transmittance spectra of nanoparticle-loaded silicone gel and silicone control gel. The graph shows the Arrhenius fit between the rate constant and the temperature. Data are expressed as mean ± SD (n = 3). (d) Pharmacodynamic effects of nanoparticle-loaded Acuvue Oasys lenses with long-term wear. The graph shows the drug release profiles of Acuvue Oasys crystal particles at approximately 37°C. Data are expressed as mean ± SD (n = 3).

Source: Copyright 2013, *Journal of Controlled Release*.[7]

contact lenses can still release the drug slowly. Figure 4(b) shows the frequency-dependent storage modulus. In addition, the lens has better light transmittance, which is a desirable characteristic of contact lenses [Fig. 4(c)]. The contact lens was shown to be safe, and the prolonged release of timolol achieved the desired pharmacodynamic effect of lowering the intraocular pressure in a glaucoma mouse model [Fig. 4(d)].[70]

3.1.2. In-situ hydrogels

In-situ hydrogels for administration at ocular surface have also attracted the interest of researchers. The *in-situ* hydrogel has the advantages of accurate dosage and long residence time on the ocular surface. After being dropped on the ocular surface, the forms of the hydrogel can be changed by pH response, temperature response, or ion response. During preparation of the hydrogel, these parameters could be controlled for the water condensation, glue's molding ability, and drug release.[71,72] This form of administration was initially dropped into the eye in liquid form, and the phase changed after instillation.[73] After the phase change, the hydrogel was stronger and more viscous than the solution state, which stayed on the ocular surface for a longer time, with a sustained and long-term release of the drug.[74]

To inhibit corneal neovascularization, Yu *et al.* developed an *in-situ* PEG-based hydrogel for the sustained release of Avastin. A four-arm PEG-maleimide (PEG-Mal) and four-arm PEG-sulfhydryl group (PEG-SH) hydrogel (1:1, 1:2, and 1:4 w/w) was developed, as shown in Fig. 5(a). Subsequently, Avastin was loaded onto the PEG hydrogel. It was found that Avastin could be released continuously *in vitro* for almost two weeks, and the integrity of Avastin remained unchanged during the entire release period without significant hydrolysis [Fig. 5(b)]. Co-culture of L-929 cells with the hydrogel demonstrated good biocompatibility. The live–dead cell assay [Fig. 5(c)] revealed that cells survived for up to 7 days in the hydrogel. The extensive pore structure of the hydrogel may be used to carry nutrients from the medium to the cells. Additionally, Fig. 5(d) shows that the PEG hydrogel has excellent mechanical characteristics.[75] Unfortunately, the appropriate animal model or animal tests were not used to confirm the therapeutic impact of the medicated hydrogel; nonetheless, it was illustrated that *in-situ* hydrogels have a longer ocular surface retention period than common eye drops.

Janga *et al.* developed a natamycin (NT) bilosome (NB) ion-sensitive *in-situ* hydrogel, which was used to treat fungal keratitis

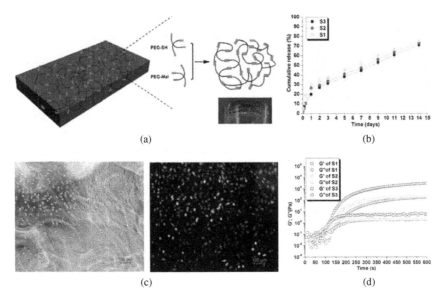

Fig. 5. (a) Formation of PEG hydrogels. (b) *In-vitro* release study. (c) Live–dead staining after co-incubation of hydrogels with L-929 cells. Dead cells are dyed red, whereas live cells are colored green. (d) Storage modulus (*G*0) and loss modulus (*G*00).

Source: Copyright 2014, *International Journal of Pharmaceutics*.[75]

and other bacterial infections. Cytotoxicity experiments and corneal histological evaluation confirmed the compatibility and safety of these preparations concerning the ocular surface. Animal experiments proved that this *in-situ* hydrogel has a longer residence time and a better therapeutic effect than the suspension. This research indicates the great potential of hydrogel slow-release system.[76]

3.1.3. *Microneedles*

Microneedles (MNs) are small-sized needles used for drug delivery. Thanks to the small size, they present a number of advantages, including being painless, minimally invasive, and convenient to operate.[77] The microneedle is composed of many microprojections with different shapes, and these microprojections are connected to a basis support.[78,79] Microneedles can directly penetrate the ocular barrier to specifically target the drug to the site where it needs to work. Since the size of the microneedles is only in the micron range, these needles can easily enter the eyes and cause less pain. Microneedles can also present sustained release of drugs, so they are suitable for various applications.[80] In recent years, with the rise of hydrogels, hydrogel-forming microneedles have been vastly reported. After the hydrogel-forming microneedles are inserted into the skin, they swell and play a hydrophilic role. Compared with other types of microneedles, hydrogel-forming microneedles have excellent drug-loading capacity and good biocompatibility, and can be molded into various shapes.[81] Currently, hydrogel-forming microneedles have not been widely reported in the ophthalmic research, which means a bright prospect in the future is exhibited.

3.2. *Promoting repair patch*

Most corneal surgeries use nylon sutures, but a series of corneal complications such as inflammation and corneal neovascularization follow. Therefore, it is necessary to replace the nylon sutures with biocompatible mesh.[82,83] Hydrogel can conform to dynamic soft tissues and withstand regular movements, so it is a good carrier for patch.[84]

Jumelle *et al.* created a photo-cross-linked hydrogel patch made up of three polymers, HA, gelatin, and PEG, for eye lesions or corneal incisions. The hydrogel solution can be administered to the injured ocular surface using a carrier such as a contact lens. After 4 min of visible light cross-linking, the hydrogel solution solidified into a solid hydrogel patch that adhered to the injury site. After curing, the contact lenses can be removed, leaving a smooth hydrogel patch [Fig. 6(a)]. The physical properties of the photo-cross-linked hydrogel can be adjusted based on the concentration ratio of the components. Due to its high transparency, water content, and stability [Fig. 6(b)], the hydrogel is suitable for use as a corneal patch. The adhesive qualities of the hydrogel patch were tested in recently transplanted pig eyeballs, and it was discovered that the patch could securely cling to the corneal surface [Fig. 6(c)].

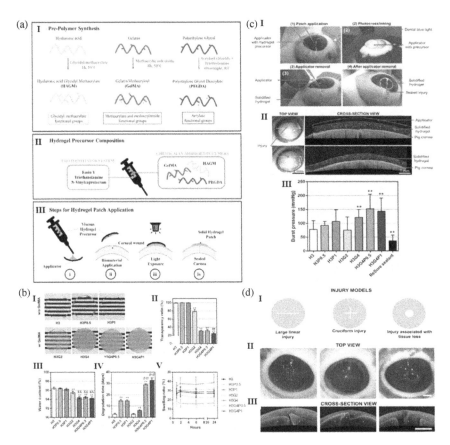

Fig. 6. (a) Synthesis and application of the hydrogel patch. (b) Physical properties of the cured hydrogels: (i), (ii) transparency, (iii) water content, (iv) degradation time, and (v) swelling ratio of the hydrogels. (c) Assessment of adhesion of the hydrogel patches to isolated porcine corneal lesions: (i) representative images, (ii) OCT images, and (iii) mean rupture pressures of the photo-cross-linked hydrogel patches. (d) Application and size evaluation of hydrogel patches in different types of open eyeball injuries: (i), (ii) damage model and (iii) OCT images after application of corneal patches to different injuries.

Source: Copyright 2014, Acta Biomaterialia.[85]

The extent of lesions that can be closed by the hydrogel patch technique was assessed by three different types of open ocular lesions (large linear lesions, cruciate lesions, and tissue loss-related lesions). It was demonstrated that the patch was able to cover the entire width of the lesion, and the hydrogel conformed well to the edges of the lesion and provided a smooth surface [Fig. 6(d)].[85]

3.3. Tissue-engineered cornea

Corneal transplantation is one of the most important ways to restore the vision. However, corneal transplantation also has disadvantages such as immune rejection, shortage of donors, and possible infections.[86] At present, there are two main sources of corneal donors—allogenic and synthetic materials. Allogenic materials are obtained from humans, and are the first choice for corneal transplantation, but they usually face a shortage of supply.[87] Hence, there was an urge for exploring new materials as corneal substitutes. Tissue-engineered corneal biomaterials are supposed to meet the requirements of transparency and mechanical elasticity. These biomaterials can be divided into three categories according to their design methods: tissue-engineered corneal epithelium, tissue-engineered corneal stroma, and tissue-engineered corneal endothelium.[88,89]

3.3.1. Tissue-engineered corneal epithelium

Tissue-engineered corneal epithelium based on amniotic membrane has been widely used for ocular surface reconstruction.[90] However, it degrades excessively, which is not useful for patients with severe ocular surface attack. Hence, there is a need for developing a novel type of tissue-engineered corneal epithelial scaffold.[91] Current materials are available to to construct tissue-engineered corneal epithelium. However, there are relatively only few studies on the tissue engineering of corneal epithelium based on hydrogel materials.

3.3.2. Tissue-engineered corneal stroma

Corneal stroma is the largest component of cornea, mainly composed of type-I collagen, with uniform arrangement and high transparency. The structure of corneal stroma is complicated, and the replacement of corneal stroma is a very difficult process.[92] However, there are some reports on the uses of tissue engineering of corneal stroma based on hydrogels. Han et al. developed a new type of tissue-engineered corneal stroma using the bacterial cellulose (BC)/ polyvinyl alcohol (PVA) hydrogel compound, which was aimed at reconstructing the cornea. It was found that the polymer exhibited good biocompatibility and tolerability, and the cornea remained almost transparent after transplantation, with no obvious side effects.[93] Furthermore, Feng et al. established a corneal scaffold built of oligoethylene glycol (OEG)-based dendritic chitosan (DC) and evaluated its viability as a corneal stroma substitute using a rabbit model [Fig. 7(a)]. The hydrogel is liquid at room temperature but forms a gel at physiological temperature, affording mechanical strength [Fig. 7(b)]. In addition, the hydrogel can promote migration of rabbit corneal stromal cells. Cells cling to the surface of the hydrogel and proliferate through the porous structure into the interior [Fig. 7(c)]. The hydrogel was proven to neither have effect on corneal transparency nor produce edema in a rabbit lamellar keratectomy corneal stromal defect model [Fig. 7(d)]. It can encourage cell migration and heal stromal defects in the cornea, making it an excellent material for corneal engineering scaffolds.[94]

3.3.3. Tissue-engineered corneal endothelium

The corneal endothelium layer forms by a single layer of cells. Endothelial cells are the most important layer of the cornea and play an important role in maintaining the transparency of the cornea. The function of endothelial cells is to prevent liquid from entering the corneal stroma. Corneal endothelial cells cannot proliferate indefinitely. Once the corneal endothelium layer is destroyed, only transplantation solely remains as the therapy.[94,95] Transplantation of endothelial cells using hydrogel as a scaffold has been studied. Liang et al. designed an in-situ biodegradable hydrogel for corneal endothelial reconstruction. This hydrogel exhibited low toxicity and was biodegradable and biocompatible. The encapsulated endothelial cells survived and maintained a normal morphology

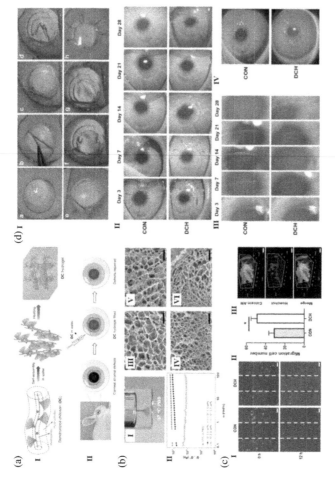

Fig. 7. (a) Schematic of hydrogel formation and application to corneal stromal defects. (b) The physical properties of hydrogels: (i) photograph of the injection of DC61 with a trace amount of rhodamine B into a PBS bath at 37°C to form a stable hydrogel; (ii) mechanical properties; and (iii)–(vi) SEM images of the hydrogel after immersion into PBS solution at 37°C. (c) Hydrogel stimulation of the migration of rabbit corneal stromal cells. (d) *In-vivo* application of hydrogels to rabbit corneal defects.

Source: Copyright 2021, *ACS Applied Materials & Interfaces*.[9]

after transplantation. This study provides an opportunity for corneal endothelial reconstruction.[96]

4. Intraocular Hydrogel Materials

4.1. *Intraocular lens*

Cataract is one of the reasons for blindness, and is caused by the change in the protein of the lens or due to a gradual loss in its transparency.[97] The lens usually needs to be replaced during cataract surgery, which makes the intraocular lens particularly important.[98] Due to the improvements in the cataract surgery, a small gap is required during the operation, which requires a foldable intraocular lens with a smaller incision. Hydrogels have many advantages of being used as intraocular lenses. For example, they are foldable, have a high water content, and can be autoclaved. They have been used as an intraocular lens for many years, and the natural lens itself is a kind of hydrogel.[99–101]

Han *et al.* evaluated the feasibility of a poloxamer hydrogel as an injectable intraocular lens, as shown in Fig. 8(a). In comparison, it was found that air vinyl filled with poloxamer hydrogel is as clear as air vinyl filled with PBS and is suitable as an intraocular lens material [Fig. 8(b)]. After phacoemulsification in rabbits, poloxamer hydrogel was injected to fill the lens [Fig. 8(c)]. One week after surgery, no active inflammation was observed in the ocular tissue and bright red reflections were seen, indicating that the refilled lens has excellent transparency. By indirect inspection, retinal blood vessels in rabbit eyes were also visible through the poloxamer in the capsule bag [Fig. 8(d)]. In conclusion, the poloxamer hydrogel demonstrates the feasibility of developing injectable intraocular lenses.[102]

4.2. *Vitreous substitute*

Retinal disease is another important reason for the vision loss.[103] The most common method to treat the retinal disease is to replace

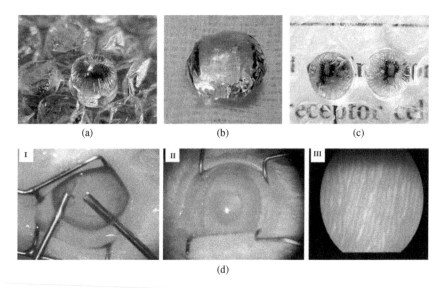

Fig. 8. (a) Air vinyl filled with poloxamer hydrogel. (b) Air vinyl filled with poloxamer hydrogel (left) and air vinyl filled with PBS (right). (c) Rabbit lenses being filled with poloxamer hydrogel post-operatively. (d) Evaluation of post-operative injection of poloxamer hydrogel in rabbits: (i) injection of poloxamer hydrogel using an 18-gauge cannula; (ii) slit-lamp photographs of rabbit eyes one week post-operatively; and (iii) the retinal blood vessels of the rabbit eyes being visualized through the poloxamer in the capsular bag during the first week after surgery.

Source: Copyright 2017, British Journal of Ophthalmology.[102]

the vitreous. The glass substitutes include silicone oil and gas that are widely used now, but each of them has its own drawbacks.[104] Gas as a substitute easily causes different refractive indices and harms the eyeball. Silicone oil is currently the most commonly used filler, but its nondegradability and toxicity to cells are a major hurdle for its use in the treatment. It is very difficult to design an optimal vitreous substitute, and various factors need to be considered, such as clinical availability, physical properties, biomechanics, and so on. Vitreous is also a kind of hydrogel to some extent, and there have been many attempts to use hydrogel as a vitreous substitute.[105–108] As a good vitreous substitute, hydrogel possesses good mechanical properties, biocompatibility, and optical properties.[109]

Tao et al. designed an injectable *in-situ* chemically cross-linked hydrogel system as a vitreous substitute. Injecting a mixture of two PEG-based components into the eye through the Michael addition route [Fig. 9(a)] formed a chemically cross-linked *in-situ* hydrogel that exhibited rheological properties similar to those of the natural vitreous. It was placed in the rabbit eyes for nine months and maintained its transparency and stability without causing any noticeable

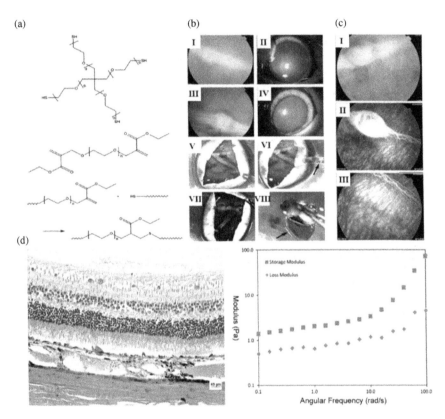

Fig. 9. (a) Schematic diagram of hydrogel formation. (b) (i)–(iv) Fundus or anterior segment photographs and (v)–(viii) enucleated eye tissue photographs after implantation in rabbit eyes. (c) (i) Fundus photographs and (ii), (iii) fluorescein angiography of rabbit eyes at 270 days post-implant of hydrogel. (d) H&E staining of rabbit eyes at 270 days post-implant of hydrogel. (e) Storage and loss moduli of hydrogels removed from surgical rabbits.
Source: Copyright 2013, *Acta Biomaterialia*.[110]

side effects [Figs. 9(b)–9(d)]. Moreover, hydrogels taken from surgical rabbits had comparable storage and loss moduli to gel samples extracted from natural porcine vitreous, indicating that the physical properties of *in-situ*-formed hydrogels were similar to those of natural animal vitreous [Fig. 9(e)]. These observations demonstrate that the hydrogel can serve as a vitreous substitute.[110]

Jiang *et al.* developed another *in-situ* hydrogel vitreous substitute, the main component of which was sodium alginate di-aldehyde cross-linked with hydroxypropyl chitosan. They showed that the hydrogel was nontoxic and exhibited no adverse effects till 90 days after vitreous replacement in the rabbit. The physical and chemical properties of hydrogel vitreous are similar to human vitreous, but its long-term safety and effectiveness need a further evaluation.[111]

5. Concluding Remarks and Future Perspectives

Hydrogels have been widely used in various fields and have attracted many researchers in ophthalmology. The bioavailability of medications on the ocular surface is considerably boosted when used as an ocular delivery system. It fits the cornea effectively as a tissue adhesive. As a tissue-engineered cornea, it can promote corneal regeneration. In these areas, hydrogels have good clinical application. Intraocular applications of hydrogels are currently relatively rare, and related research could lead to considerable advancements in this field. However, hydrogel materials also face certain issues, such as sterilization problems, shelf-life problems, and poor degradation. Most of the tests conducted in the literature were of duration less than one year; hence, the long-term safety and effectiveness need to be further explored. If possible, some clinical experiments can be carried out after the animal experiments are completed. Furthermore, this research is just in the experimental stage, and while the efficacy is excellent, there are few commercial items available, so it cannot be adequately promoted to the commercialization phase. All in all, hydrogel is an ideal material for ophthalmic applications, although it has some disadvantages.

References

1. S. Sharma and S. Tiwari, *Int. J. Biol. Macromol.*, 2020, **162**, 737–747.
2. A. Presland and J. Myatt, *Anaesth. Intensive Care Med.*, 2010, **11**, 438–443.
3. S. Kirchhof, A. M. Goepferich and F. P. Brandl, *Eur. J. Pharm. Biopharm.*, 2015, **95**, 227–238.
4. Y. Han, L. Jiang, H. Shi, C. Xu, M. Liu, Q. Li, L. Zheng, H. Chi, M. Wang, Z. Liu, M. You, X. J. Loh, Y.-L. Wu, Z. Li and C. Li, *Bioact. Mater.*, 2022, **9**, 77–91.
5. R. DeVolder and H. J. Kong, *Wiley Interdiscip. Rev. Syst. Biol. Med.*, 2012, **4**, 351–365.
6. Y. Qiu and K. Park, *Adv. Drug Deliv. Rev.*, 2001, **53**, 321–339.
7. C. Torres-Luna, X. Fan, R. Domszy, N. Hu, N. S. Wang and A. Yang, *Eur. J. Pharm. Sci.*, 2020, **154**, 105503.
8. S. J. Buwalda, K. W. Boere, P. J. Dijkstra, J. Feijen, T. Vermonden and W. E. Hennink, *J. Control. Release*, 2014, **190**, 254–273.
9. L. Yahia, *J. Biomed. Sci.*, 2015, **4**, 13:1–13:23.
10. M. C. Catoira, L. Fusaro, D. Di Francesco, M. Ramella and F. Boccafoschi, *J. Mater. Sci., Mater. Med.*, 2019, **30**, 115.
11. E. Caló and V. V. Khutoryanskiy, *Eur. Polym. J.*, 2015, **65**, 252–267.
12. G. Chen, W. Tang, X. Wang, X. Zhao, C. Chen and Z. Zhu, *Polymers (Basel)*, 2019, **11**, 1420.
13. M. W. Ansari and A. Nadeem, *Atlas of Ocular Anatomy*, Springer, Cham, 2016, pp. 11–27.
14. I. K. Gipson, *Invest. Ophthalmol. Vis. Sci.*, 2013, **54**, ORSF48–ORSF53.
15. A. G. Y. Chiou, G. J. Florakis and M. Kazim, *Surv. Ophthalmol.*, 1998, **43**, 19–46.
16. R. C. Cooper and H. Yang, *J. Control. Release*, 2019, **306**, 29–39.
17. P. C. Nicolson and J. Vogt, *Biomaterials*, 2001, **22**, 3273–3283.
18. A. K. Agrawal, M. Das and S. Jain, *Expert Opin. Drug Deliv.*, 2012, **9**, 383–402.
19. T. R. Thrimawithana, S. Young, C. R. Bunt, C. Green and R. G. Alany, *Drug Discov. Today*, 2011, **16**, 270–277.
20. H. Chen, Y. Jin, L. Sun, X. Li, K. Nan, H. Liu, Q. Zheng and B. Wang, *Colloid Interface Sci. Commun.*, 2018, **24**, 54–61.

21. T. R Thrimawithana, S. Young, C. R. Bunt, C. R. Green and R. G. Alany, *Drug Deliv. Lett.*, 2011, **1**, 40–44.
22. X. Zhao, I. Seah, K. Xue, W. Wong, Q. S. W. Tan, X. Ma, Q. Lin, J. Y. C. Lim, Z. Liu, B. H. Parikh, K. N. Mehta, J. W. Lai, B. Yang, K. C. Tran, V. A. Barathi, K. H. Cheong, W. Hunziker, X. Su and X. J. Loh, *Adv. Mater.*, 2022, **34**, e2108360.
23. R. Gaudana, J. Jwala, S. H. Boddu and A. K. Mitra, *Pharm. Res.*, 2009, **26**, 1197–1216.
24. A. Urtti, *Adv. Drug Deliv. Rev.*, 2006, **58**, 1131–1135.
25. P. M. Hughes, O. Olejnik, J.-E. Chang-Lin and C. G. Wilson, *Adv. Drug Deliv. Rev.*, 2005, **57**, 2010–2032.
26. M. Karayilan, L. Clamen and M. L. Becker, *Biomacromolecules*, 2021, **22**, 223–261.
27. J. U. Prause and M. A. Saornil, *Eye Pathology*, Springer-Verlag, Berlin, 2015, pp. 1–39.
28. M. F. Ramos, J. Baker, E.-A. Atzpodien, U. Bach, J. Brassard, J. Cartwright, C. Farman, C. Fishman, M. Jacobsen, U. Junker-Walker, F. Kuper, M. C. R. Moreno, S. Rittinghausen, K. Schafer, K. Tanaka, L. Teixeira, K. Yoshizawa and H. Zhang, *J. Toxicol. Pathol.*, 2018, **31**, 97S–214S.
29. D. M. Maurice, *J. Physiol.*, 1957, **136**, 263–286.
30. N. R. Galloway, W. M. K. Amoaku, P. H. Galloway and A. C. Browning, *Common Eye Diseases and Their Management*, Springer, Cham, 2016, pp. 7–16.
31. A. Muroo and J. Pospisil, *Acta Univ. Palacki. Olomuc., Fac. Rer. Nat., Phys.*, 2000, **39**, 87–95.
32. A. Tousimis and B. S. Fine, *AMA Arch. Ophthalmol.*, 1959, **62**, 974–976.
33. N. A. Delamere, *Advances in Organ Biology*, ed. J. Fischbarg, Elsevier, 2005, pp. 127–148.
34. E. R. Tamm and E. Lütjen-Drecoll, *Microsc. Res. Tech.*, 1996, **33**, 390–439.
35. H. Kolb, *Webvision: The Organization of the Retina and Visual System [Internet]*, University of Utah Health Sciences Center, Salt Lake City, 1995, pp. 11–33.
36. G. D. Hildebrand and A. R. Fielder, *Pediatric Retina*, Springer, Berlin, 2011, pp. 39–65.

37. N. Evangelou and O. S. Alrawashdeh, *Optical Coherence Tomography in Multiple Sclerosis*, Springer, Cham, 2016, pp. 3–19.
38. H. Kolb, *Am. Sci.*, 2003, **91**, 28–35.
39. W. Chen, X. Tan and X. Chen, *Pediatric Lens Diseases*, Springer, Singapore, 2017, pp. 21–28.
40. D. G. Cogan, *Exp. Eye Res.*, 1962, **1**, 291–295, IN1–IN3.
41. G. Thin and J. Ewart, *J. Anat. Physiol.*, 1876, **10**, i3–i230.
42. A. T. Chin and C. R. Baumal, *Ocular Fluid Dynamics: Analtomy, Physiology, Imaging Techniques, and Mathematical Modeling*, Birkhauser, Cham, 2019, pp. 289–302.
43. Y. Song, N. Nagai, S. Saijo, H. Kaji, M. Nishizawa and T. Abe, *Mater. Sci. Eng. C*, 2018, **88**, 1–12.
44. Y. Han, C. Xu, H. Shi, F. Yu, Y. Zhong, Z. Liu, X. J. Loh, Y.-L. Wu, Z. Li and C. Li, *Chem. Eng. J.*, 2021, **421**, 129734.
45. J.-Y. Lai, L.-J. Luo and D. D. Nguyen, *Chem. Eng. J.*, 2020, **402**, 126190.
46. G. Chouhan, R. J. A. Moakes, M. Esmaeili, L. J. Hill, F. deCogan, J. Hardwicke, S. Rauz, A. Logan and L. M. Grover, *Biomaterials*, 2019, **210**, 41–50.
47. Z. Shi, S. K. Li, P. Charoenputtakun, C.-Y. Liu, D. Jasinski and P. Guo, *J. Control. Release*, 2018, **270**, 14–22.
48. L. Feng, S. K. Li, H. Liu, C.-Y. Liu, K. LaSance, F. Haque, D. Shu and P. Guo, *Pharm. Res.*, 2014, **31**, 1046–1058.
49. X. Rong, J. Yang, Y. Ji, X. Zhu, Y. Lu and X. Mo, *J. Drug Deliv. Sci. Technol.*, 2019, **49**, 556–562.
50. Z. Zhong, X. Deng, P. Wang, C. Yu, W. Kiratitanaporn, X. Wu, J. Schimelman, M. Tang, A. Balayan and E. Yao, *Biomaterials*, 2021, **267**, 120462.
51. D. H. Geroski and H. F. Edelhauser, *Invest. Ophthalmol. Vis. Sci.*, 2000, **41**, 961–964.
52. E. Eljarrat-Binstock, J. Pe'er and A. J. Domb, *Pharm. Res.*, 2010, **27**, 530–543.
53. Y. E. Choonara, V. Pillay, M. P. Danckwerts, T. R. Carmichael and L. C. du Toit, *J. Pharm. Sci.*, 2010, **99**, 2219–2239.
54. Y. Yu, L. C. M. Lau, A. C.-Y. Lo and Y. Chau, *Transl. Vis. Sci. Technol.*, 2015, **4**, 5.
55. S. L. Fialho and A. da Silva Cunha, *Drug Deliv.*, 2005, **12**, 109–116.

56. J. L. Bourges, C. Bloquel, A. Thomas, F. Froussart, A. Bochot, F. Azan, R. Gurny, D. BenEzra and F. Behar-Cohen, *Adv. Drug Deliv. Rev.*, 2006, **58**, 1182–1202.
57. J. K. Gandhi, F. Mano, R. Iezzi, Jr., S. A. LoBue, B. H. Holman, M. P. Fautsch, T. W. Olsen, J. S. Pulido and A. D. Marmorstein, *PLOS ONE*, 2020, **15**, e0227641.
58. V. H. L. Lee and J. R. Robinson, *J. Ocul. Pharmacol. Ther.*, 1986, **2**, 67–108.
59. J. Hao, X. Wang, Y. Bi, Y. Teng, J. Wang, F. Li, Q. Li, J. Zhang, F. Guo and J. Liu, *Colloids Surf. B, Biointerfaces*, 2014, **114**, 111–120.
60. S. Ding, *Pharm. Sci. Technol. Today*, 1998, **1**, 328–335.
61. J. C. Imperiale, G. B. Acosta and A. Sosnik, *J. Control. Release*, 2018, **285**, 106–141.
62. Y. Yu, R. Feng, J. Li, Y. Wang, Y. Song, G. Tan, D. Liu, W. Liu, X. Yang, H. Pan and S. Li, *Asian J. Pharm. Sci.*, 2019, **14**, 423–434.
63. M. Dubald, S. Bourgeois, V. Andrieu and H. Fessi, *Pharmaceutics*, 2018, **10**, 10.
64. R. Moreddu, D. Vigolo and A. K. Yetisen, *Adv. Healthc. Mater.*, 2019, **8**, e1900368.
65. X. Li, Y. Cui, A. W. Lloyd, S. V. Mikhalovsky, S. R. Sandeman, C. A. Howel and L. Liao, *Cont. Lens Anterior Eye*, 2008, **31**, 57–64.
66. J. Kim, A. Conway and A. Chauhan, *Biomaterials*, 2008, **29**, 2259–2269.
67. X. Hu, L. Hao, H. Wang, X. Yang, G. Zhang, G. Wang and X. Zhang, *Int. J. Polym. Sci.*, 2011, **2011**, 814163.
68. A. Childs, H. Li, D. M. Lewittes, B. Dong, W. Liu, X. Shu, C. Sun and H. F. Zhang, *Sci. Rep.*, 2016, **6**, 34905.
69. J. F. Huang, J. Zhong, G. P. Chen, Z. T. Lin, Y. Deng, Y. L. Liu, P. Y. Cao, B. Wang, Y. Wei, T. Wu, J. Yuan and G. B. Jiang, *ACS Nano*, 2016, **10**, 6464–6473.
70. H. J. Jung, M. Abou-Jaoude, B. E. Carbia, C. Plummer and A. Chauhan, *J. Control. Release*, 2013, **165**, 82–89.
71. A. Chowhan and T. K. Giri, *Int. J. Biol. Macromol.*, 2020, **150**, 559–572.
72. U. Laddha and H. Mahajan, *Int. J. Adv. Pharm.*, 2017, **06**, 31–40.

73. H. Gupta, S. Jain, R. Mathur, P. Mishra, A. K. Mishra and T. Velpandian, *Drug Deliv.*, 2007, **14**, 507–515.
74. K. Rathore, *Int. J. Pharm. Sci.*, 2010, **2**, 30–34.
75. J. Yu, X. Xu, F. Yao, Z. Luo, L. Jin, B. Xie, S. Shi, H. Ma, X. Li and H. Chen, *Int. J. Pharm.*, 2014, **470**, 151–157.
76. K. Y. Janga, A. Tatke, S. P. Balguri, S. P. Lamichanne, M. M. Ibrahim, D. N. Maria, M. M. Jablonski and S. Majumdar, *Artif. Cells Nanomed. Biotechnol.*, 2018, **46**, 1039–1050.
77. G. Ma and C. Wu, *J. Control. Rel.*, 2017, **251**, 11–23.
78. R. F. Donnelly, T. R. R. Singh and A. D. Woolfson, *Drug Deliv.*, 2010, **17**, 187–207.
79. A. J. Guillot, A. S. Cordeiro, R. F. Donnelly, M. C. Montesinos, T. M. Garrigues and A. Melero, *Pharmaceutics*, 2020, **12**, 569.
80. P. Gupta and K. S. Yadav, *Life Sci.*, 2019, **237**, 116907.
81. J. G. Turner, L. R. White, P. Estrela and H. S. Leese, *Macromol. Biosci.*, 2021, **21**, 2000307.
82. M. W. Grinstaff, *Biomaterials*, 2007, **28**, 5205–5214.
83. L. Koivusalo, M. Kauppila, S. Samanta, V. S. Parihar, T. Ilmarinen, S. Miettinen, O. P. Oommen and H. Skottman, *Biomaterials*, 2019, **225**, 119516.
84. J. Wang, L. Wang, C. Wu, X. Pei, Y. Cong, R. Zhang and J. Fu, *ACS Appl. Mater. Interfaces*, 2020, **12**, 46816–46826.
85. C. Jumelle, A. Yung, E. S. Sani, Y. Taketani, F. Gantin, L. Bourel, S. Wang, E. Yuksel, S. Seneca, N. Annabi and R. Dana, *Acta Biomater.*, 2022, **137**, 53–63.
86. A. Shah, J. Brugnano, S. Sun, A. Vase and E. Orwin, *Pediatr. Res.*, 2008, **63**, 535–544.
87. C. E. Ghezzi, J. Rnjak-Kovacina and D. L. Kaplan, *Tissue Eng. B, Rev.*, 2015, **21**, 278–287.
88. L. S. Wray and E. J. Orwin, *Tissue Eng. A*, 2009, **15**, 1463–1472.
89. E. S. Gil, B. B. Mandal, S.-H. Park, J. K. Marchant, F. G. Omenetto and D. L. Kaplan, *Biomaterials*, 2010, **31**, 8953–8963.
90. S. Tseng, *Biosci. Rep.*, 2001, **21**, 481–489.
91. H. Zhao, M. Qu, Y. Wang, Z. Wang and W. Shi, *PLOS ONE*, 2014, **9**, e111846.
92. S. Matthyssen, B. Van den Bogerd, S. N. Dhubhghaill, C. Koppen and N. Zakaria, *Acta Biomater.*, 2018, **69**, 31–41.

93. Y. Han, C. Li, Q. Cai, X. Bao, L. Tang, H. Ao, J. Liu, M. Jin, Y. Zhou, Y. Wan and Z. Liu, *Biomed. Mater.*, 2020, **15**, 035022.
94. L. Feng, R. Liu, X. Zhang, J. Li, L. Zhu, Z. Li, W. Li and A. Zhang, *ACS Appl. Mater. Interfaces* 2021, **13**, 49369–49379.
95. S. Proulx and I. Brunette, *Exp. Eye Res.*, 2012, **95**, 68–75.
96. Y. Liang, W. Liu, B. Han, C. Yang, Q. Ma, F. Song and Q. Bi, *Colloids Surf. B, Biointerfaces*, 2011, **82**, 1–7.
97. J. H. de Groot, F. J. van Beijma, H. J. Haitjema, K. A. Dillingham, K. A. Hodd, S. A. Koopmans and S. Norrby, *Biomacromolecules*, 2001, **2**, 628–634.
98. R. J. Olson, L. Werner, N. Mamalis and R. Cionni, *Am. J. Ophthalmol.*, 2005, **140**, 709–716.
99. S. Percival and A. Jafree, *Eye*, 1994, **8**, 672–675.
100. J. H. de Groot, C. J. Spaans, R. V. van Calck, F. J. van Beijma, S. Norrby and A. J. Pennings, *Biomacromolecules*, 2003, **4**, 608–616.
101. M. K. Yoo, Y. J. Choi, J. H. Lee, W. R. Wee and C. S. Cho, *J. Drug Deliv. Sci. Technol.*, 2007, **17**, 81–85.
102. Y. K. Han, J. W. Kwon, J. S. Kim, C.-S. Cho, W. R. Wee and J. H. Lee, *Br. J. Opthalmol.*, 2003, **87**, 1399–1402.
103. R. C. Heath Jeffery, S. A. Mukhtar, I. L. McAllister, W. H. Morgan, D. A. Mackey and F. K. Chen, *Ophthalmic Genet.*, 2021, **42**, 431–439.
104. Q. Lin, Z. Liu, D. S. L. Wong, C. C. Lim, C. K. Liu, L. Guo, X. Zhao, Y. J. Boo, J. H. M. Wong, R. P. T. Tan, K. Xue, J. Y. C. Lim, X. S and X. J. Loh, *Biomaterials*, 2022, **280**, 121262.
105. K. Januschowski, S. Schnichels, J. Hurst, C. Hohenadl, C. Reither, A. Rickmann, L. Pohl, K. U. Bartz-Schmidt and M. S. Spitzer, *PLOS ONE*, 2019, **14**, e0209217.
106. Q. Lin, J. Y. C. Lim, K. Xue, X. Su and X. J. Loh, *Biomaterials*, 2021, **268**, 120547.
107. N. K. Tram, P. Jiang, T. C. Torres-Flores, K. M. Jacobs, H. L. Chandler and K. E. Swindle-Reilly, *Macromol. Biosci.*, 2020, **20**, e1900305.
108. H. Barth, S. Crafoord, S. Andreasson and F. Ghosh, *Graefe's Arch. Clin. Exp. Ophthalmol.*, 2016, **254**, 697–703.
109. S. Feng, H. Chen, Y. Liu, Z. Huang, X. Sun, L. Zhou, X. Lu and Q. Gao, *Sci. Rep.*, 2013, **3**, 1838.

110. Y. Tao, X. Tong, Y. Zhang, J. Lai, Y. Huang, Y. R. Jiang and B. H. Guo, *Acta Biomater.*, 2013, **9**, 5022–5030.
111. X. Jiang, Y. Peng, C. Yang, W. Liu and B. Han, *J. Biomed. Mater. Res. A*, 2018, **106**, 1997–2006.

Green Nanotechnology and Phytosynthesis of Metallic Nanoparticles: The Green Approach, Mechanism, Biomedical Applications and Challenges*

Abdulrahman Alomar, Tabarak Qassim, Yusuf AlNajjar, Alaa Alqassab and G. Roshan Deen[†]

Materials for Medicine Research Group, School of Medicine
The Royal College of Surgeons in Ireland (RCSI)
Medical University of Bahrain
Busaiteen 228, Kingdom of Bahrain
[†]*rdeen@rcsi.com*

The synthesis of nanoparticles is generally divided into bottom-up and top-down approaches which involve physical, chemical, and biological methods. The physical and chemical methods are associated with issues such as high cost, non-ambient reaction conditions, and toxicity. Biological methods or green-methods, using plants, bacteria, algae, and fungi have been developed in recent years to overcome the issues associated with conventional methods. The synthesis of nanoparticles using plants and plant-products as chemical reducing and stabilizing agents from metal precursors is termed phytosynthesis. Plants are available in plenty, safe to handle and contain a wide variety of water-soluble metabolites such as anthocyanins, flavonoids, polyphenols, alkaloids, and terpenoids, that act as excellent chemical reducing and stabilizing

---

[†]Corresponding author.
*This review article is dedicated to Prof. Fryad Z Henari, Emeritus Professor, on his retirement.
To cite this article, please refer to its earlier version published in the *World Scientific Annual Review of Functional Materials*, Volume 2, 2430001 (2024), DOI: 10.1142/S2810922824300010.

agents. These metabolites reduce the metal precursors to metal nanoparticles in a much shorter time as compared to bacteria and fungi. Furthermore, both bacteria and fungi require much longer incubation time for the chemical reduction process. A wide variety of plants have been used to synthesize nanoparticles, oxides, and alloys of gold, silver, titanium, platinum, palladium, copper, cobalt, selenium, zinc, titanium, and iron for various biomedical applications. The types of plants and parts used such as stem, leaf, flower, fruit, pods, and peel have significant effects on the size and shape of the synthesized nanoparticle. Although the phytosynthesis method is advantageous in many aspects of synthesis, there are challenges associated with scale-up process for larger scale production which could be overcome in the future. This review summarizes the phytosynthesis process, green nanotechnology, characterization methods, mechanisms, various biomedical applications, and challenges.

Keywords: Nanoparticles; nanotechnology; phytosynthesis; plants; biomedical applications.

1. Introduction

Nanotechnology is an interdisciplinary science that deals with the synthesis and structural manipulation of particles at the nanometer length scale. The concept of nanotechnology was first introduced by Richard P Feynman (Nobel Laureate in physics) in 1959 in his famous lecture entitled, "There's plenty of room at the bottom" at the American Physical Society meeting.[1] The ultra-fine particles in the size range of 1–100 nm are called nanoparticles that are obtained either by scaling up from single groups of atoms (bottom-up approach) or by reducing bulk materials (top-down approach). The nanoparticles can be composed of one or more species of atoms (or molecules) and can exhibit a wide range of properties that are dependent on the size of the particles. Nanotechnology is now widely used in translational research including diagnostics and therapeutics, drug development, drug delivery systems, water decontamination, information and communication technologies, non-linear optical devices, etc.[2] Bulk materials (at macro or micro level) when reduced to nanoscale (nanoparticle), often exhibit novel and interesting properties in strength, chemical reactivity, optical, catalytic, superparamagnetic, and electrical conductivity.

Nanoparticles are synthesized by two approaches, (i) the bottom-up approach and (ii) the top-down approach. In the bottom-up approach, nanoparticles are synthesized or fabricated from the atomic level using chemical (chemical reduction, sol–gel, reverse micelles, electrochemistry) and biological (algae, virus, bacteria, plants) processes. In the top-down approach, bulk materials are broken down into smaller and ultra-fine particles using physical methods such as laser ablation, ultrasonication, lithography, plasma radiations, and spray pyrolysis.[3] The various approaches of nanoparticle synthesis are illustrated in Fig. 1.

Physical and chemical methods are widely used in the development of various nanoparticles for a wide range of applications. The use of toxic chemicals such as organic solvents, chemical reducing agents, organic stabilizers, high temperature, and acidic pH conditions limits the biological and clinical applications of nanoparticles that are developed by physical and chemical methods. Although

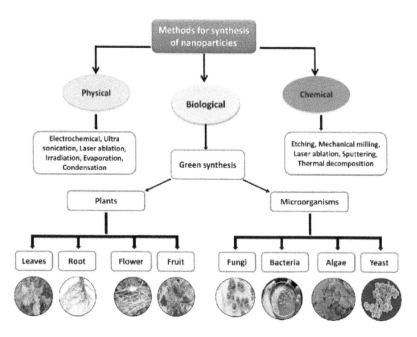

Fig. 1. Illustration of various approaches in the synthesis of nanoparticles. Adapted from Ref. 3 based on collective common agreement 4.0.

these chemicals provide stability to the nanoparticles against agglomeration, they render them toxic and unsuitable for any biomedical applications.[4]

For successful clinical and biological applications, the nanoparticles must be free of toxic chemicals (non-toxic) and stable with prolonged shelf-life. In this regard, biological or green methods using bacteria, viruses, yeasts, and plant extracts offer great promise in the development of reliable, non-toxic, and stable nanoparticles.[5] These methods do not employ hazardous chemicals, extreme temperature and pH conditions thus making the biological methods an attractive and safe alternative to chemical and physical methods.

In this paper, a review of the synthesis of various metallic nanoparticles using the aqueous extract of different types of plants as chemical and stabilizing agents, reaction mechanisms, factors affecting the synthesis, and biomedical applications are presented. The first part of this review briefly covers the general features of nanoparticles, followed by phytosynthesis, factors affecting the synthesis, biomedical applications, and challenges in the subsequent sections.

2. Classification of Nanoparticles

Nanoparticles are classified based on dimensions, morphology, composition, agglomeration, and uniformity.[6] In terms of dimensions, they are classified as one-dimensional (1D), two-dimensional (2D), and three-dimensional materials (3D).

Materials with one dimension on the nanometre length scale are called 1D nanomaterials, and these are mainly surface coatings or thin films. Thin films have been widely used in various fields such as chemical and biological sensors, electronic devices, information storage systems, and magneto-optic and optical devices. Thin films in the form of monolayers can be deposited by physical methods and developed in a controlled manner by chemical methods.[7] Materials with two dimensions on the nanometre length scale are called 2D nanomaterials, such as nanowires, nanotubes, dendrimers, nanofibers, patterned surfaces (including metallic patterned surfaces) with plasmonic resonances for applications in nanophotonics.[8-10]

Particles that are in this size range but have a larger aspect ratio are also classified as 2D nanomaterials. Materials that have nanoscale length scales on all three dimensions are called 3D nanomaterials such as nanoparticles, nanocrystals, colloids, precipitates, fullerenes (buckyballs), and 3D printed photonic crystals which are developed using two-photon lithography.[11-13]

An illustration of nanomaterials based on dimensions is shown in Fig. 2.

Nanoparticles are classified based on their surface morphology in terms of aspect ratio, flatness, and spatial position of elements (hybrid nanoparticles). Nanoparticles with high aspect ratio include nanotubes and nanowires and nanoparticles with small aspect ratio include shapes such as prism, oval, cube, spheres, and helices. In recent years much progress has been made in the field of magnetic nanospheres and ferrofluids. Polymeric nano/microspheres containing magnetic nanoparticles have been developed for various biological applications particularly, magnetic field-driven targeted drug delivery.[14] The magnetic nanoparticles for *in vitro* applications are superparamagnetic colloids and contain appropriate stabilizing agents (coatings) which makes them stable and biocompatible.

In terms of chemical composition, the nanoparticles can be composed of a single component material or a composite of several materials. The three main types of nanoparticles based on chemical composition, mixed nanoparticles, core–shell nanoparticles, and layered nanoparticles, and these are illustrated in Fig. 3.[15]

Mixed nanoparticles can be either ordered or randomized to correspond to ordered nanoalloys or mixed alloys. Core–shell nanoparticles consist of a core made of one type of atom and a shell of another type of atom. The core–shell types also include multishell or "onion-like" and alternating shell types of nanoparticles. These types of nanoparticles are synthesized by modifying the

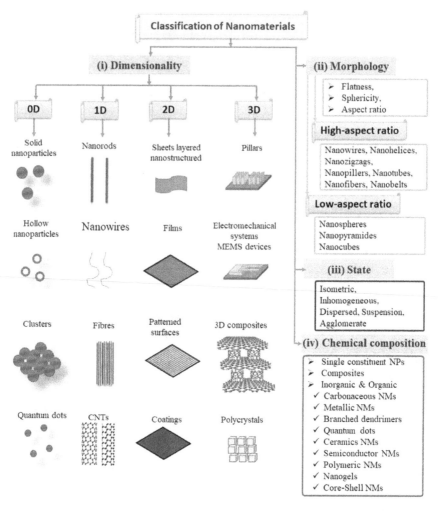

Fig. 2. Illustration of nanoparticles based on dimensions. Adapted from Ref. 7 with permission from Elsevier.

experimental conditions. Layered nanoparticles are referred to as Janus or dumb-bell and they consist of two types of nanoparticles that have a common interface and heterojunction.

Nanoparticles can exist as aerosols, suspensions or agglomerates based on their chemistry and electromagnetic properties. Forces such as van der Waals (acting in short distances), magnetic interactions, electrostatics, and adhesive cause agglomeration or coagulation of

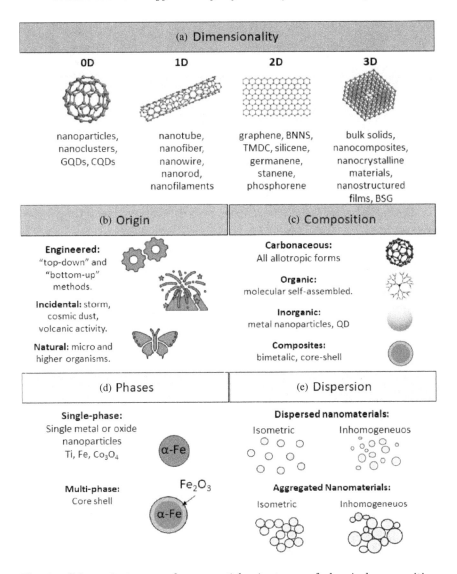

Fig. 3. Schematic images of nanoparticles in terms of chemical composition, phase, and dispersion adapted from Ref. 15 based on collective common agreement 4.0.

nanoparticles. The formation of agglomerates is usually avoided by coating the nanoparticles with other inorganic or organic substances leading to stabilization either through electrostatic or steric interactions.

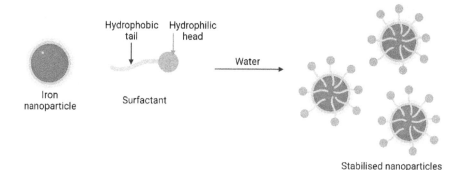

Fig. 4. An illustration of the stabilization process applied to magnetic nanoparticles using sodium dodecyl sulphate as the stabilizing agent.

Magnetic nanoparticles tend to form clusters easily and are usually coated with a non-magnetic material such as sodium dodecyl sulphate (SDS) to avoid the formation of aggregates as illustrated in Fig. 4.

3. Characterization of Nanoparticles and Environmental Implications

The unique physicochemical properties of nanoparticles are electrical and thermal conductivity, catalytic activity, light absorption, and scattering. The physicochemical properties of nanoparticles can be influenced by manipulating the size and shape of the particles. The physicochemical properties of nanoparticles are characterized by using various methods, such as X-ray diffraction (XRD), X-ray photoelectron spectroscopy (XPS), Infrared spectroscopy (IR), Scanning electron microscopy (SEM), Transmission electron microscopy (TEM), Atomic Force microscopy (AFM), and Brunauer–Emmet–Teller (BET).[16,17] These methods allow the characterization of nanoparticles in terms of size, morphology, optical properties, crystallography, and surface properties that are vital for a wide range of engineering and biomedical applications.

XRD is mainly used to study the crystallography of nanomaterials.[16,18,19] It is the study of the arrangement of atoms and molecules in crystalline solids.[19] The method uses X-ray scattering which gives rise to diffraction patterns that reveal the arrangement of atoms in a crystalline structure.[20] XPS is a sensitive method that also characterizes the elemental ratio and the nature of bonds on the surface of the nanoparticle.[21,22] Infra-red spectroscopic (IR) studies are used to identify the functional groups on the surface of the nanoparticles based on their vibrational absorption frequencies.[23,24] The composition of the material can also be quantified using IR spectroscopy.[25] Scanning electron microscopy (SEM) is used to generate high-resolution images of individual nanoparticles (size and morphology) through the back-scattering of electrons from the sample.

Similarly, transmission electron microscopy (TEM) uses an electron beam to generate a high-resolution image of nanoparticles beyond the limitations of SEM. The main difference is that in TEM the transmitted electrons are detected as opposed to the reflected electrons in SEM. BET test is used to determine the surface area of nanoparticles. The BET method utilizes gas adsorption to nanoparticles to determine their surface area. Typically, nitrogen gas is adsorbed to saturation on the nanoparticles and after saturation by the gas, the nanoparticles are removed from the nitrogen environment and heated to remove the adsorbed gas. Based on the amount of nitrogen adsorbed, the total surface area of the nanoparticle is then determined.

The impact of nanoparticles on human health and the environment (water, soil, and air) is still a matter of debate. Nanoparticles pose a threat to the environment and living organisms due to several factors, such as their size, density, and chemical composition.[27] Nanoparticles of heavy metals such as cadmium, lead, chromium, iron, copper, nickel, platinum, and silver accumulate in the environment posing a great threat to human health and environment. For example, cadmium selenide (CdS) nanoparticles release Cd^{2+} ions, a heavy metal, which is extremely toxic to humans leading to kidney and skeletal damage, liver disease and lung cancer.

Furthermore, cadmium and other heavy metals are also hazardous to plants, microorganisms, and aquatic organisms above a threshold concentration.[28,29] The size of nanoparticles also poses a risk, as this allows them to enter cells or attach to cellular membrane which impairs the functions of the cells. In addition to the size, the shape of nanoparticles also has a significant impact on cellular functions. For example, carbon nanotubes can pierce through the cell membranes leading to the damage of cells.[27] Studies have presented evidence that some metallic nanoparticles with the size of 50 nm perforate the membranes of type-1 alveolar cells leading to cell death.[29]

4. Synthesis of Nanoparticles

Various methods of nanoparticle synthesis exist; however, they are generally divided into two main categories: (i) Top-down synthesis and (ii) Bottom-up synthesis.[4,28] Both categories include physical, chemical, or biological methods. Chemical and physical methods have been associated with environmental pollution due to the use of toxic solvents and very high temperatures. To overcome the effects of chemical and physical methods, eco-friendly biological or green methods have been developed. The biological methods utilize microorganisms, plants, enzymes, fungi, and proteins to synthesize the nanoparticles from metal salt precursors.[29]

4.1. Top-down method

The top-down method is also known as a destructive method, as it involves the breaking down of bulk material into nanosized materials using physical methods. Top-down synthesis methods include nanolithography, mechanical milling, thermal decomposition, sputtering, and laser ablation.[11,30]

Nanolithography involves different procedures, such as photolithography, electron beam lithography, focused ion beam lithography, soft lithography, nanoimprint, and dip-pen lithography.[31] Overall, nanolithography utilizes a laser beam to generate clusters of nanoparticles with specific size, shape, and interlayer spacings.[32,33] Nanolithography generates nanoparticles from bulk material which is broken down into nanoparticles using a laser beam. This method is not cost-effective and requires sophisticated equipment.

Mechanical milling is arguably the most used top-down synthesis method due to its convenience.[34,35] In mechanical milling, a high energy mill with a suitable medium produces a powder (charged), which contains a mixture of elements. The milling process works by reducing the size of the charged powder into nanosized particles accompanied by annealing.[34] High energy mills such as attrition jet, planetary, oscillating, and vibration mills are used in this milling process.[35] Advantages of mechanical milling include reduced environmental waste production, controlled degradation process, and the ability to combine immiscible material blends.[36]

Thermal decomposition is a chemical top-down synthesis method that is used to generate monodispersed nanoparticles.[37,38] Thermal decomposition utilizes heat to break the chemical bonds in compounds. The production of nanoparticles is carried out by heating metals at a specific temperature leading to the formation of a secondary product, the nanoparticles.[38] The benefit of thermal decomposition is the ability to produce large quantities of stable, monodispersed nanoparticles.[39] The thermal stability of compounds in thermal decomposition depends on the metal cation and the ligand and bulky ligands produce nanoparticles that are highly stable.[40] Sputtering is another top-down nanoparticle synthesis method, wherein nanoparticles are ejected from a desired material. The formation and ejection of nanoparticles are brought about by the bombardment of the surface of a target material with gas plasma or energetic ions of a gas. This results in an exchange of momentum between atoms and ions causing sputtering and ultimately leading to the cause it was used twice in the deposition of a thin layer of nanoparticles.[41,42]

Laser ablation is a common top-down synthesis method that employs a high-energy laser beam to produce nanoparticles from target materials. A metal is submerged in a liquid and irradiated by a high-energy laser concentrated at a particular point on the metal to produce the nanoparticles.[43,44] The laser ablation method works by removing surface atoms and multiphoton excitation to form the nanoparticles. The advantage of this method is the ability to generate nanoparticles of high purity and is predetermined by the choice of the target material. The only disadvantage of this method is the size distribution and crystal structure of the material cannot be controlled precisely.[44]

4.2. Bottom-up method

The bottom-up method, also known as the self-assembly approach, is a synthesis method that builds nanoparticles from atoms or basic units into larger units, i.e. nanoparticles. Materials scientists have developed various bottom-up approaches for nanoparticle synthesis, such as biosynthesis, sol–gel, spinning, chemical vapor deposition (CVD), and pyrolysis.[11,45] Biosynthesis methods have been developed to overcome the environmental and stability issues associated with conventional synthesis methods. Biosynthesis allows to produce non-toxic and biodegradable nanoparticles in a green and environmentally friendly way.[46,47] Biosynthesis utilizes natural elements such as typically plants, fungi, or bacteria to produce stable nanoparticles as opposed to those produced by conventional means. The size and stability of the biosynthesized nanoparticles plays a crucial role in biomedical applications.[48]

Sol–gel synthesis method is a highly preferred bottom-up method due to its simplicity and ability to form a large quantity of nanoparticles. This method involves a colloidal solution of solid particles dispersed in a liquid (sol), and a macromolecule submerged in a liquid (gel).[49,50] The sol–gel method is known as the wet-chemical process as it involves a solution containing precursors such as metal oxides and chlorides. The dispersion of the precursors in a liquid is achieved either by mechanical forces (stirring, shaking) or sonication. The two-phase system undergoes phase separation from which the nanoparticles are isolated.[49–51] Phase separation is performed by various methods, such as sedimentation, filtration, and centrifugation followed by drying to remove any moisture.[45] The sol–gel method offers various advantages as it does not require high temperatures and results in the bulk production of pure nanoparticles.[52]

The spinning method utilizes a spinning disc reactor (SDR) that causes atoms or molecules to fuse and to form the nanoparticles.[53] The SDR contains an inert gas filled and temperature-controlled chamber with a rotating disc. Liquids containing the nanoparticle precursors and water are pumped onto the rotating disc at a controlled speed. As the disc spins, the atoms or molecules fuse forming the nanoparticles. The nanoparticles are precipitated, isolated, and dried. The advantage of this method is that the characteristics of the nanoparticles can be controlled by controlling the SDR operating parameters such as disc rotation speed, liquid flow rate, location of feed, disc surface, liquid/precursor ratio, etc.[54]

CVD is a chemical method for nanoparticle synthesis that involves the deposition of a thin film of gaseous reactants onto a substrate.[55,56] The gaseous reactants used involves a single component or a mixture of components. The reaction takes place in a chamber containing the gas mixtures at ambient temperature and a substrate that is heated. Once the gases and the heated substrate come into contact, a chemical reaction occurs, producing a thin film of nanoparticles on the substrate which is then collected.[57] The advantage of this method is that the nanoparticles produced are pure with low polydispersity. However, the disadvantages are the high toxicity of the gaseous byproducts and the need for sophisticated equipment.

Pyrolysis is a widely used method to produce nanoparticles on a large scale. This method involves spraying a precursor solution in the form of tiny droplets onto a heat source. The heating of the droplets results in the evaporation of the solvent leading to the formation of nanoparticles.[58,59] Typically, in a large-scale nanoparticle production facility, the precursor solution is sprayed through a small opening at high pressure into a furnace. The byproduct gases of the ignition are then collected and filtered to gather the nanoparticles.[60] The main advantage of this method is the ability to produce large quantities of material, however, the main drawback is the use of high heat source.

This method is widely used to achieve crystallographic control over the nucleation and growth of noble-metal nanoparticles. The metal salts are chemically reduced with strong chemical reducing agents such as sodium borohydride, sodium citrate and hydrazine followed by surface modification using polymers or surfactants. Nanoparticles produced by this colloidal approach result in high shape polydispersity. The development of reliable experimental protocols for the synthesis of nanoparticles with definite chemical compositions and high monodispersity still remains a challenge.

5. Green Synthesis and Phytosynthesis of Metallic Nanoparticles

Synthesis of nanoparticles using natural materials as chemical reducing and stabilizing agents have been explored in recent years to overcome the toxicity issues associated with physical and chemical methods. The biosynthesis of nanoparticles is a bottom-up approach and the main chemical reaction in this approach is chemical reduction/oxidation and surface modification (covalent and non-covalent interactions). Bacteria, actinomycetes, fungi and plants have been used in the biosynthesis of nanoparticles. As organic solvents are not used in this method, biosynthesis is regarded the "green" and environmental-friendly method. The general concept of green synthesis of nanoparticles is illustrated in Fig. 5.

In this section, the synthesis and experimental protocols of a few important metallic nanoparticles using extracts of plants (phytosynthesis) as chemical reducing and stabilizing agents is presented and the process of phytosynthesis is illustrated in Fig. 6.

Plants contain a wide variety of metabolites, readily available and are safe to handle. These factors make them the material of choice in the synthesis of nanoparticles. A wide variety of plants have been investigated for their role in the synthesis of nanoparticles through chemical reduction and stabilization processes. The water-soluble phytochemicals present in plants reduce the metal ions to metal nanoparticles in a much shorter time as compared to

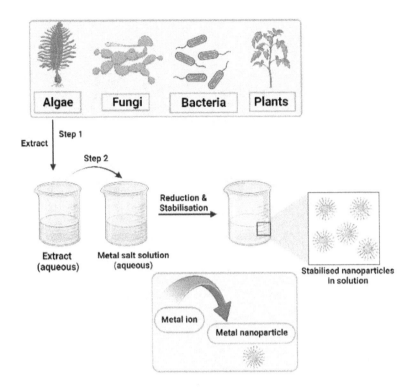

Fig. 5. Illustration of general concept of green synthesis using natural resources.

fungi and bacteria. Both fungi and bacteria require longer incubation time for the reduction of metal ions to metal nanoparticles.[61] By combining the techniques of plant tissue culture and downstream processing procedures synthesis of metallic and metallic oxide nanoparticles on an industrial scale could be made possible.

By using infrared (IR) spectroscopy, the main phytochemicals responsible for immediate chemical reduction have been identified as terpenoids, flavones, ketones, amide, carboxylic acids, and aldehydes. The phytochemicals present in xerophytes (*Bryophyllum* sp.), mesophytes (*Cyprus* sp.), and hydrophytes (*Hydrilla* sp.) have been studied for their role in the synthesis of silver nanoparticles. In the case of xerophytes, an anthraquinone viz. emodin has been identified as the main chemical constituent that is involved in tautomerization and formation of silver nanoparticles.

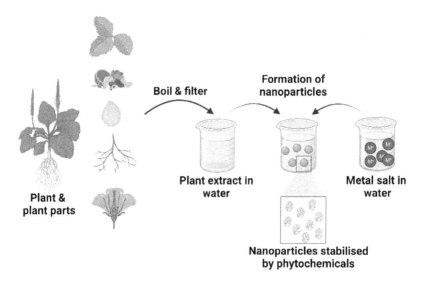

Fig. 6. Phytosynthesis of metal nanoparticles using plants. Original image created using Biorender.com.

Benzoquinones such as cyperoquinone, dietchquinone, and remirin are present in the mesophytes and these are involved in the chemical reduction of metal salts to metal nanoparticles. Hydrophytes contain catechol and protocatechualdehyde and these are converted into protocatechuic acid under alkaline conditions which play in the chemical reduction process.[61] The chemical structure of some important phytochemicals present in plants and the mechanism of chemical reduction of metal salts and stabilization of nanoparticles are shown in Fig. 7.

5.1. *Green synthesis of Titanium nanoparticles (TiO$_2$)*

The phytosynthesis of titanium nanoparticles has drawn immense research attention due to its quantum efficiency, chemical inertness and stability and has found applications in medicine and environmental remediation. The synthesis of TiO$_2$ nanoparticles on a large scale using green methods has gained momentum due to its low cost, and environmental friendliness. A wide range of plants and

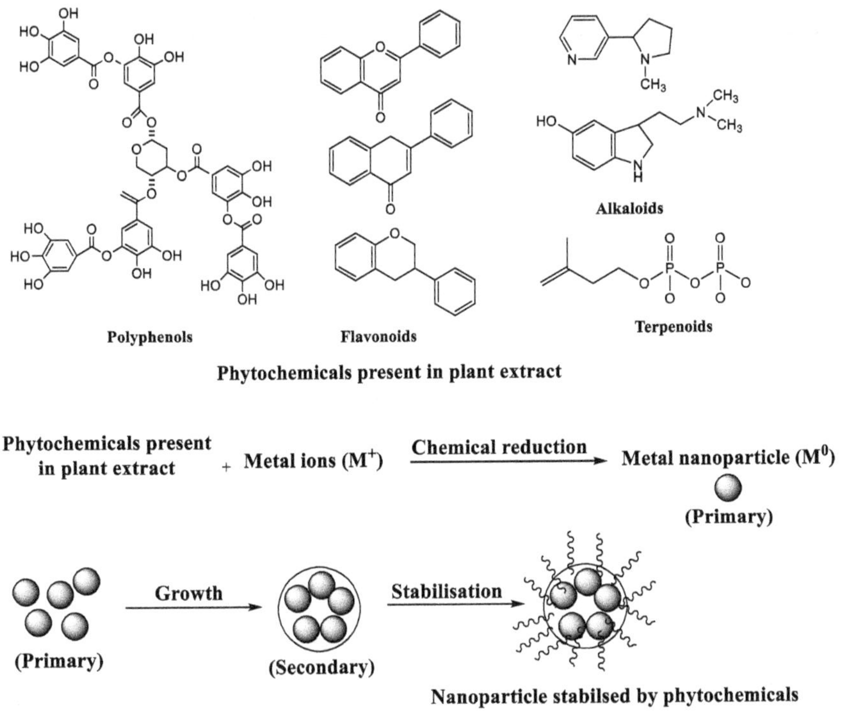

Fig. 7. Chemical structure of major phytochemicals present in plant extract and the mechanism of chemical reduction and stabilization processes.

their parts have been used in the synthesis of titanium nanoparticles of various shapes and sizes. The aerial parts of the gum Arabic tree (*Acacia nilotica*) have been used to synthesis TiO_2 nanoparticles for cancer therapy from tetra-butyl orthotitanate $(Ti(C_4H_9O)_4)$. The resulting nanoparticles were spherical in shape with a size range of 20–40 nm.[62] Extracts of African custard apple or wild soursop (*Ananos seneglensisa*) have also been used to synthesize titanium nanoparticles with a range of 20–40 nm from Titanium tetrachloride $(TiCl_4)$.[64] Interconnected and spherical nanoparticles with an average size of 120 nm for photocatalytic applications have been synthesized using neem leaves (*Azadirachta indica*).[65] The plant extract of Curcuma longa resulted in an anatase form of TiO_2 nanoparticles and the reduction is attributed to the high levels of

Table 1. Phytosynthesis of titanium nanoparticles.

Name of plant	Parts used	Size (nm)	Shapes	References
Acacia nilota	Leaves	10	Spherical	63
Azadirachta indica	Leaves	15–50	Spherical	65
Syzygium cumini	Leaves	22	Spherical round	68
Psidium Guajava	Leaves	32.58	Spherical	69
Mentha arvensis	Leaves	20–70	Spherical	70
Nyctanthes arbor-tristis	Leaves	100–150, 100–200	Cubic, crystalline, spherical	71
Solanum trilobatum	Leaves	70	Spherical, oval	72
Jatropha curcas & citrus aurantium	Leaves	25–50	Spherical	73
Ocimum basilicum	Leaves	50	Hexagonal	74
Cassia auriculata	Leaves	38	Spherical	75
Euphorbia prostrata	Leaves	81–84	Spherical	76
Ledebrouria revoluta	Bulb	47	Tetragonal	77
Calotopis gigantea	Flower	10–52	Crystalline, Spherical, oval	78
Hibiscus-rosa-sinensis	Petals	7–24	Spherical	79
Jatropha curcas	Latex	25–50	Spherical, uneven	80

terpenoid and flavonoid content of the plant. Leaves rich in alcohol and primary and aromatic amines such as Psidium guajava and Echinacea purpurea have also been used in the synthesis of TiO_2 nanoparticles for cancer therapy and photocatalysis.[66,67] A few important plants used in the synthesis of TiO_2 nanoparticles are summarized in Table 1.

5.2. *Platinum nanoparticles*

Platinum nanoparticles and complexes are widely used in anti-cancer drugs, oncological disease therapy and dentistry. It is used as a catalyst in various industries such as electroplating, coating, and jewelry. Synthesis of platinum nanoparticles from plants for various applications has been reported. Indian stinging nettle (*Tragia involucrate*) has been used to reduce hexachloroplatinic acid ($H_2PtCl_6 \cdot 6H_2O$) to platinum nanoparticles. The nanoparticles showed a strong surface plasmon resonance (SPR) peak at 252 nm and were spherical in shape with an average size of 10 nm.[66] Brahmi or water hyssop leaves (*Bacopa monnieri*) have also been used to synthesize platinum nanoparticles in the size range of 5–20 nm. A red shift in SPR peak wavelength was observed with an increase in the size of the nanoparticles.[67] Polyphenols present in plant leaf extracts are good chemical reducing and stabilizing agents. Tea leaves (*Camellia sinensis*), which are a rich source of polyphenols, have been used in the green synthesis of highly stable platinum nanoparticles in the size range of 30–60 nm.[81] Platinum nanoparticles synthesized using holy basil leaves (*Ocimum sanctum*) have been used in the electrolysis of water to produce hydrogen.

5.3. *Palladium nanoparticles*

Palladium nanoparticles have been synthesized using plant materials as a cost-effective and eco-friendly process. The various amounts of primary and secondary metabolites (phytochemicals) present in the plants reduce the palladium salt to palladium nanoparticles. Various parts of the plant such as leaves, seeds, flowers, peels, fruits, and roots have been explored in the synthesis of palladium nanoparticles and a few important ones are listed in Table 2. Spherical and porous palladium nanoparticles of size 100 nm have been synthesized using extracts of Chaga mushroom (*Inonotus obliquus*). The phytochemicals on the surface of the nanoparticles imparted near infrared features to the nanoparticles.[82] White tea extract (*Camellia sinensis*), and fern leaves extract (*Filicium decipiens*) have also been used in the synthesis of spherical palladium nanoparticles.[83,84]

Table 2. Phytosynthesis of palladium nanoparticles.

Name of plant	Part of plant	Size (nm)	Shape	References
Inonotus obliquus	Entire mushroom	100	Spherical	82
Filicium decipiens	Leaves	2–22	Spherical	84
Coleus amboinicus	Leaves	40–50	Spherical	85
Cinnamomum camphora	Leaves	3–6	Variable	86
Catharanthus roseus	Leaves	38	Spherical	87
Pulicariaglutinosa	Leaves	3–5	Spherical	88
Glycine max	Leaves	15	Spherical	89
Cinnamom zeylanicum	Bark	15–20	Crystalline	90
Musa paradisica	Peel	50	Crystalline	91
Curcuma longa	Tuber	10–15	Spherical	92

5.4. Silver nanoparticles

Due to their distinctive conducting and antimicrobial properties, silver nanoparticles have attracted much research interest in biomedical and industrial applications. In recent years, the synthesis of silver nanoparticles by green methods has gained importance as plants contain unique secondary metabolites such as alkaloids, sugars, phenolic acids, terpenoids, and flavonoids that are involved in the chemical reduction of silver salts and in-situ stabilization of silver nanoparticles. Pods of okra plant (*Abelmoschus esculentus*), olive leaves, and rock garden plant (*Onosma sericeum*) have been widely used in the synthesis of silver stable silver nanoparticles.[93–95]

Different parts of the plant contain different amounts of phytochemicals and this effect on the formation of silver nanoparticles was studied using extract of the bark and powder of *Curcuma longa*. It was found that the bark extract produced higher concentrations of silver nanoparticles than the powder. Some important

Table 3. Name and parts of plants used in the phytosynthesis of silver nanoparticles of various sizes and shapes.

Name of plant	Part of plant	Size (nm)	Shape	References
Mentha piperita	Leaves	35	Spherical	96
Acalypha indica	Leaves	20–30	Spherical	97
Morinda cirtrifolia	Leaves, fruit, seeds	3–11	Spherical	98
Nymphae odorata	Leaves	15	Spherical	99
Capparis zeylanica	Leaves	23	Spherical	100
Cycas	Leaves	2–6	Spherical	101
Tribulus terrestris	Fruit	16–28	Spherical	102
Tanacetum vulgare	Fruit	16	Spherical	103
Jatropha Curcas	Latex	10–20	Cubic	104
Rhododedendron-dauricam	Flower	25–50	Spherical	105

plants used in the synthesis of silver nanoparticles are listed in Table 3.

5.5. *Gold nanoparticles*

Gold nanoparticles have been widely used in gene expression, catalysis, nanoelectronics, non-linear optics and cancer therapy. Plants such as rock rose (*Cistus incancus*), sea beans (*Salicornia brachiate*), and Deccan hemp (*Hibiscus cannabinus*) have been used to synthesize spherical gold nanoparticles with a size range of 10–35 nm. The leaves of mangosteen tree (*Garcinia mangostana*) contain rich amounts of anthocyanins, benzophenones, phenols, and flavonoids. The extract of these leaves has been used to synthesize spherical gold nanoparticles with a size of 32 nm.

Triangular and hexagonal gold nanoparticles were synthesized using a dilute extract of the plant *Phyllanthus amarus* that contains phyllanthin as the active chemical reducing agent.[106] A wide variety

Table 4. Some important plants that are used in the phytosynthesis of gold nanoparticles.

Plant	Parts used	Size (nm)	Shape	References
Lawsonia inermis	Leaves	20	Spherical	107
Croton Caudatus Geisel	Leaves	20	Spherical	108
Aegle marmelos	Leaves	4–10	Spherical	109
Tamarind	Leaves	20–40	Triangle	110
Cinnamon	Bark	35	Spherical	111
Aloe vera	Plant Extract	50–350	Crystalline	112
Rosa hybrid	Rose petals	6–60	Anisotropic	113
Cuminum cyminum	Seeds	1–10	Spherical	114
Tanacetum vulgare	Fruit	11	Triangular	103
Nyctanthes arbortristis	Flower extract	20	Spherical, hexagonal	115

of plants used in the phytosynthesis of gold nanoparticles are summarized in Table 4.

5.6. *Copper nanoparticles*

Copper oxide nanoparticles exhibit anti-oxidant, anti-bacterial, and antimicrobial properties. The copper nanoparticles have antibacterial properties against common pathogens such as *Escherichia coli* and *Staphylococcus aureus*.[116] The decontaminating properties of copper and copper oxide nanoparticles have been widely reported. Copper nanoparticles synthesized using fruits of *Citrus medica* called as finger citron or Buddha's fingers exhibit excellent antimicrobial properties. The change in color from blue to yellow upon the addition of the fruit extract to 100 mm copper sulfate solution indicated the formation of copper nanoparticles. A strong SPR peak at 631 nm indicated the formation of the nanoparticle in the solution. The nanoparticles were spherical in shape with a size range of 10–60 nm.[117]

Copper nanoparticles in the size range of 10–50 nm have also been synthesized using the leaves of *Hegenia abyssinica* (African redwood). This plant widely found in many African countries is used in herbal medicines to treat intestinal parasites (tapeworms and ringworms) by the tribes.[118,119] The nanoparticles produced were polydisperse with a wide range of shapes such as spherical, cylindrical, triangular and prisms. Clove (*Syzgium aromaticum*) oil has also been used in the synthesis of copper nanoparticles of spherical nanoparticles with an average size of 20 nm.[120] Copper nanoparticles synthesized using the stem of *Euphoria nivulia* were reported to be toxic to human lung cancer cells (A549) showing promise in cancer therapy applications.[121]

5.7. Selenium nanoparticles

Selenium is an essential trace element involved in the normal functioning of immune systems, iodine and peroxide metabolism, and prevention of degenerative diseases.[4] Selenium deficiency disturbs the cellular equilibrium between oxidants and antioxidants that can alleviate the risks associated with oxidative stress, cardiac and muscular disturbances. Selenium nanoparticles have gained research interest in recent years as food additives and as anticancer agents.

Various plant species (extracts of root, shoot, bark, stem, flowers, and leaves) have been explored for the synthesis of selenium nanoparticles. The fruit extract of *Emblica officinalis* (Indian gooseberry) has been used to reduce sodium selenite (Na_2SeO_3) to selenium nanoparticles at room temperature. A change in color from colorless to brick-red, and the observance of SPR peak at 270 nm indicated the formation of selenium nanoparticles. The SPR peak shifted to longer wavelength with an increase in the size of the nanoparticle. The nanoparticles produced were mainly spherical in shape with an average size range of 15–70 nm.[116] Extracts of Hawthorn fruit, garlic (*Allivum sativum*), *Dillenia indica*, ginger (*Zingiber officinale*), latex of papaya (*Carica papaya*), lemon, and cinnamon barks (Cinnamomum zeylanicum) have also been used to synthesize selenium nanoparticles of spherical shape in the size range of 15–20 nm.[4]

5.8. Iron nanoparticles

Various types of iron and iron oxide nanoparticles (Fe-FeO) have been synthesized using various plants for applications in photocatalysis, medical imaging and degradation of toxic dyes in water. Iron nanoparticles have been synthesized using tea leaves extract, *Moringa oleifera* seeds, *Trigonella-foenum graecum* seeds, papaya leaves, and hibiscus flowers. The shape of nanoparticles formed using papaya leaf extract was highly polydisperse indicating the lower stabilizing effect of the phytochemicals present in papaya leaf. In all cases, a strong SPR peak at 420 nm indicated the formation of iron nanoparticles.[123,124]

6. Biomedical Applications of Nanoparticles

6.1. Titanium nanoparticles (TiO_2)

The current conventional cancer treatment methods include surgery, chemotherapy, and radiation.[125] These therapies are associated with considerable side-effects such as triggering of metastasis, mutations, and systemic side effects.[126-128] There is always a need for methods that minimize these side-effects, and using titanium oxide nanoparticles for cancer therapy has been explored.[93] These nanoparticles lead to the formation of oxidative stress and cytokine production in lung cancer cells (LL2) and colon cancer cells (CT26).[94] The anticancer activity of titanium oxide nanoparticles on two types of human colorectal cancer cells (HCT116 and HT29) showed dose-dependent and cell-dependent activities.[93]

The use of TiO_2 doped with silver, as an anticancer agent, synthesized using *Acacia nilotica*, has been demonstrated.[73] The nanoparticles were tested for anticancer activity against human breast cancer cell line (MCF-7). The MTT assay and a histogram showed a concentration-dependent cell viability of these nanoparticles.[129]

Further, the depletion of intracellular glutathione (GSH) clearly showed that the anticancer activity was due to oxidative stress (Karger-Basel hypothesis).[129,130] The oxidative stress mechanism of nanoparticles poses a risk as it can also damage healthy tissues. One way to limit the unwanted damage is by administering the nanoparticles intratumorally using a gel plug, however, this does not eliminate the risk.[131]

Bacterial infections in all forms are a huge threat to human health and global health care systems.[132] The conventional way to treat such infections is by antibiotics therapy. However, in recent years, the antibiotic resistance has increased due to the overuse and misuse of antibiotics. Antibiotic resistance further complicates the treatment of bacterial infections, and there is an urgent need to develop new methods of treatment.[133] Titanium oxide nanoparticles have been studied as potential bactericidal agents due to their oxidizing potential properties. TiO_2 produces reactive oxygen species (ROS) which results in damage of bacterial cell wall/membrane. This makes the bacteria osmotically fragile thereby leading to its destruction.[73,134] The antibacterial activity of TiO_2 nanoparticles synthesized using the extracts of *Hypheae thelbiecea* (fungi) and *Ananos seneglensisa* (African custard apple) showed much higher antibacterial activity against *Bacillus subtilis*, *Staphylococcus aureus*, and *Escherichia coli*, compared to the commercially available materials.[64] The results showed that the activity of TiO_2 nanoparticles was synergistic due to the various phytochemicals that decorate the surface of the particles.[135]

Titanium oxide nanoparticles have been exploited as photocatalysts in environmental remediation applications. The high surface-to-volume ratio of the nanoparticles plays a key role as it increases the surface area for photo-induced reactions. This enhances the rate of light absorption, increasing the density of the photo-induced carrier. This allows for increased photo-reduction and surface activity.[136] In the field of medicine, the properties of photocatalysis are utilized for disinfection of materials surfaces in hospitals.[137,138] Paints containing these nanoparticles were able to destroy a wide range of microorganisms using natural light. The TiO_2 nanoparticles developed using *Azadirachta indica* plant[65] were effective in the degradation of the dye, methyl orange. However, the photocatalysis was influenced by the size, shape, and charge of the nanoparticles.[65,138] The mechanism involved in photocatalytic disinfection is through lipid peroxidation, which deteriorates the integrity of the cell membrane of the microorganisms resulting in their death.[139]

6.2. *Platinum nanoparticles*

Platinum nanoparticles have been tested for possible bactericidal activity on both gram-positive and gram-negative bacteria. This increase in interest in these materials, as antibacterial agents is due to the rising bacterial resistance against conventional antibiotics.[140,141] The antibacterial mechanism of platinum nanoparticles is hypothesized due to the loss of cell membrane integrity, and the production of ROS.[66,142-144] These were confirmed by a study that detected the presence of cytosolic protein leakage following the administration of the nanoparticles due to damage of the cell membrane.[66]

The platinum nanoparticles synthesized using the plant, *Tragia involucrate* (Indian stinging nettle) showed concentration-dependent antibacterial activity against *S. aureus* and *E. Coli*.[145,146] Platinum nanoparticles synthesized using dates (Zahidi) also showed concentration-dependent antibacterial activity against *P. aeruginosa* and *S. pyogens*.[147]

The accumulation of ROS in cells leads to a phenomenon known as oxidative stress. ROS are normal byproduct of oxygen metabolism, and under normal conditions these species are removed through various biochemical reactions in the body. However, certain stress factors such as radiations, pollutants, and certain types of medicines cause an imbalance in the production and accumulation of ROS. Increased ROS accumulation results in the damage of cellular components.

Exogenous antioxidants are molecules that minimize the effects of ROS.[148-151] Recent studies have suggested the use of platinum nanoparticles as a viable antioxidant, that can reduce the accumulation of ROS.[66,148,152,153] Platinum nanoparticles synthesized using the plant, Bacopa monnieri (Brahmi) displayed good antioxidant properties on Zebrafish.[66] The photocatalytic property of the nanoparticle is responsible for its antioxidant property leading to the enhanced production of superoxide dismutase, catalase, and glutathione peroxidase.[66,154–156] On the contrary, studies have also shown that the antioxidant property of platinum nanoparticles is concentration-dependent, and high concentration can lead to disruption of normal cell functions and viability.[157]

Platinum-based cancer therapeutics, such as cisplatin, carboplatin, and oxaliplatin, have been used since the 1960s due to their anticancer effects on testicular and ovarian cancers.[158,159] However, the conventional platinum based anticancer drugs lack specificity in differentiating between the healthy and diseased cells.[152,154] The lack of specificity leads to systemic side effects.[160]

The PI-stained, Pt nanoparticles when treated with cells showed a high rate of nuclear fragmentation, confirming the induction of apoptosis. Furthermore, a cell cycle analysis was performed to assess the effects of the nanoparticles on the cell cycle using SiHa cells. Flow cytometry was used to analyze the cell cycle. The results showed a significant decrease in the proportion of cells in the G0/G1 and S-phase, in addition to a significant increase in the proportion of cells in the G2/M phase. This finding indicated that the nanoparticles induce cell cycle arrest caused by damage to the DNA.

Bendale et al.[127] conducted a similar study on various cancer cell lines such as human lung adenocarcinoma, ovarian teratocarcinoma, and pancreatic cancer cells. The results showed significant inhibition of the cancer cells, with the highest effect on the ovarian teratocarcinoma cells. Furthermore, they also studied the cytotoxic effect of nanoparticles on normal cells. The platinum nanoparticles were ︎d with peripheral blood mononuclear cells (PBMC) at a ︎tion (200 μg/mL) for 48 h. The test showed no signifi-effects of the PBMCs, thus suggesting the selectivity of platinum nanoparticles for cancer cells. The exceptional characteristics of these nanoparticles in cancer therapy shown by these studies make these materials as choice materials in oncology.[161]

6.3. Palladium nanoparticles

Photothermal therapy (PTT) has recently been spotlighted as a promising application in cancer treatment due to it being minimally invasive, highly selective, and target-specific.[162] PTT utilizes light-absorbing nanoparticles that convert photon energy to thermal energy leading to thermal ablation of cancer cells.[162,163] The nanoparticles embedded in tumor are irradiated by a near-infrared laser (NIR) with a wavelength range of 750–2500 nm. To achieve the effects of photothermal ablation, the nanoparticles must have a maximal absorption wavelength ranging in the near-infrared light range.[162,164]

Palladium nanoparticles are excellent candidates for PTT due to their high absorption in the NIR region and high photothermal conversion efficiency. The exceptional chemical and physical properties of these nanoparticles, in addition to their size heterogeneity, have been exploited in the photothermal ablation of cancer cells.[163,166,167] The nanoparticles were infused with human lung adenocarcinomas for 48 h and exposed to NIR with a wavelength of 808 nm at an intensity of 4 W/cm^2 for 20 min. At the end of the exposure time, the temperature increased significantly by 25°C. The laser irradiation was followed by fluorescent microscopy, which revealed a localized ablation of the cancer cells.[161] Similar results were achieved in a study that utilized palladium nanoparticles synthesized using the Chaga mushroom (*Inonotus obliquus*).[82] The presence of the surface coating (due to phytochemicals of the mushroom extract) is of great importance as it increases cancer cell selectivity, increases the porosity of the nanoparticles, and increases the NIR absorbance. The photothermal conversion effect of the nanoparticles was tested on HeLa cervical cancer cells. The cells were incubated with 40 μg/ml of palladium nan irradiated using an 808 nm NIR diode laser, which is widely used in biomedical applications due to its high tissue penetration.[82]

The irradiation was performed at 4 W/cm^{-2} for 5 min, followed by fluorescence microscopy that confirmed the photoablation of the cancer cells. Despite the advantages, PTT also has its limitations in cancer treatment, as it is restricted by the depth of the tumor. The therapy utilizes light to cause the ablation of cancer cells, and the depth of the cancer cells limits the light penetration deep into the tumors.[136] In addition, the use of palladium-based nanoparticles poses a slight risk, as unbound palladium is toxic to humans.[167] Furthermore, the chaga mushroom-based palladium nanoparticles exhibit cancer cell selectivity that depends on the amount of surface coatings. Interestingly, the nanoparticles with no coatings of the phytochemicals were significantly cytotoxic to MCF-10A (normal cell line).[168]

Recent studies showed promising potential for palladium nanoparticles as antiproliferative agents on cancer cells.[163] The nanoparticles were successful anti-cancer agents and induced apoptosis of cancer cells without affecting the normal cells.[83,163] Azizi et al.[83] utilized palladium nanoparticles synthesized using white tea (Camellia sinensis) extract, as an anticancer agent on the MOLT-4 cell line. MOLT-4 cells are a human lymphocytic leukemia cell line.[169] The study showed that incubation of the MOLT-4 cells with the green synthesized nanoparticles resulted in apoptosis leading to damage to DNA and arrest of G2/M cell cycle. The exact mechanism of action of palladium nanoparticles as anti-cancer agent is still debated; however, it is suggested that the nanoparticles exhibit a physicochemical interaction with the functional groups of cellular proteins, nitrogen bases, and phosphate groups of the DNA, thus inducing apoptosis.[83]

To further confirm that palladium nanoparticles are responsible for inducing apoptosis, a MOLT-4 cells sample was incubated with the green synthesized nanoparticles and neat white tea extract. The results showed that the MOLT-4 cells incubated with the nanoparticles had a decreased population after 48 h, while the cells sample incubated with neat white tea extract showed no decrease in population, confirming the anti-cancer effect of the nanoparticles.[83] The selectivity of the palladium nanoparticles towards cancer cells is theorized to be due to the phytochemicals, in particular, flavonoids present in the white tea extract.[23,83] The study concluded that palladium nanoparticles are 3000 times more toxic to cancer cells than conventional doxorubicin, however, this does not have any clinical evidence.[77]

Delivery of therapeutics in traditional drug delivery systems suffers from many issues such as poor solubility of drugs, weak targeting capabilities, nonspecific distribution, systemic toxicity, and low therapeutic index.[29,170] These uses can be improved using nanoparticles as drug delivery devices/systems to overcome these issues and to improve the bioavailability and biodistribution of therapeutics.[170-172] The unique properties of nanoparticles encourage extensive research on the use of nanoparticles for targeted delivery of anticancer drugs.[173,174] PEGylation of nanoparticles using poly(ethylene glycol) has been studied in the development of biocompatible palladium-based nanoparticles for the delivery of doxorubicin (DOX).

A drug release profile test was performed *ex-vivo* under different pH conditions, pH 7.4 (blood pH), pH 6.8 (tumor tissue pH), and pH 5.5 (pH of mature endosomes in a tumor cell). The results indicated that the drug release was significantly higher at pH 5.5, followed by a slightly lower release at pH 6.8, and lowest at pH 7.4. The drug release ratio was 89%, 68%, and 19%, respectively. The pH sensitive, drug release profile allows for targeted drug delivery into cancer cells, minimizing systemic side effects.[174]

The DOX conjugated PEGylated palladium nanoparticles were tested in mice carrying the cervical carcinoma HeLa cells.[48,160] The mice were divided into four groups, the first group was injected with DOX-palladium nanoparticles, the second group was injected with free DOX, the third group was injected with PEGylated palladium nanoparticles, and the fourth group was injected with saline solution. The tumor volume was recorded over the course of 18 days. The mice injected with normal saline and PEGylated palladium nanoparticles showed no tumor growth suppression, and the tumors had a volume of 1102.27–1050.03 mm^3, respectively.

On the other hand, tumor growth suppression was recorded in the groups that were injected with free DOX and DOX-palladium nanoparticles. Tumor volumes were 500.02 mm^3 and 164.74 mm^3, respectively.[174] The results proved that using palladium nanoparticles for the delivery of anticancer drugs is due to its high specificity with cancer cells, high drug releasing capabilities, and minimal side-effects. In another study,[82] porous spherical palladium nanoparticles synthesized from chaga mushrooms were used to deliver doxorubicin. The presence of phytochemicals of the mushroom on the surface of the palladium nanoparticles increased their specificity.

The incubation was performed at room temperature and in the dark to eliminate any unwanted reactions. After incubation, the mixture was centrifuged at 8000 rpm for 15 min to remove the unbound doxorubicin. Maximum DOX loading of about 110 μg was achieved after 6 h of incubation. The exceptional DOX loading capability of the palladium nanoparticles was attributed to the electrostatic interactions between the positively charged doxorubicin and negatively charged nanoparticle surface. The release profile of the DOX-palladium nanoparticles was evaluated in two solutions at pH 7.4 and 5.6. An average release profile of 93.6% at pH 5.6 and 29% at pH 7.4 was observed. The higher release of the drug in an acidic environment is more beneficial and can lead to the destruction of cancer cells.[82,168]

Furthermore, the drug-loaded nanoparticles were infused in the HeLa cells to observe the intracellular distribution of the loaded nanoparticles by using fluorescence microscopy. After 2 h of incubation, signals were observed in the cytoplasm and perinuclear area, and after 6 h, signals were observed within the nuclei, and after 12 h of incubation, strong signals were observed from within the nucleus and cytoplasm. These results clearly indicated a sustained release profile in comparison to the free drug.[82]

6.4. Silver nanoparticles

Silver nanoparticles have demonstrated promising activity in tackling gram-positive (*S. aureus*) and gram-negative (*K. pneumoniae, E. coli*) bacteria. The exact action of action is till debatable, however, one accepted mechanism (specifically for *E. coli*) is through increased permeability of the cell membrane by silver nanoparticles and thereby impeding the transport throughout the membrane, eventually leading to the destruction of the bacterial cell. Other mechanisms are related to the intercalation and condensation of DNA bacteria which hinders cell proliferation.[20,175]

Three possible antibacterial mechanisms of silver nanoparticles have been outlined by Kuberan[176,177] as (i) inhibition of bacterial cell membrane synthesis by silver ions, (ii) disruption of thiol groups of bacterial cell membrane by silver nanoparticles, and (iii) penetration of silver ions into the cell leading to DNA damage and obstructed protein synthesis. Morones *et al.*[178] suggested that silver behaves as soft acid and interferes with the sulfur and phosphorus bases of the DNA thereby inhibiting its replication.

The antibacterial activity of silver nanoparticles was examined by agar well diffusion assays and measuring the zones of inhibition.[175] In this study, silver nanoparticles were tested against three gram-positive bacteria (*Bacillus subtilis, Staphylococcus aureus and Streptomyces griseous*) and one gram-negative bacteria (*E. Coli*). The zones of inhibitions (in mm) were 22 ± 0.5 for *Bacillus subtilis*, 30 ± 0.5 for *E. Coli*, 34 ± 0.5 for *Staphylococcus aureus*, and 16 ± 0.5 for *Streptomyces griseous*.

Silver nanoparticles synthesized using plant extracts showed excellent antibacterial and antifungal properties. The effect was more synergistic due to the phytochemicals on the surface of the nanoparticles.[185] However, an opposite effect showing a decrease in antibacterial activity was demonstrated by Kim *et al.*[180] The subtle differences are attributed to the differing nature of bacterial colonies (genetic differences) and the type of plant used.[181]

Anticancer properties of silver nanoparticles have not been widely investigated, however, there exist a few reports.[182] The anticancer properties of green synthesized nanoparticles at various concentrations were studied using MCF7 cell lines. Based on the percentage cell viability and IC50 values, it was shown that silver nanoparticles exhibit concentration-dependent cytotoxicity, through the generation of ROS leading to oxidative stress.[183] However, this effect was observed after 48 h of reaction time. The size of silver nanoparticles was also found to affect the cytotoxicity, and particles in the size range of 10–30 nm exhibited the highest value.

6.5. Gold nanoparticles

Gold nanoparticles are known to exhibit excellent antibacterial and anticancer activities.[184] Phytosynthesized gold nanoparticles have been tested against several pathogenic bacteria such as *Pseudomonas aeruginosa*, *Salmonella typhi*, *Escherichia coli*, and *Staphylococcus aureus*. The variations in antibacterial activity against these pathogens are theorized to be due to differences in bacterial cell wall composition and the surface coverage of nanoparticles with phytochemicals. The formation of reactive oxygen species is one factor associated with the antibacterial activity of these nanoparticles. The accumulation of reactive oxygen species causes oxidative stress and release of the enzyme lactate dehydrogenase within bacterial cells leading to its death.[185]

Gold nanoparticles are considered an excellent candidate in cancer treatment, due to their unique features such as fluorescence, high surface coverage, conductivity, variable shape and size. These features have been utilized to target specific cells in cancer diagnosis and therapy.[186,187] The anticancer potential of gold nanoparticles against lung (A549) and liver cancer cells (Hep-G-2) has been reported.[188]

The cancer cells were treated with various concentrations of gold nanoparticles (1, 10, 25, 50, and 100 μg/ml) for 24 h. The percentage of cell viability was recorded by the MTT assay. The approximate percentage cell viability for the HepG-2 cell line after treatment was 95%, 89%, 76%, 75%, and 48% for each respective concentration. As for the A549 cell line, the approximate percentage of cell viability was 95%, 90%, 87%, 77%, and 50%. The concentration-dependent anticancer activity of gold nanoparticles using plant extracts is demonstrated in this study. Oxidative stress, cytokine release, mitochondrial toxicity, DNA damage, apoptosis and necrosis are some of the mechanisms that are related to the anticancer effect of gold nanoparticles.[189–192]

6.6. Copper nanoparticles

In recent years, antimicrobial resistance poses a significant threat in healthcare as it leads to increased medical costs, prolonged hospital admissions, and increased mortality. This issue continues to grow due to misuse, and overuse of conventional antimicrobials.[133] This calls for an urgent intervention either to reduce the growth of resistance or develop new antimicrobials against multi-drug resistant pathogens. Studies have shown that copper nanoparticles are good antimicrobial materials against many species of pathogens. Hindawi et al.[118] investigated the effect of phytosynthesized copper nanoparticles against two gram-positive bacteria (S. aureus and B. subtilis) and two gram-negative bacteria (E. coli and P. aeruginosa). The results showed excellent antibacterial effect of copper nanoparticles in comparison to ampicillin which was used as the positive control. The copper nanoparticles are believed to strongly adsorb onto the surface of the microbe thereby destabilizing the cell membrane.[193] In addition, copper ions undergo redox reactions on the surface of certain microbes, which lead to the generation of hydrogen peroxide that damages the cell membrane and causes cell death.[194]

The binding capacity of nanoparticles containing drugs to receptors on cancer cells is one of the crucial factors in cancer therapy.[195] Copper nanoparticles containing phytochemicals on the surface have been investigated for anti-cancer and cytotoxicity against MCF-7 breast cancer cells.[196,197] The study revealed an inhibition of the growth of the cancer cells by more than 93%, with an IC_{50} value of 61.25 μg.ml^{-1}. The inhibition was associated with DNA-fragmentation that leads to cell death. The copper nanoparticles were also effective against human colon cancer cells (Caco-2) and human hepatic cancer cells (HepG2).[198,199] The MTT assay results revealed a concentration-dependent growth inhibition of the cancer cells. These studies indicate that phytosynthesized copper nanoparticles are synergistically effective against cancer cells.

6.7. Selenium nanoparticles

The antioxidant properties of selenium nanoparticles are dependent on the size, shape, and surface coating of the nanoparticles. Selenium plays a major role in the production of glutathione peroxidase which is a free-radical scavenger.[200–203] The selenium nanoparticles synthesized using the fruit extract of Indian gooseberry (E. officinalis) exhibited excellent antimicrobial and antioxidant properties.[122] The antioxidant potent of phytosynthesized selenium nanoparticles was much higher in comparison to ascorbic acid various free-radical scavenging assays.

A recent study evaluated the anticancer properties of phytosynthesized selenium nanoparticles against human hepatic cancer cells (HepG2).[204] HepG2 cells are characterized by an epithelial-like morphology with high proliferation rates. In this study, the anticancer effects of selenium nanoparticles synthesized using hawthorn fruit extract were evaluated by MTT assays and flow cytometry. To examine the anti-proliferative effects of nanoparticles using the MTT assays, HepG2 cells were cultured in 96-well microplates at a concentration of 1×10^5 cells per well for 24 h. The cells were incubated with the nanoparticles at concentrations of 5, 10, 20, and 40 µg/mL, for 24 h. The results revealed promising antiproliferative effects, wherein proliferation was inhibited by 50% after 24 h of incubation with nanoparticles.

Furthermore, the apoptotic potential of selenium nanoparticles was evaluated using the Annexin V-FITC/PI staining assay (a type of flow cytometry). The cell cultures were then incubated for 24 h with selenium nanoparticles at concentrations of 5, 10, and 20 µg/mL. The results showed significant rates of apoptosis, thus confirming that the anticancer properties of selenium nanoparticles are associated with the induction of apoptosis.[204]

6.8. Iron nanoparticles

Iron nanoparticles synthesized using papaya plant leaves showed dose-dependent excellent antibacterial properties against *Klebsiella spp.*, *E. coli spp.*, *Pseudomonas spp.*, and *S. aureus*. The antibacterial activity of iron nanoparticles is attributed to the formation of reactive oxygen species which causes oxidative stress leading to cell death.[205,207] The electrostatic interactions between the bacterial cell wall and nanoparticles also play a role in the destruction of the peptidoglycan cell wall of the bacteria. In addition to antibacterial properties, the iron nanoparticles also exhibited cytotoxic properties. The cytotoxic effect was studied using the human cervical cancer cell line (BHK-21) and Vero cells (a non-tumorigenic kidney epithelial cell line derived from the African green monkey). The phytosynthesized iron nanoparticles were highly cytotoxic and selective to the BHK-21 cells lines with a survival rate of 5%, after 48 h of incubation. The results indicate that phytochemicals-coated iron nanoparticles exhibit anticancer properties causing oxidative stress within the cells.[208,209]

The biomedical applications of the various nanoparticles synthesized using green methods are schematically illustrated in Fig. 8.

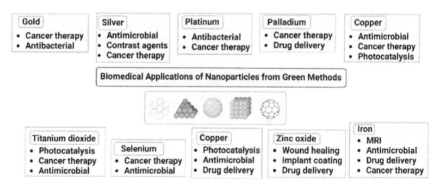

Fig. 8. Summary of biomedical applications of nanoparticles synthesized using green methods.

7. Major Challenges and Future Perspectives

Numerous studies have been reported on the phytosynthesis of metal nanoparticles for a wide range of biomedical applications. The large-scale production of nanoparticles using plant extracts is challenging due to the optimization of reaction processes. The size and shape of nanoparticles produced using plant extracts depend on optimized process parameters such as pH, temperature, and stirring speed. If these parameters are not optimized, the resulting nanoparticles might not have the properties for the intended applications. The influence of each metabolite of plant extract and their toxicity should be analyzed completely before the scale-up of nanoparticle production. The separation and purification of nanoparticles from large-scale reactors are not an energy-consuming process.

The futuristic approach is to address the major challenges and optimize the reaction processes such that the method becomes comparable to or better than conventional methods. This requires a strong interdisciplinary collaboration between academia and industries in basic and engineering research. Further, metabolite analysis and toxicology protocols need to be developed and updated constantly for process optimization and to keep in line with advances in technology. If these steps are undertaken carefully, the phytosynthesis of nanoparticles will become a sustainable approach leading to enormous economic opportunity. These important considerations are summarized in Fig. 9.

8. Conclusion

In the past decade, nanotechnology has become a highly attractive research niche in the biomedical field. Nanoparticles and nanomaterials have proven to possess a vast array of properties that can be utilized for targeted biomedical applications such as cancer diagnosis and therapy, antimicrobials, photocatalysis, photothermal therapy, antioxidants, and imaging. Phytosynthesis is a cost-effective, non-toxic, and environmentally friendly method that is used to

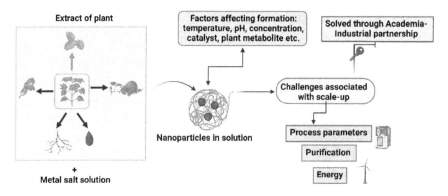

Fig. 9. Concept map of phytosynthesis and associated challenges for large-scale synthesis.

synthesize nanoparticles using extracts from plants. Various metals are used to create nanoparticles, commonly titanium, platinum, palladium, silver, gold, copper, selenium, and iron. Currently, nanoparticles are yet to be used extensively in medicine as it is still a relatively new discipline. Preliminary studies suggest the high potential and efficacy of nanoparticles in medicine; however, further extensive research is needed to safely start implementing them in human trials.

References

1. R. P. Feynman, *Eng. Sci.*, 1960, **23**, 22–36.
2. D. Nath and P. Banerjee, *Env. Toxicol. Pharmacol.*, 2013, **36**, 97–1014.
3. F. Khan, M. Sharif, M. Asif, M. Siddiqui, P. Malan and F. Ahmad, *Nanomaterials*, 2022, **12**, 673–691.
4. M. Ikram, B. Javed, N. I. Raja and Z. R. Mashwani, *Int. J. Nanomed.* 2021, **16**, 249–268.
5. S. Z. Karazhavanov and P. Rareendram, *J. Am. Chem. Soc.*, 2003, **125**, 13940–13945.
6. C. Bvzea, I. Pacheco-Blandino and K. Robbbie, *Biointerphases*, 2007, **2**, MR17–MR172.
7. T. A. Salah, *Env. Tech. Innov.*, 2020, **20**, 101067.
8. T. Min *et al.*, *Nature Mat.* 2021, **20**, 93–99.

9. Z. Dong, T. Wang, X. Chi, J. Ito, C. Tserkezis, S. L. K. Yap, A. Rusydi, F. Tjiptoharsono, D. Thian, N. A. Mortensen and J. K. W. Yang, *Nano Lett.*, 2019, **19**, 8040–8048.
10. S. S. Kruk, L. Wang, B. Sain, Z. Dong, J. Yang, T. Zentgraf and Y. Kivshar, *Nature Photonics*, 2022, **16**, 561–565.
11. P. Li, S. Chen, H. Dai, Z. Yang, Z. Chen, Y. Wang, Y. Chen, W. Peng, W. Shan and H. Duan, *Nanoscale*, 2021, **13**, 1529–1565.
12. C. Jung, G. Kim, M. Jeong, J. Jang, Z. dong, T. Badloe, J. K. W. Yang and J. Rho, *Chem. Rev.*, 2021, **121**, 13013–13050.
13. K. Salaita, Y. Wang and G. A. Mirkin, *Nature Nanotech.*, 2007, **2**, 145–155.
14. K. M. Krishnan, *IEEE Trans.*, 2010, **46**, 2123–2158.
15. A. Barhoum, M. L. Garcia-Betancourt, J. Jeevanandam, E. A. Hussien, S. A. Mekkawy, M. Mostafa, M. M. Omran, M. S. Abdullla and M. Bechelany, *Nanomaterials*, 2022, **2**, 177–194.
16. I. Khan, K. Saeed and I. Khan, *Arab. J. Chem.*, 2019, **12**, 908–931.
17. A. M. Ealias and S. M P, *IOP Conf. Ser. Mat. Sci. Eng.*, 2017, **263**, 032019.
18. F. Yano, A. Hiraoka, T. Itoga, H. Kojima, K. Kanehori and Y. Mitsui, *Appl. Surf. Sci.* 1996, **100**, 138–142.
19. S. S. Dhilip Kumar and H. Abrahamse, *Int. J. Mol. Sci.* 2020, **21**, 6752–2766.
20. R. Kaiser and G. Miskolczy, *J. Appl. Phys.* 1970, **41**, 1064–1072.
21. C. P. Bean and J. D. Livingstone, *J. Appl. Phys.* 1959, **30**, 120–129.
22. O. Sublemontier, C. Nicolas, D. Aureau, M. Patanen, H. Kintz, X. Liu, M.-A. Gaveau, J.-L. Le Garrec, E. Robert, F.-A. Barreda, A. Etcheberry, C. Reynaud, J. B. Mitchell and C. Miron, *Phys. Chem. Lett.*, 2014, **5**, 3399–3403.
23. Á. I. López-Lorente and B. Mizaikoff, *Trends in Anal. Chem.*, 2016, **84**, 97–106.
24. J. Kiefer, J. Grabow, H.-D. Kurland and F. A. Müller, *Anal. Chem.*, 2015, **87**, 12313–12317.
25. M.-I. Baraton and L. Merhari, *Proceedings of SPIE*, Boston, United States, 2007.
26. C. J. Barnett, E. Gowenlock, A. O. White and A. R. Barron, *Nanoscale Adv.*, 2021, **3**, 631–640.

27. G. Bystrzejewska-Piotrowska, J. Golimowski and P. L. Urban, *Waste Manag.*, 2009, **29**, 2587–2595.
28. U. N. E. Programme, *United Nations Environement Programme, Chemicals Branch*, 2010.
29. H. Ali, E. Khan and I. Ilahi, *J. Chem.*, 2019, **3**, 1–14.
30. I. Ijaz, S. Ezaz Gilani, A. Nazir and A. Bukhari, *Green Chem. Lett. Rev.*, 2020, **13**, 59–81.
31. A. Pimpin and W. Srituravanich, *Eng. J.*, 2012, **16**, 37–56.
32. J. C. Hulteen, D. A. Treichel, M. T. Smith, M. L. Duval, T. R. Jensen and R. P. Van Duyne, *J. Phys. Chem. B*, 1999, **103**, 3854–3863.
33. E. Grabowska, M. Marchelek, M. Paszkiewicz-Gawron and A. Zaleska-Medynska, *Metal Oxide-Based Photocatalysis*, ed. A. Zaleska-Medynska, Elsevier, 2018, pp. 51–209.
34. T. Yadav, R. M. Yadav and D. Singh, *Nanosci. Nanotechnol.*, 2012, **2**, 22–48.
35. R. Arbain, M. Othman and S. Palaniandy, *Miner. Eng.*, 2011, **24**, 1–9.
36. G. Gorrasi and A. Sorrentino, *Green Chem.*, 2015, **17**, 31–39.
37. A. Odularu, *Bioinorg. Chem. Applic.*, 2018, **20**, 1–6.
38. M. Salavati-Niasari, F. Davar and N. Mir, *Polyhedron*, 2008, **27**, 3514–3518.
39. T. Huy, N. Quy and A.-T. Le, *Adv. Nat. Sci. Nanosci. Nanotechnol.*, 2013, **4**, 033001.
40. M. Lalia-Kantouri, *J. Therm. Anal. Calor.* 2005, **82**, 75–82.
41. P. Shah and A. Gavrin, *J. Magn. Magn. Mat.*, 2006, **301**, 118–123.
42. E. Lugscheider, S. Bärwulf, C. Barimani, M. Riester and H. Hilgers, *Surf. Coat. Technol.*, 1998, **108–109**, 398–402.
43. V. Amendola and M. Meneghetti, *Phys. Chem. Chem. Phys.*, 2009, **11**, 3805–3821.
44. M. Kim, S. Osone, T. Kim, H. Higashi and T. Seto, *KONA Powder Part. J.*, 2017, **34**, 80–90.
45. R. C. Thiruvalam, J. C. Pritchard, N. Dimitratos, J. A. Lopez-Sanchez, J. K. Edwards, A. F. Carley, G. J. Hutchins and C. J. Keily, *Faraday Discuss.*, 2011, **152**, 63–86.
46. N. Sheoran and P. Kaur, *Biotechnol. Res. Inno.*, 2018, **2**, 45–55.
47. P. Kuppusamy, M. M. Yusoff, G. P. Maniam and N. Govindan, *Saudi Pharm. J.*, 2016, **24**, 473–484.

48. N. Rabiee, M. Bagherzadeh, M. Kiani and A. M. Ghadiri, *Adv. Powder Technol.* 2020, **31**, 1402–1411.
49. S. Ramachandran and R. Sivasamy, *Int. J. Nanosci.*, 2018, **17**, 1760008.
50. A. J. Kora and L. Rastogi, *Arab. J. Chem.* 2018, **11**, 1097–1106.
51. S. Mann, S. L. Burkett, S. A. Davis, C. E. Fowler, N. H. Mendelson, S. D. Sims, D. Walsh and N. T. Whilton, *Chem. Mat.*, 1997, **9**, 2300–2310.
52. M. D'Arienzo, R. Scotti, B. Di Credico and M. Redaelli, *Stud. Surf. Sci. Catal.* 2017, **177**, 477–540.
53. S. Mohammadi, A. Harvey and K. V. K. Boodhoo, *Chem. Eng. J.*, 2014, **258**, 171–184.
54. C. Tai, C.-T. Tai, M.-H. Chang and H.-S. Liu, *Ind. Eng. Chem. Res*, 2007, **46**, 5536–5541.
55. L. Xia, *Bioceramics*, ed. A. Osaka and R. Narayan, Elsevier, 2021, pp. 5–24.
56. S. Bhaviripudi, E. Mile, S. A. Steiner, A. T. Zare, M. S. Dresselhaus, A. M. Belcher and J. Kong, *J. Am. Chem. Soc.*, 2007, **129**, 1516–1517.
57. M. Adachi, S. Tsukui and K. Okuyama, *Jap. J. App. Phys.*, 2003, **42**, 4–7.
58. H. K. Kammler, L. Mädler and S. E. Pratsinis, *Chem. Eng. Technol.*, 2001, **24**, 583–596.
59. H. Zhang and M. T. Swihart, *Chem Mat.*, 2007, **19**, 1290–1301.
60. R. D'Amato, M. Falconieri, S. Gagliardi, E. Popovici, E. Serra, G. Terranova and E. Borsella, *J. Analy. App. Pyrol.*, 2013, **104**, 461–469.
61. D. Patra and R. El Kurdi, *Green Chem. Lett. Rev.* 2021, **14**, 474–487.
62. A. K. Jha, K. Prasad and K. Prasad, *Biochem. Eng. J.*, 2009, **43**, 303–306.
63. T. N. Rao, Riyazuddin, P. Babji, N. Ahmad, R. A. Khan, I. Hassan, S. A. Shahzad and F. M. Husain, *Saudi J. Biol. Sci.*, 2019, **26**, 1385–1391.
64. H. Hassan, K. I. Omoniyi, F. G. Okibe, A. Nuhu and E. G. Echioba, *J. Appl. Sci. Environ. Manag.*, 2019, **23**, 1795–1804.
65. R. Sankar, K. Rizwana, K. S. Shivashangari and V. Ravikumar, *App. Nanosci.*, 2015, **5**, 731–736.

66. A. M. Selvi, S. Palanisamy, S. Jeyanthi, M. Vinosha, S. Mohandoss, M. Tabarsa, S. You, E. Kannapiran and N. M. Prabhu, *Process Biochem.*, 2020, **98**, 21–33.
67. J. Nellore, C. Pauline and K. Amarnath, *J. Neurodegener. Dis.*, 2013, **2013**, 972391.
68. N. K. Sethy, Z. Arif, P. K. Mishra and P. Kumar, *Green Process. Synth.*, 2020, **9**, 171–181.
69. T. Santhoshkumar, A. A. Rahuman, C. Jayaseelan, G. Rajakumar, S. Marimuthu, A. V. Kirthi, K. Velayutham, J. Thomas, J. Venkatesan and S.-K. Kim, *Asian Pac. J. Trop. Med.*, 2014, **7**, 968–976.
70. W. Ahmad, K. K. Jaiswal and S. Soni, *Inorg. Nano-Metal Chem.*, 2020, **50**, 1032–1038.
71. S. Basu, P. Maji and J. Ganguly, *Appl. Nanosci.*, 2015, **6**, 1–5.
72. G. Rajakumar, A. A. Rahuman, C. Jayaseelan, T. Santhoshkumar, S. Marimuthu, C. Kamaraj, A. Bagavan, A. A. Zahir, A. V. Kirthi, G. Elango, P. Arora, R. Karthikeyan, S. Manikandan and S. Jose, *Parasitol. Res.*, 2013, **113**, 469–479.
73. A. C. Nwanya, P. E. Ugwuoke, P. M. Ejikeme, O. U. Oparaku and F. I. Ezema, *Int. J. Electrochem. Sci.*, 2012, **7**, 11219–11235.
74. H. A. Salam, R. Sivaraj and R. Venckatesh, *Mater. Lett.*, 2014, **131**, 16–18.
75. D. A. C. Mohan, *World J. Pharm. Res.*, 2017, **7**, 1058–1065.
76. A. A. Zahir, I. S. Chauhan, A. Bagavan, C. Kamaraj, G. Elango, J. Shankar, N. Arjaria, S. M. Roopan, A. A. Rahuman and N. Singh, *Antimicrob. Agents Chemother.*, 2015, **59**, 4782–4799.
77. R. Aswini, S. Murugesan and K. Kannan, *Int. J. Environ. Anal. Chem.*, 2020, **14**, 1–11.
78. S. Marimuthu, A. A. Rahuman, C. Jayaseelan, A. V. Kirthi, T. Santhoshkumar, K. Velayutham, A. Bagavan, C. Kamaraj, G. Elango, M. Iyappan, C. Siva, L. Karthik and K. V. B. Rao, *Asian Pac. J. Trop. Med.*, 2013, **6**, 682–688.
79. V. Verma, M. Al-Dossari, J. Singh, M. Rawat, M. G. M. Kordy and M. Shaban, *Environ. Nanotechnol.*, 2014, **3**, 73–81.
80. M. Hudlikar, S. Joglekar, M. Dhaygude and K. Kodam, *Mater. Lett.*, 2012, **75**, 196–199.
81. A. A. Alshatwi, J. Athinarayanan and P. V. Subbarayan, *J. Mater. Sci. Mater. Med.*, 2015, **26**, 5330.
82. Y.-G. Gil, S. Kang, A. Chae, Y.-K. Kim, D.-H. Min and H. Jang, *Nanoscale*, 2018, **10**, 19810–19817.

83. S. Azizi, M. M. Shahri, H. S. Rahman, R. A. Rahim, A. Rasedee and R. Mohamad, *Int. J. Nanomed.*, 2017, **12**, 8841–8853.
84. G. Sharmila, M. F. Fathima, S. Haries, S. Geetha, N. Manoj Kumar and C. Muthukumaran, *J. Mol. Struc.*, 2017, **1138**, 35–40.
85. C. Bathula, S. K, A. Kumar K, H. Yadav, S. Ramesh, S. Shinde, N. K. Shrestha, K. Mallikarjuna and H. Kim, *Colloids Surf. B*, 2020, **192**, 111026.
86. X. Yang, Q. Li, H. Wang, J. Huang, L. Lin, W. Wang, D. Sun, Y. Su, J. B. Opiyo, L. Hong, Y. Wang, N. He and L. Jia, *J. Nanoparticle Res.*, 2009, **12**, 1589–1598.
87. A. Kalaiselvi, S. M. Roopan, G. Madhumitha, C. Ramalingam and G. Elango, *Spectrochim. Acta A*, 2015, **135**, 116–119.
88. M. Khan, M. Khan, M. Kuniyil, S. F. Adil, A. Al-Warthan, H. Z. Alkhathlan, W. Tremel, M. N. Tahir and M. R. H. Siddiqui, *Dalt. Trans.*, 2014, **43**, 9026–9031.
89. R. K. Petla, S. Vivekanandhan, M. Misra, A. K. Mohanty and N. Satyanarayana, *J. Biomater. Nanobiotechnol.*, 2012, **3**, 14–19.
90. M. Sathishkumar, K. Sneha, I. S. Kwak, J. Mao, S. J. Tripathy and Y. S. Yun, *J. Hazard. Mater.*, 2009, **171**, 400–404.
91. A. Bankar, B. Joshi, A. R. Kumar and S. Zinjarde, *Mater. Lett.*, 2010, **64**, 1951–1953.
92. M. Sathishkumar, K. Sneha and Y. S. Yun, *Mater. Lett.*, 2013, **98**, 242–245.
93. A. M. N. Jassim, S. Farhan and R. Dadoosh, *Adv. Environ. Biol.*, 2016, **10**, 51–67.
94. M. M. H. Khalil, E. H. Ismail, K. Z. El-Baghdady and D. Mohamed, *Arab. J. Chem.*, 2014, **7**, 1131–1139.
95. S. Dogan and M. Gundogan, *Turk. J. Chem.*, 2020, **44**, 1587–1600.
96. A. Khatoon, F. Khan, N. Ahmad, S. Shaikh, S. M. D. Rizvi, S. Shakil, M. H. Al-Qahtani, A. M. Abuzenadah, S. Tabrez, A. B. F. Ahmed, A. Alafnan, H. Islam, D. Iqbal and R. Dutta, *Life Sci.*, 2018, **209**, 430–434.
97. C. Krishnaraj, E. G. Jagan, S. Rajasekar, P. Selvakumar, P. T. Kalaichelvan and N. Mohan, *Colloids Surf. B*, 2010, **76**, 50–56.
98. V. Morales-Lozoya, H. Espinoza-Gómez, L. Z. Flores-López, E. L. Sotelo-Barrera, A. Núñez-Rivera, R. D. Cadena-Nava, G. Alonso-Nuñez and I. A. Rivero, *Appl. Surf. Sci.*, 2021, **537**, 147855.
99. A. Gudimalla, J. Jose, R. J. Varghese and S. Thomas, *J. Polym. Environ.*, 2020, **29**, 1412–1423, doi:10.1007/s10924-020-01959-6.

100. M. Nilavukkarasi, S. Vijayakumar and S. P. Kumar, *Mater. Sci. Energy Technol.*, 2020, **3**, 371–376.
101. A. K. Jha and K. Prasad, *Int. J. Green Nanotechnol.*, 2010, **1**, P110—P117.
102. V. Gopinath, D. MubarakAli, S. Priyadarshini, N. M. Priyadharsshini, N. Thajuddin and P. Velusamy, *Colloids Surf. B*, 2012, **96**, 69–74.
103. S. P. Dubey, M. Lahtinen and M. Sillanpää, *Process Biochem.*, 2010, **45**, 1065–1071.
104. H. Bar, D. K. Bhui, G. P. Sahoo, P. Sarkar, S. P. De and A. Misra, *Colloids Surf. A*, 2009, **339**, 134–139.
105. A. K. Mittal, A. Kaler and U. C. Banerjee, *Nano Biomed. Eng.*, 2012, **4**, 75–81.
106. F. Khan, M. Shariq, M. Asif, M. A. Siddiqui, P. Malan and F. Ahmad, *Nanomaterials*, 2022, **12**, 673–685.
107. P. Kumari and A. Meena, *Colloids Surf. A*, 2020, **606**, 125447.
108. P. V. Kumar, S. M. J. Kala and K. S. Prakash, *Mater. Lett.*, 2019, **236**, 19–22.
109. A. K. Jha and K. Prasad, *Int. J. Green Nanotechnol.*, 2011, **3**, 92–97.
110. B. Ankamwar, M. Chaudhary and M. Sastry, *Synth. React. Inorganic, Met. Nano-Metal Chem.*, 2005, **35**, 19–26.
111. O. S. ElMitwalli, O. A. Barakat, R. M. Daoud, S. Akhtar and F. Z. Henari, *J. Nanoparticle Res.*, 2020, **22**, 651–672.
112. S. P. Chandran, M. Chaudhary, R. Pasricha, A. Ahmad and M. Sastry, *Biotechnol. Prog.*, 2006, **22**, 577–583.
113. M. Noruzi, D. Zare, K. Khoshnevisan and D. Davoodi, *Spectrochim. Acta A*, 2011, **79**, 1461–1465.
114. K. Sneha, M. Sathishkumar, S. Y. Lee, M. A. Bae and Y.-S. Yun, *J. Nanosci. Nanotechnol.*, 2011, **11**, 1811–1814.
115. R. K. Das, N. Gogoi and U. Bora, *Bioprocess Biosyst. Eng.*, 2011, **34**, 615–619.
116. M. El Zowalaty, N. A. Ibrahim, M. Salama, K. Shameli, M. Usman and N. Zainuddin, *Int. J. Nanomed.*, 2013, **38**, 4467.
117. S. Shende, A. P. Ingle, A. Gade and M. Rai, *World J. Microbiol. Biotechnol.*, 2015, **31**, 865–873.
118. H. C. A. Murthy, T. Desalegn, M. Kassa, B. Abebe and T. Assefa, *J. Nanomat.*, 2020, **2020**, 3924081.
119. B. Assefa, G. Glatzel and C. Buchmann, *J. Ethnobiol. Ethnomed.*, 2010, **6**, 20–31.

120. B. Desta, *J. Ethnopharmacol.*, 1995, **45**, 27–33.
121. M. Valodkar, P. S. Nagar, R. N. Jadeja, M. C. Thounaojam, R. V. Devkar and S. Thakore, *Colloids Surf. A*, 2011, **384**, 337–344.
122. L. Gunti, R. S. Dass and N. K. Kalagatur, *Front. Microbiol.*, 2019, **10**, 931.
123. S. Paul, T. Aka, M. Ashaduzzaman, M. M. Rahaman and O. Saha, *Heliyon*, 2020, **6**, e04603.
124. H. A. F. Buarki, F. Al Hannan, F. Z. Henari,, *J. Nanotechnol.*, 2022, **14**, 758–769.
125. M. I. Damyanov CA, Pavlov VS, Avramov L, *Ann. Complement Altern. Med.*, 2018, **1**, 75–83.
126. S. Tohme, R. L. Simmons and A. Tsung, *Cancer Res.*, 2017, **77**, 1548–1552.
127. S. C. Formenti and S. Demaria, *Lancet Oncol.*, 2009, **10**, 718–726.
128. N. Gegechkori, L. Haines and J. J. Lin, *Med. Clin. North Am.*, 2017, **101**, 1053–1073.
129. N. R. Kukia, Y. Rasmi, A. Abbasi, N. Koshoridze, A. Shirpoor, G. Burjanadze and E. Saboory, *Asian Pac. J. Cancer Prev.*, 2018, **19**, 2821–2829.
130. R. Fujiwara, Y. Luo, T. Sasaki, K. Fujii, H. Ohmori and H. Kuniyasu, *Pathobiology*, 2015, **82**, 243–251.
131. S. Çe∆meli and C. B. Avci, *J. Drug Target.*, 2019, **27**, 762–766.
132. S. Doron and S. L. Gorbach, *International Encyclopedia of Public Health*, ed H. K. Heggenhougen, Academic Press, Oxford, 2008, pp. 273–282.
133. C. L. Ventola, *J. Formul. Manag.*, 2015, **40**, 277–283.
134. M. Nadeem, D. Tungmunnithum, C. Hano, B. H. Abbasi, S. S. Hashmi, W. Ahmad and A. Zahir, *Green Chem. Lett. Rev.*, 2018, **11**, 492–502.
135. P. S. M. Kumar, *J. Environ. Nanotechnol.*, 2014, **3**, 73–81.
136. Y. Lan, Y. Lu and Z. Ren, *Nano Energy*, 2013, **2**, 1031–1045.
137. J. G. McEvoy and Z. Zhang, *Minerology Petrology*, 2020, **114**, 1–13.
138. R. Bucureşteanu, R. Apetrei, M. Ioniţă, L.-O. Cinteză, L.-M. Diţu and M. Husch, *E3S Web Conf.*, 2019, **111**, 04046.
139. C. Aprile, A. Corma and H. Garcia, *Phys. Chem. Chem. Phy.*, 2008, **10**, 769–783.
140. A. L. Ruiz, C. B. Garcia, S. N. Gallón and T. J. Webster, *Int. J. Nanomed.*, 2020, **15**, 169–179.

141. N. Y. Lee, W. C. Ko and P. R. Hsueh, *Front. Pharmacol.*, 2019, **10**, 1153.
142. K. Bloch, K. Pardesi, C. Satriano and S. Ghosh, *Front. Chem.*, 2021, **9**, 624344.
143. K. B. Ayaz Ahmed, T. Raman and V. Anbazhagan, *RSC Adv.*, 2016, **6**, 44415–44424.
144. X. Yu, L. Yuan, N. Zhu, K. Wang and Y. Xia, *J. Photochem. Photobiol. B*, 2019, **195**, 27–32.
145. A. D. Selvan, D. Mahendran, R. Senthil Kumar and A. Kalilur Rahman, *J. Photochem. Photobiol. B Biol.* 2018, **180**, 243–252.
146. A. K. Dhara, S. Pal and A. K. N. Chaudhuri, *Phytother. Res.*, 2002, **16**, 326–330.
147. N. H. Ali and A. M. Mohammed, *Biotechnol. Rep.* 2021, **30**, e00635.
148. S. Onizawa, K. Aoshiba, M. Kajita, Y. Miyamoto and A. Nagai, *Pulm. Pharmacol. Ther.*, 2009, **22**, 340–349.
149. V. Lobo, A. Patil, A. Phatak and N. Chandra, *Pharmacogn. Rev.*, 2010, **4**, 118–126.
150. E. B. Kurutas, *Nutr. J.*, 2016, **15**, 71–83.
151. G. Pizzino, N. Irrera, M. Cucinotta, G. Pallio, F. Mannino, V. Arcoraci, F. Squadrito, D. Altavilla and A. Bitto, *Oxid. Med. Cell Lonyev.*, 2017, **2017**, 8416763.
152. S. A. Fahmy, E. Preis, U. Bakowsky and H. M. E. Azzazy, *Molecules*, 2020, **25**, 1132–1142.
153. J. Kim, M. Takahashi, T. Shimizu, T. Shirasawa, M. Kajita, A. Kanayama and Y. Miyamoto, *Mech. Ageing. Dev.*, 2008, **129**, 322–331.
154. F. Giacco and M. Brownlee, *Circ. Res.*, 2010, **107**, 1058–1070.
155. S. Reuter, S. C. Gupta, M. M. Chaturvedi and B. B. Aggarwal, *Free Radic. Biol. Med.*, 2010, **49**, 1603–1616.
156. A. Singh, R. Kukreti, L. Saso and S. Kukreti, *Molecules*, 2019, **24**, 654–662.
157. S. Gunes, Z. He, D. van Acken, R. Malone, P. J. Cullen and J. F. Curtin, *Nanomedicine*, 2021, **36**, 102436.
158. S. Dilruba and G. V. Kalayda, *Cancer Chemother. Pharmacol.*, 2016, **77**, 1103–1124.
159. L. Kelland, *Nat. Rev. Cancer*, 2007, **7**, 573–584.
160. X. Wang and Z. Guo, *Chem. Soc. Rev.*, 2013, **42**, 202–224.
161. Y. Bendale, V. Bendale and S. Paul, *Integr. Med. Res.*, 2017, **6**, 141–148.

162. W. Yang, H. Liang, S. Ma, D. Wang and J. Huang, *Sust. Mater. Technol.*, 2019, **22**, e00109.
163. T. T. V. Phan, T. C. Huynh, P. Manivasagan, S. Mondal and J. Oh, *Nanomaterials*, 2019, **10**, 1275–1277.
164. S. Bharathiraja, N. Q. Bui, P. Manivasagan, M. S. Moorthy, S. Mondal, H. Seo, N. T. Phuoc, T. T. Phan, H. Kim, K. D. Lee and J. Oh, *Sci. Rep.*, 2018, **8**, 500–509.
165. S. Mallidi, G. P. Luke and S. Emelianov, *Trends Biotechnol.*, 2011, **29**, 213–221.
166. S. Gurunathan, E. Kim, J. W. Han, J. H. Park and J. H. Kim, *Molecules*, 2015, **20**, 22476–22498.
167. B. Rubio-Ruiz, A. M. Pérez-López, T. L. Bray, M. Lee, A. Serrels, M. Prieto, M. Arruebo, N. O. Carragher, V. Sebastián and A. Unciti-Broceta, *ACS Appl. Mat. Interf.*, 2018, **10**, 3341–3348.
168. X. Deng, Z. Shao and Y. Zhao, *Adv. Sci.*, 2021, **8**, 3625–3633.
169. E. Bump, S. Hoffman and W. Foye, *J. Med. Chem.*, 1975, **18**, 803–812.
170. T. Sun, Y. S. Zhang, B. Pang, D. C. Hyun, M. Yang and Y. Xia, *Angew. Chem. Int. Ed Engl.*, 2014, **53**, 12320–12364.
171. P. Ruenraroengsak, P. Novak, D. Berhanu, A. J. Thorley, E. Valsami-Jones, J. Gorelik, Y. E. Korchev and T. D. Tetley, *Nanotoxicology*, 2012, **6**, 94–108.
172. W. Park and K. Na, *Wiley Interdiscip. Rev. Nanomed. Nanobiotechnol.*, 2015, **7**, 494–508.
173. M. Ferrari, *Nat. Rev. Cancer*, 2005, **5**, 161–171.
174. S. Kundu and U. Nithiyanandham, *RSC Adv.* 2013, **3**, 25278–25290.
175. D. Sharma, L. Ledwani and N. Bhatnagar, *New Front. Chem.*, 2015, **24**, 121–135.
176. R. Kuberan, K. Sudha, S. Gnanasekar and S. Sivaramakrishnan, *Spectro. Acta Part A Mole. Biomol. Spect.*, 2014, **136**, 1126–1132.
177. X.-F. Zhang, W. Shen and S. Gurunathan, *Int. J. Mol. Sci.*, 2016, **17**, 1603.
178. J. R. Morones, J. L. Elechiguerra, A. Camacho, K. Holt, J. B. Kouri, J. T. Ramírez and M. J. Yacaman, *Nanotechnology*, 2005, **16**, 2346–2353.
179. L. Wang, C. Hu and L. Shao, *Int. J. Nanomed.*, 2017, **12**, 1227–1249.

180. I. X. Yin, J. Zhang, I. S. Zhao, M. L. Mei, Q. Li and C. H. Chu, *Int. J. Nanomed.*, 2020, **15**, 2555–2562.
181. I. Sondi and B. Salopek-Sondi, *J. Colloid. Interface Sci.*, 2004, **275**, 177–182.
182. I. Camarillo, F. Xiao, S. Madhivanan, T. Salameh, M. Nichols, L. Reece, J. Leary, K. Otto, A. Natarajan, A. Ramesh and R. Sundararajan, *Biomedicines* 2020, **1**, 498–505.
183. E. Barcińska, J. Wierzbicka, A. Zauszkiewicz-Pawlak, D. Jacewicz, A. Dabrowska and I. Inkielewicz-Stepniak, *Oxid. Med. Cell. Longev.*, 2018, **2018**, 8251961.
184. K. B. A. Ahmed, S. Subramanian, A. Sivasubramanian, G. Veerappan and A. Veerappan, *Spectro. Acta Part A Mole. Biomol. Spectr.*, 2014, **130**, 54–58.
185. M. Mohamed, S. Fouad, H. Elshoky, G. Mohammed and T. Salaheldin, *Int. J. Vet. Sci. Med.*, 2017, **5**, 117–138.
186. K. Sztandera, M. Gorzkiewicz and B. Klajnert-Maculewicz, *Mole. Pharm.*, 2019, **16**, 1–23.
187. R. Shukla, V. Bansal, M. Chaudhary, A. Basu, R. Bhonde and M. Sastry, *Langmuir*. 2005, **21**, 10644–10654.
188. S. Rajeshkumar, *J. Genet. Eng. Biotechnol.*, 2016, **14**, 195–202.
189. H. K. Patra, S. Banerjee, U. Chaudhuri, P. Lahiri and A. K. Dasgupta, *Nanomedicine*, 2007, **3**, 111–119.
190. S. Ito, N. Miyoshi, W. G. Degraff, K. Nagashima, L. J. Kirschenbaum and P. Riesz, *Free Radic. Res.*, 2009, **43**, 1214–1224.
191. Y. Pan, A. Leifert, D. Ruau, S. Neuss, J. Bornemann, G. Schmid, W. Brandau, U. Simon and W. Jahnen-Dechent, *Small*, 2009, **5**, 2067–2076.
192. B. Kang, M. A. Mackey and M. A. El-Sayed, *J. Am. Chem. Soc.*, 2010, **132**, 1517–1519.
193. M. Raffi, S. Mehrwan, T. Bhatti, J. Akhter, A. Hameed, W. Yawar and M. Hassan, *Ann. Microbiol.*, 2010, **60**, 75–80.
194. N. Hoshino, T. Kimura, A. Yamaji and T. Ando, *Free Radic. Biol. Med.*, 1999, **27**, 1245–1250.
195. N. T. K. Thanh and L. A. W. Green, *Nano Today*, 2010, **5**, 213–230.
196. S. V. P. Ramaswamy, S. Narendhran and R. Sivaraj, *Bull. Mat. Sci.*, 2016, **39**, 361–364.
197. U. Jinu, M. Gomathi, I. Saiqa, N. Geetha, G. Benelli and P. Venkatachalam, *Microb. Pathog.*, 2017, **105**, 86–95.

198. R. Hassanien, D. Z. Husein and M. F. Al-Hakkani, *Heliyon*, 2018, **4**, e01077.
199. D. Xu, E. Li, B. Karmakar, N. S. Awwad, H. A. Ibrahium, H.-E. H. Osman, A. F. El-kott and M. M. Abdel-Daim, *Arab. J. Chem.*, 2022, **15**, 103638.
200. W. Y. Qiu, Y. Y. Wang, M. Wang and J. K. Yan, *Colloids Surf. B Biointerfaces*, 2018, **170**, 692–700.
201. S.-Y. Zhang, J. Zhang, H.-Y. Wang and H.-Y. Chen, *Mat. Lett.*, 2004, **58**, 2590–2594.
202. M. Singh, N. Sharma, H. S. Paras, N. S. Hans, N. P. Singh and A. Sarin, *Env. Prog. Sustainable Energy*, 2019, **38**, 721–726.
203. J. T. Rotruck, A. L. Pope, H. E. Ganther, A. B. Swanson, D. G. Hafeman and W. G. Hoekstra, *Science*, 1973, **179**, 588–590.
204. D. Cui, T. Liang, L. Sun, L. Meng, C. Yang, L. Wang, T. Liang and Q. Li, *Pharm. Biol.*, 2018, **56**, 528–534.
205. S. Chatterjee, A. Bandyopadhyay and K. Sarkar, *J. Nanobiotechnol.*, 2011, **9**, 34.
206. L. Gabrielyan, A. Hovhannisyan, V. Gevorgyan, M. Ananyan and A. Trchounian, *Appl. Microbiol. Biotechnol.*, 2019, **103**, 2773–2782.
207. M. B. Sathyanarayanan, R. Balachandranath, Y. Genji Srinivasulu, S. K. Kannaiyan and G. Subbiahdoss, *ISRN Microbiol.*, 2013, **2013**, 272086.
208. Z. Vardanyan, V. Gevorkyan, M. Ananyan, H. Vardapetyan and A. Trchounian, *J. Nanobiotechnol.*, 2015, **13**, 69.
209. H. Lu, S. Yang, B. Wilson, S. McManus, C. Chen and R. Prud'homme, *Appl. Nanosci.*, 2017, **7**, 83–93.

© 2025 World Scientific Publishing Company
https://doi.org/10.1142/9789819806409_0008

Functional Coatings for Built Environments Based on Nanotechnology*

Pin Jin Ong [†], Suxi Wang [†], Ady Suwardi [†,ǁ], Jing Cao [†],
Fuke Wang [†], Xuesong Yin [†], Pei Wang [†], Fengxia Wei [†],
Dan Kai [†,‡,§], Enyi Ye [†,‡,§], Jianwei Xu [†,‡,¶], Ming Hui Chua [†,‡],
Warintorn Thitsartarn [†], Qiang Zhu [†,‡,§,**,‡‡] and Xian Jun Loh [†,‡,ǁ,††,‡‡]

[†]*Institute of Materials Research and Engineering (IMRE),*
*Agency for Science Technology and Research (A*STAR),*
2 Fusionopolis Way Innovis #08-03,
Singapore 138634, Singapore
[‡]*Institute of Sustainability for Chemicals,*
Energy and Environment (ISCE²)
*Agency for Science, Technology and Research (A*STAR)*
1 Pesek Road, Jurong Island, Singapore 627833, Singapore
[§]*School of Chemistry, Chemical Engineering and Biotechnology*
Nanyang Technological University
21 Nanyang Link, Singapore 637371, Singapore
[¶]*Department of Chemistry, National University of Singapore*
3 Science Drive 3, Singapore 117543, Singapore
[ǁ]*Department of Materials Science and Engineering*
National University of Singapore, 9 Engineering Drive 1 #03-09 EA
Singapore 117575, Singapore
[**]*zhuq@imre.a-star.edu.sg*
[††]*lohxj@imre.a-star.edu.sg*

[‡‡]Corresponding authors.
*To cite this article, please refer to its earlier version published in the *World Scientific Annual Review of Functional Materials*, Volume 2, 2330004 (2024), DOI: 10.1142/S2810922823300040.

Coatings provide underlying item surfaces or bulk materials with protection, enhancement, and/or additional usefulness and attributes. Since materials can be modified or enhanced to possess different properties such as mechanical, thermal, or chemical to improve urban built environment functions, nanotechnologies have been widely included in functional coatings in recent years. Recent studies on functional coatings for green and smart buildings are summarized in this review paper. These comprised phase change materials, photocatalytic, hydrophilic, hydrophobic, solar reduction, and solar utilization coatings. This study offers a thorough overview of functional coating methods, from the production of raw materials through their application to building components.

Keywords: Coating; nanotechnology; energy conversion; solar; phase transition.

1. Introduction

In daily life, coatings are everywhere. Several coating materials have been developed for food, medicines, wearables, consumer goods, industries, machinery, automobiles, and architectural components.[1-7] To safeguard, enhance, and/or give new functions and characteristics to the surfaces of different materials, coating materials are frequently applied as an exterior coating. Coating methods can stop or lessen corrosion and deterioration of exterior surfaces, which are frequently caused by exposure to elements such as humidity, ultraviolet (UV) rays, and weather.[8-12] The stability, longevity, and lifetime of the things or surfaces they are put on are improved as a result. Surfaces can be made antimicrobial, superhydrophobic, and superhydrophilic using coating methods.[13-15] Functional coatings are used in food and medicine to mask flavor and odor, protect and stabilize in physiological settings, target bodily release, and other purposes. Functional coating materials and technologies have received a lot of focus and effort due to their wide range of applications.

Depending on the desired purpose, different deposition methods and materials can be acquired through coating technologies. These coating materials ranged from organic, inorganic, to hybrid polymers/composites, depending on their features and utility.[16-21] The recent rapid growth of nanotechnology has sparked the manufacture of coatings that use this technology. The manipulation of matter at the nanoscale is known as nanotechnology (0.1–100 nm). It uses nanostructured materials including carbon nanotubes (CNTs), fullerenes, quantum dots, inorganic nanoparticles (NPs), etc. Because they have a greater surface area-to-volume ratio, higher surface energy, and fewer flaws than bulk materials, nanomaterials exhibit various attributes and performance.[22–24] Electronics, energy, and other industries all employ nanotechnology.[25–33]

NPs are commonly used in coatings. Surface coatings using NPs can change stability, wettability, and performance. Due to their large surface area-to-volume ratio and free surface energy, NPs have low activity or dispensability because they prefer to clump together in solutions. They may be covered or coated with polymer to prevent this. Accordingly, encapsulation has been widely investigated.[34–39] By creating a thin layer of core–shell using TiO_2 NP around the Au particle, Seh et al. created core–shell NPs with superior dispensability and reactivity.[40] The TiO_2 shell prevents the Au core from aggregating while nitro-phenol was decreased by core–shell TiO_2@Au NPs. After five recyclings, the core–shell NPs produced identical catalytic results. Shah et al. invented unique encapsulating techniques utilizing water as the solvent for various NPs since the core–shell NP is essential for catalytic reactions and coating.[41] Silica shell thickness is controlled by a variable coating time. Ag NPs in Ag@SiO_2–SH NPs may be preserved by the silica shell under the circumstances of silica coating in an aqueous solution.

Functional coatings including fire-retardant and anti-corrosion coatings have significantly improved due to nanotechnology. Researchers have investigated the use of these coatings on textiles based on different materials and evaluated their flame retardation performance.[42–45] Polymer nanocomposite coatings with anti-fouling, anti-corrosion, and self-healing properties for a myriad of applications were reviewed recently.[46–52] Similarly, Adak et al. summarized nanotechnology-based coating for solar energy-based devices such as photovoltaic (PV) panels and concentrated solar power (CSP) reflectors. The basic principles, the deposition methods' cost-effectiveness, and the coatings' long-term utility were discussed.[53]

Surfaces may be coated with functional nanocomposite materials. Deposition techniques like UV-assisted nucleation, roll-to-roll processing, pulsed UV laser irradiation, physical vapor deposition (PVD), and chemical vapor deposition (CVD) are capable of generating or modifying nanoscale coating surface structures.[54-60] On the other hand, coating methods include spraying, casting, spinning, dipping, dropping, and taping.[61-64]

Due to global warming and climate change, researchers are exploring various ways to create a circular economy by focusing on sustainable materials or designs.[65-73] In this regard, "green" buildings and the associated technologies have been designated as a significant infrastructure aim. Their engineering and architectural design can result in less waste, pollution, and carbon footprint, more effective use of energy and water, and greater protection of the health and wellness of inhabitants. In this area, the conversion of heat through various approaches has been extensively studied, such as the thermoelectric conversion of waste heat into power, solar panels, energy-efficient lighting systems, and energy-chromic smart windows.[74-82] Additionally, phase change material (PCM) for thermal storage could be used for cooling to address such issues.[83-85] The most basic coating is probably paint, which is now used for more than just esthetic purposes and has solar reflectance and antimicrobial properties.[86-89] Other functional coatings, particularly those used on the exterior walls, roofs, and windows of buildings, can add useful features like indoor temperature regulation, solar reduction, self-cleaning, etc.

The research on functional coating materials that use nanotechnology for enhanced features and performance in relation to the design and construction of green buildings is summarized in this paper. These include PCM, photocatalytic, hydrophilic, hydrophobic, solar reduction, and utilization coatings. To reduce their carbon footprint, experts from across the world have been experimenting with functional coatings on green building materials. Superhydrophobic coatings provide materials used in civil engineering and the exteriors of buildings with the ability to reject water, which promotes self-cleaning, anti-corrosion, and anti-icing. As a result, less time, effort, and resources are required to maintain building components. They can increase wall breathability, lower thermal conductivity, and offer biological agents and anti-fouling resistance. Increasing the surface area of water droplets by spreading them or by using photocatalysis to break down dirt and pollutants when exposed to sunlight, hydrophilic coatings may also enable building exteriors to self-clean, saving resources and labor. Additionally, hydrophilic coatings are antimicrobial and humidity-regulating.[90,91] Solar heat gain in buildings is decreased by solar reduction coatings, which results in less energy being used for thermal control within structures. Through its ability to store thermal energy, PCM coatings save energy while cooling or heating a building's interior.[92-94] We'll talk about the components, methods, purposes, and overall effectiveness of these functional coatings. The future of these functional coating materials' research and development will also be covered, with a focus on evaluating the likelihood of their widespread commercial application in green buildings.

2. Coatings with Phase Change Material

During phase transition which occurs at a specific and constant temperature, PCM may store a significant quantity of latent heat that is released.[95-98] The performance of PCM is impacted by factors such as stability, flammability, and thermal conductivity. Buildings, textiles, and composite materials use PCM for energy-saving and thermal regulation.

To lower the indoor temperature and enhance the passive cooling performance of buildings envelopes, Ling *et al.* integrated heat-reflective coatings, PCMs, and insulation layers into test rooms and investigated the effects on thermal regulation performance.[99] Two types of PCMs, RT31/fumed SiO_2 and RT31/expanded graphite (RT31/EG), were used and two different structures of the rooms, single and double cavity walls, were employed as well. The results indicated that RT31/SiO_2 can limit the maximum temperature rise

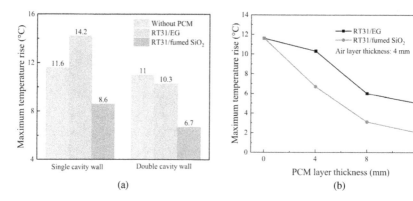

Fig. 1. (a) Maximum temperature rise comparisons between the different PCMs in different structures. (b) Maximum temperature rise comparisons among the different thicknesses of PCMs.

Source: Reproduced with permission from Ref. 99. Copyright 2021 Elsevier.

to 6.7°C as compared to RT31/EG's 8.6°C due to SiO_2 having a lower thermal conductivity [Fig. 1(a)]. The effect was also more prominent for the single cavity walls as the decrease in insulation resulted in a higher sensitivity toward thermal conductivity. A thicker PCM layer resulted in a smaller maximum temperature rise due to larger thermal storage capacity and thermal resistance [Fig. 1(b)]. However, an optimum thickness of 8 mm was noted as the PCM does not undergo a complete phase change when the layer is too thick.

Meanwhile, Chen *et al.* used a microencapsulated PCM and developed a coating for temperature regulation in buildings.[100] $CaCO_3$ and SiO_2 are commonly used as shell materials for encapsulating PCMs but they have poor corrosion resistance and therefore have limited applications. To address this issue, the authors used $BaSO_4$ as the shell material and encapsulated *n*-eicosane via emulsion-templated controllable precipitation. The $BaSO_4$ shell endows the microcapsules with high acid and alkali resistance, rapid thermal transfer, high thermal conduction, and good thermal cycling stability, with an overall latent heat of over 100 J/g. A simulated study using house models revealed that the indoor temperature was reduced by 3.2°C compared to latex coating, in addition to having high corrosion resistance, scratch resistance, and hardness.

Instead of applying PCM coating on the exterior of buildings, Nam et al. investigated the effects of PCM coating on wood-based furniture to reduce their sensitivity to temperature changes and provide heat storage capability.[101] The n-eicosane was vacuum-impregnated into three types of wood panel materials: medium-density fiberboard (MDF), particle board (PB), and plywood (PW), and a low-pressure melamine (LPM) coating was applied on the surface to prevent leakage (Fig. 2). Results indicated that the improved thermal conductivity of the wood materials enabled the PCM to store and release heat efficiently and PB has the highest enthalpy and latent heat, followed by PW and MDF. A reduction in peak temperature and time lag effect was confirmed by dynamic

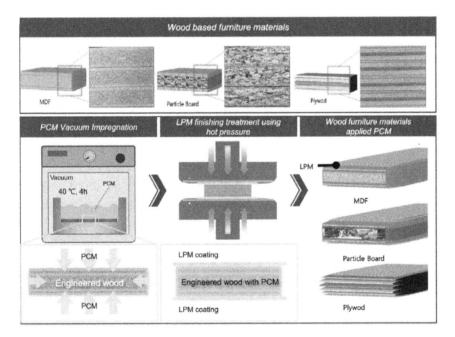

Fig. 2. Manufacturing process of wood-based furniture materials characterized by PCMs.

Source: Reproduced with permission from Ref. 101. Copyright 2022 Elsevier.

heat transfer analysis and therefore reduced the risk of water vapor condensation.

Besides using PCM coatings for building applications, they have been extensively studied for other purposes as well. Yilbas et al. used PCM coating to study the reversible change of a hydrophobic surface's wetting state.[102] Micro/nano-sized spherules and fibrils were dispersed on the surface via solution crystallization. The n-octadecane was coated after functionalized silica particles were deposited. A 140° water contact angle (WCA) was produced when functionalized silica particles were exposed to n-octadecane solid flakes. The n-octadecane formed a liquid layer over the silica particles at higher temperatures, producing a hydrophilic surface. By hardening and liquefying n-octadecane film, reversible surface wetting may be achieved. This work advances systems with many uses, such as microfluidic devices.

To enhance the anti-icing performance, Shamshiri et al. fabricated an icephobic coating using PCM microcapsules and poly(ethylene glycol)–poly(dimethylsiloxane) (PEG–PDMS) copolymers.[103] Using the in-situ polymerization method, n-dodecane and n-tetradecane are encapsulated within urea–formaldehyde shells and mixed with self-lubricated PDMS coating containing 2.5-wt.% hydroxyl-terminated PEG–PDMS copolymer. Differential scanning calorimetry (DSC) results revealed that the PCM-embedded coatings have lower ice nucleation temperatures compared to PDMS copolymer as the latent heat release by the PCM microcapsule preserved the liquid-like layer, known as the "quasi-liquid layer" (QLL), hence prolonging ice nucleation [Figs. 3(a) and 3(b)]. The QLL itself can reduce surface ice adhesion strength but incorporating PCM into the matrix further enhanced this property. As shown in Fig. 3(c), the ice adhesion strength decreased from 380 kPa to 27 kPa as the concentration of the PCM microcapsules increased and this is due to the preservation of the QLL for a longer duration attributing to the release of latent heat. In addition, there was no significant change in the wettability of the PCM-embedded coatings after 10 icing/de-icing cycles [Fig. 3(d)]. Nevertheless, it was noted that a high concentration of PCM microcapsules

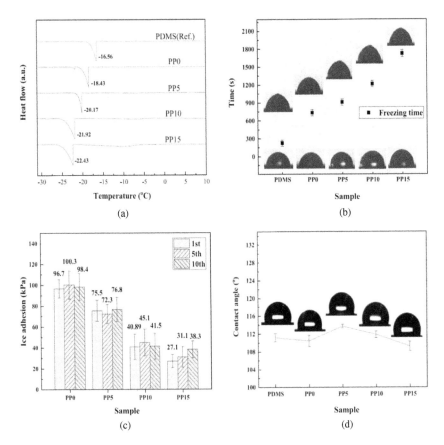

Fig. 3. (a) Delayed ice nucleation temperature, evaluated using DSC. (b) Freezing times on the prepared samples with the images of the water droplets at $t = 0$ s and at the instant of complete freezing. (c) The ice adhesion strength of the PCM-containing samples over 10 icing/de-icing cycles; the ice adhesion strengths of PDMS coating corresponding to the 1st, 5th, and 10th cycles are measured as 379.4 ± 49.3, 385.4 ± 40.2, and 404.2 ± 45.1 kPa, respectively. (d) WCAs of the prepared samples after 10 icing/de-icing cycles.

Source: Reproduced with permission from Ref. 103. Copyright 2023 Elsevier.

weakens the mechanical properties of the coating and hence affects its long-term utility.

Chalco-Sandoval et al. explored the possibility of using PCM coating to develop heat management materials for food packaging applications.[104] Polystyrene (PS) trays were coated with electrospun

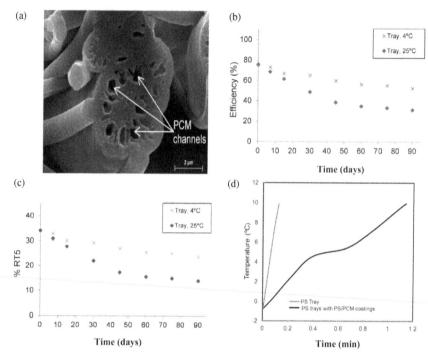

Fig. 4. (a) Cross-section SEM image of the PS tray with the ultrathin fiber-structured PS/PCM coating. (b) Encapsulation efficiencies at 4°C and 25°C. (c) Calculated amounts of RT5 (%) encapsulated in the PS-tray systems at 4°C and 25°C. (d) Surface temperature as a function of time for PS tray with and without the ultrathin fiber-structured PS/PCM coating.

Source: Reproduced with permission from Ref. 104. Copyright 2017 Elsevier.

PS/PCM (RT5) fibers and thereafter annealed to enhance their adhesion. The scanning electron microscopy (SEM) image indicated successful encapsulation of the PCM inside the fibrillar PS matrix, consisting of multiple channels [Fig. 4(a)]. Encapsulation efficiency and loading of the PCM were found to be 78% and ~34 wt.%, respectively, and it was noted that these parameters were affected by storage temperature and time [Figs. 4(b) and 4(c)]. The decrease in encapsulation efficiency at 25°C could be attributed to the different PCM–PS interactions during coating formation while the porous structure of the PS fibers led to leakage of the PCM over time. The PS/PCM-coated tray delayed the time required to raise the temperature above the chilling temperature owing to the latent heat of the PCM and the PS/PCM coating insulation properties [Fig. 4(d)].

PCM-based coatings have potential in green construction materials. Their properties can be enhanced via nanotechnology, enhancing their capacity for heat storage. Some of the techniques include electrospinning, vacuum impregnation, NP addition, and microencapsulation. Although PCM coating technology has advanced, commercial uses are still constrained. Problems with PCM leakage and form stability in additives or matrices exist.[105–109] Repeated cycles may result in the leakage of PCM from the matrix. Since microcapsules can degrade and reduce thermal integrity, this is particularly true for encapsulation technologies. Second, these materials are likely to be exposed to intense sunlight after being deposited onto building components, which might result in photodegradation and lower their performance. Researchers are always attempting to enhance the functional performance of PCM materials in light of these challenges to make them a workable green material for the future.

3. Photocatalytic Coatings

Photocatalytic coatings comprising semiconductor nanomaterials such as TiO_2, ZnO, NiO, CuO, and WO_3 can impede and decrease the growth of microorganisms on the surface of materials. These nanomaterials are easy to prepare, stable, of low cost, and possess antimicrobial and self-cleaning properties and hence have broad applications in areas such as buildings, solar cells, water decontamination, and textile.[110–113] In the presence of a light source, the photocatalysis process produces highly reactive species such as hydrogen peroxide, super-oxide radical anions, singlet oxygen, and hydroxyl radicals. The working principle that underpins the functionalities of heterogeneous and heterojunction photocatalysts in the construction industry was reviewed by Meda *et al.* The synthesis procedures, the advantages and disadvantages of these photocatalysts, and how to incorporate these materials into construction materials and as coatings on buildings were covered as well.[114]

Appasamy *et al.* investigated the potential of nitrogen-doped TiO$_2$/single-wall CNT nanocomposite as a self-cleaning and photocatalytic coating for solar PV cells using the sol–gel method.[115] Degradation studies [Figs. 5(a) and 5(b)] using methylene blue (MB) revealed a degradation rate (C/C_0) of 1.93 mg/L and an enhanced photocatalytic degradation efficiency of 72.43%, as well

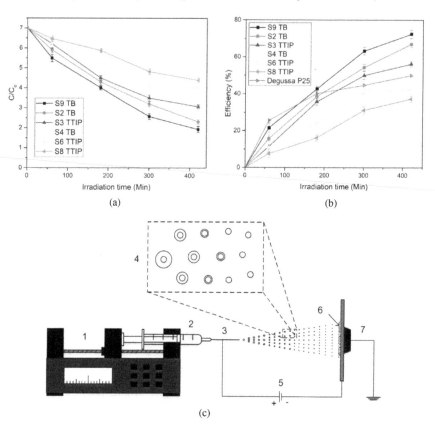

Fig. 5. (a) Photocatalytic degradation rate. (b) Photocatalytic degradation efficiency. Error bars: standard deviations of triplicate experiments. (c) Scheme of the electrospray setup used for ZnO NP coating preparation: 1: syringe pump, 2: ZnO NP suspension, 3: capillary needle, 4: aerosol formation and solvent evaporation, 5: high-voltage power supply, 6: glass substrate, and 7: grounded collector.

Source: Panels (a) and (b) reproduced with permission from Ref. 115. Copyright 2020 Elsevier. Panel (c) reproduced with permission from Ref. 116. Copyright 2019 Elsevier.

as wettability as high as 94.3 ± 2°. More importantly, not only does the voltage output of the PV cells was not negatively affected by the coating, but it also resulted in a higher voltage after the dust deposition test and therefore can maintain the optimum performance of the PV.

To produce antimicrobial functionalized surfaces on glass substrates, Valenzuela et al. created electrospray coatings using sol–gel ZnO suspensions and the electrospray setup is shown in Fig. 5(c).[116] The ZnO-functionalized microspheres have sizes in the 100–300-nm range and surface densities of up to 0.30 mg·cm^{-2} were observed. Under different UV irradiation times to simulate either winter or summer conditions, photoinduced hydrophilicity was noted and a wettability of <10° was recorded for the WCA. The coatings displayed excellent photocatalytic properties (>95% MB degradation) and antibacterial activity against *Staphylococcus aureus* (>99.5% reduction), which is attributed to both bioavailable Zn^{2+} ions and photogenerated reactive oxygen species from the dissolution of ZnO.

Through the hydrolysis and precipitation of titanyl sulfate ($TiOSO_4$), Wang et al. fabricated SiO_2–TiO_2 coatings on SiO_2 microspheres via spray coating.[117] These coatings achieved a WCA of 0° within 0.35 s [Fig. 6(a)] and exhibited excellent photocatalytic degradation performance against methyl orange (MO), MB, and rhodamine B (RhB) under UV and sunlight irradiation. In addition, these coatings were applied to the surfaces of different substrates such as concrete, wood, bricks, and glass slides, and the photocatalytic degradation of RhB on these surfaces was investigated under sunlight for 10 min [Fig. 6(b)]. The results revealed that the RhB droplets were entirely decolorized, indicative of the great adaptability of the coatings.

Similarly, TiO_2–SiO_2 nanocomposites were recently developed by Luna et al. using titania nanosheet (TNS) to investigate their self-cleaning and anti-pollution properties in urban buildings.[118] As the synthesized nanosheet has a high fluorine content on its surface, agglomeration of the nanosheets occurred and dispersion was difficult. To circumvent this issue, the adsorbed residual fluoride

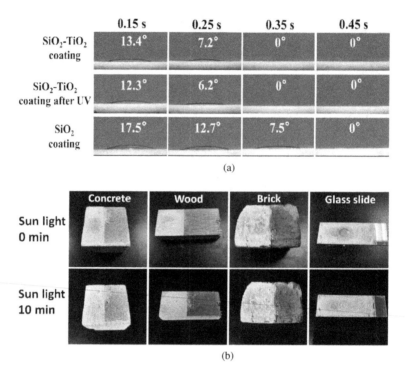

Fig. 6. (a) WCAs of the SiO_2–TiO_2 coating, the SiO_2–TiO_2 coating after UV irradiation, and the SiO_2 coating at different points in time. (b) The adaptability of the SiO_2–TiO_2 coating to the surfaces of concrete, wood, brick, and glass slide.
Source: Reproduced with permission from Ref. 117. Copyright 2022 Elsevier.

was removed via treatment with NaOH and it was revealed that the photocatalytic performance improved as compared to the untreated samples. As compared to the untreated surface and using only commercial P25 TiO_2 in silica sol (S4P25), the treated TNS (S4NSNaOH) demonstrated an effective reduction of MB and NO^x concentrations under simulated sunlight irradiation. Furthermore, it demonstrated higher photocatalytic oxidation of soot on limestone surfaces.

As a green synthesis approach for obtaining a photocatalytic film for waste treatment, Wen *et al.* used green tea extract as a capping and reducing agent to immobilize Ag NPs onto a 2,2,6,6-tetramethylpiperidine-1-oxyl (TEMPO)-mediated tea cellulose–PVA (Ag–TTC–PVA) film.[119] The carboxylic acid groups present in TTC facilitated the immobilization of Ag ions on the cellulose backbone and TTC itself serves as an inhibition for the fast release of cytotoxic Ag NPs. Compared to the TTC–PVA film, the transparent and flexible Ag–TTC–PVA composite film has enhanced thermal and mechanical properties. Photocatalytic studies involving the degradation of MO under UV irradiation were carried out and results indicated that Ag–TTC–PVA film has a higher degradation rate than the TTC–PVA film and uncoated film in the dark and under UV irradiation, with the latter having an especially prominent effect.

Traditionally, CVD, PVD, and the sol–gel process are used to prepare photocatalytic coatings. Even though the CVD and PVD techniques can create coatings with uniform and tunable thickness and chemical composition, they are expensive, energy-intensive, and not feasible for large-scale applications. On the other hand, the sol–gel process is cost-effective but time-consuming. Instead, thermal spraying can be employed to obtain highly efficient and mechanically resistant photocatalytic coatings, with the added advantage of the ease of application on different substrates of different sizes.

4. Hydrophilic Coatings

In hydrophilic coatings, water droplets are spread across the surface to form a film, thus conferring photocatalytic properties that remove dirt and impurities when exposed to light. A hydrophilic surface has a WCA ≤ 90° while a WCA < 5° exhibits superhydrophilicity. UV protection, among other things, can be provided by hydrophilic coatings that lack photocatalytic properties. The creation of hydrophilic coatings sparked a lot of curiosity decades ago.

Hydrophilic materials are essential components for self-cleaning functions, and it has increasingly gained much popularity as a research topic. Self-cleaning coatings were examined by Ganesh *et al.*, who concentrated on well-liked and investigated materials such as ZnO, TiO_2, silicone, etc.[120] They also emphasized their manufacturing process and commercial applications. The mechanism of hydrophilic coatings, their manufacturing processes, and commercially available self-cleaning products were also covered by the authors. The growing significance of self-cleaning coatings was noted by Ragesh et al.[121] This was because more coatings were being applied to glass, paints, cement, and other construction materials, which made them a more significant part of our everyday lives. The paper also discussed several synthetic techniques for producing coatings that are hydrophilic, hydrophobic, oleophobic, and possess other multifunctional properties.

da Silva Cardoso et al. investigated superhydrophilic transparent thin films by doping erbium (Er^{3+}) and yttrium (Y^{3+}) into TiO_2 to remove oleic acid from the surfaces.[122] Glass substrates were dipped in both pure TiO_2 sol and TiO_2 sol that had been doped with Y^{3+} or Er^{3+}, then dried three times. The effects of the drying temperature, the number of layers, and the loading of Y^{3+} or Er^{3+} on the apparent WCA for thin films were looked at. The zero CA that all coated glass samples display suggests superhydrophilicity. On the other hand, Altınışık et al. prepared a series of carbazole-based conjugated backbone polymers with different lengths of hydrophilic PEG as coatings with anti-fogging and UV-absorbing properties.[123] The conjugated main chain contributed to strong UV absorption while the bathochromic shift of the main chain absorption band was influenced by the electron-donating behavior of PEG. An optimum PEG length could reduce the WCA from 60° to 29° due to the highest surface roughness as longer PEG moieties could result in agglomeration, irregular, and amorphous surfaces, thus lowering surface roughness.

Hydrophilic surfaces can confer anti-biofouling properties as the surface hydration layer can serve as a physical barrier to resist biofouling processes such as bacterial and protein attachments.[124] Due to their ability to form hydration layers via electrostatic attraction and hydrogen bonding, hydrophilic polymers such as PEG and zwitterion-containing polymers have been used to develop anti-biofouling coatings.[125,126] However, due to their poor mechanical stability and water solubility, hydrophilic polymers need to be used in conjunction with other types of surfaces or materials before they can be coated. For example, a hydrophilic polymer brush can be constructed to functionalize a hydrophobic surface via photo-initiation or atom transfer radical polymerization. Aguilar-Sanchez et al. reported the use of TEMPO-oxidized cellulose nanofibrils/poly(vinyl alcohol) (T-CNF/PVA) to coat a polyethersulfone (PES) substrate for the development of ultrafiltration membranes.[127] The surface modification is shown in Fig. 7(a). PVA promotes the interfacial interaction between hydrophobic PES and hydrophilic T-CNF

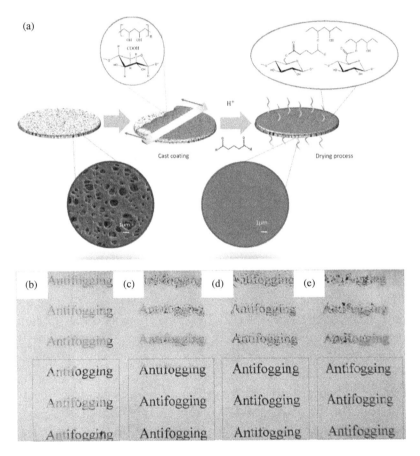

Fig. 7. (a) Schematic illustration of the surface modification method of the commercial PES membranes by coating. Anti-fogging performances of (b) S1, (c) S2, (d) S3, and (e) S4.

Source: Panel (a) reproduced with permission from Ref. 127. Copyright 2021 The Royal Society of Chemistry. Panels (b)–(e) reproduced with permission from Ref. 133. Copyright 2021 Elsevier.

and crosslinks T-CNF while glutaraldehyde was added to improve the mechanical properties and stability. The coated membranes have a significant reduction in bovine serum albumin fouling due to a decrease in surface roughness and an increased negative surface charge, resulting in electrostatic repulsions between the surface of the membrane and the foulants. In addition, the membranes demonstrated an anti-biofouling effect against *Escherichia coli* (*E. coli*) due to the aldehyde and carbonyl groups. An alternative way to prepare hydrophilic coatings involves physical mixing or copolymerization with hydrophobic polymers such as PDMS, polyurethane (PU), or methyl methacrylate.[128–131]

TiO_2 NPs that have been exposed to UV or visible light have self-cleaning properties. Without photocatalytic components like TiO_2 or ZnO particles, Thompson *et al.* described a superhydrophilic film with self-cleaning capabilities utilizing commercial SNOWTEX® aqueous colloidal silica NP solution.[132] The particles are 50 nm in size, and the film is 180-nm thick. According to experiments, the WCA was 10°, and in light rain, bare glass was twice as effective at removing water. Li *et al.* also used colloidal silica with different morphologies to create superhydrophilic coatings on glass substrates via the dip coating method.[133] It was noted that between spherical (S1) and dendritic silica (S2–S4), the latter exhibited better antifogging performance [Figs. 7(b)–7(e)] as well as good mechanical and self-cleaning performance. An optimum branch length of 60-nm dendritic silica NPs has a WCA of 1.7° and increased the maximum light transmittance by 2.2%. On the other hand, Yang *et al.* prepared a hydrophilic silicon dioxide film on an acrylate PU surface using silicon dioxide sol by hydrolyzing tetraethoxysilane (TEOS).[134] The effect of concentration of the silicon dioxide sol on the wettability of the film was investigated and results indicated that as the ratio of the water/silicon dioxide sol increased from 0:1 to 3:1, the WCA decreased from 35° to 16°.

Hydrophilic coatings are applied in a variety of methods due to the various needs and applications. To achieve the desired coating quality, each method calls for different materials, deposition techniques, densities, and thicknesses. Each approach has advantages and disadvantages. Sol–gel coating is slow and inefficient for large volumes, but dip coating is a cost-effective process utilized for high-volume manufacturing. Instead of spraying or spinning to coat, which may result in uneven spraying or uncoated areas, dip coating may be better for ensuring uniformity and better quality.

5. Coatings That Resist Water

The idea of hydrophobic coatings arises due to the lotus blossom. In Ref. 135, Ward found that the lotus flower maintains cleanliness in its muddy surroundings and thereafter, a self-cleaning phenomenon known as the "Lotus Effect" has been thoroughly studied. In contrast to hydrophilic coatings, hydrophobic coatings can repel water, and other functionalities like antibacterial and self-cleaning have been widely investigated. This discovery led to the adaptation of hydrophobic coatings for packaging, buildings, and textiles.[136–138]

By choosing particles with smaller dispersion forces between them and low-surface-energy materials, Pirich investigated hydrophobic coatings with self-cleaning properties.[139] Strong stain resistance and self-cleaning qualities would be present on these surfaces. Hydrophobic coatings can be used for anti-icing applications as well. Studies on superhydrophobic and icephobic PU coating materials were summarized by Rabbani *et al.*[140] Although this coating was created around 50 years ago, performance has greatly increased during the past 10 years. According to studies, a hydrophobic surface makes it harder for ice to adhere.[141] To mass-produce low-cost, high-performance hydrophobic coatings, several challenges and technical constraints must be overcome. In addition, the development of graphene-based nanocomposite hydrophobic coatings for anti-corrosion was also studied by others.[142–144]

To meet industrial demands, a variety of methods are utilized to produce hydrophobic surfaces with acceptable wettability. These comprise lithography, polymer replica, etching, casting, sol–gel, self-assembly, and electro-deposition. The methods first result in rough surfaces, then coatings of low surface energy such as silanes, thiols, azides, or long-chain fatty acids are applied. Among various polymer systems, silicon-based materials and polymers are widely investigated for their hydrophobic effect.[145–151] Based on functionalized

SiO$_2$ NPs, Schaeffer et al. created a superhydrophobic coating with optical transparency.[152] The coating has low roughness, a WCA higher than 160°, and is mechanically durable. Such a coating technique may be used to lessen dust accumulation and window and mirror fogging. A melamine sponge with functionalized silica/graphene oxide (GO) wide ribbon was used by Mao et al. to generate a multifunctional coating.[153] It is extremely pliable, flame-resistant, and has a 130.3° WCA. In a two-step process, a paper substrate was dip-coated in fumed SiO$_2$ and carbon black/CNT/methyl cellulose suspensions. This paper was developed by Li et al. for high-performance paper-based strain sensors.[154] The coating was self-cleaning, anti-corrosive, and had a 154° WCA. Following cyclic stress, the paper-based strain sensor demonstrated reliable and repeatable strain-detecting behavior while maintaining its hydrophobic coating. Multilayered GO coatings and silicon coatings with a 156° WCA are examples of other coating compositions.[155]

To make hydrophobic coatings, a variety of silane precursors have also been used. To produce superhydrophobic and structurally robust PDMS foams, Zhang et al. used an ultrafast flame treatment.[156] Using GO and ammonium polyphosphate (APP) to create a coating on PU foam via the electrostatic spray method, Guo et al. produced very hydrophobic and flame-resistant coatings.[157] After silane functionalization, the water had a CA of 158.4°. PDMS-coated 3D graphene nanoribbons were produced by Cao et al. and Qiang et al. and they are electrically conductive, stable structurally, and hydrophobic.[158,159] Through spraying and co-curing, Xie et al. produced a superhydrophobic coating using ZnO and acrylic PU.[160] This method strived to address the issue of the typical way of preparing nanocomposite coatings, whereby surface NPs are trapped in the matrix resin, thus impacting their anti-fouling abilities. The coating demonstrated good stability as well as abrasion resistance. It can potentially be used in marine environments since fewer Chlorella were absorbed by the coating as compared to untreated.

Using a polymethylhydrosiloxane (PMHS)/TEOS hybrid material, Tai et al. fabricated a hydrophobic ceramic hollow fiber membrane (CHFM) for membrane distillation (MD).[161] This hybrid technology separates vaporized solutes from feed solutions via a hydrophobic porous membrane at moderately high temperatures. The CHFM was first dip-coated with PMHS/TEOS hybrid solution, followed by post-coating spinning to promote pore formation and enhance the gas permeability [Fig. 8(a)]. With increasing PMHS concentrations, higher hydrophobicity was achieved with WCA ranging from 108.2° to 124.1°, attributing to more $-CH_3$

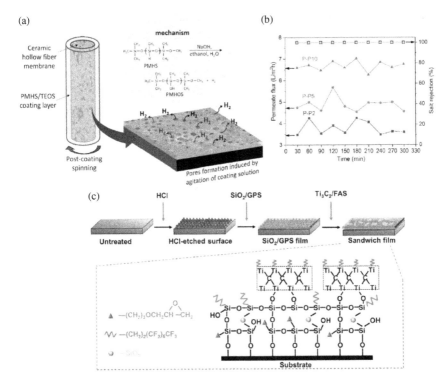

Fig. 8. (a) Schematic formation of pores on the membrane coating layer via post-coating spinning. (b) MD performances of CHFM coated with hybrid solutions of different PMHS concentrations at 35,000-ppm saline feed water (DCMD conditions: feed temperature: 80°C; permeate temperature: 10°C; feed crossflow velocity: 0.018 m/s; and permeate crossflow velocity: 0.022 m/s). (c) Schematic illustration of the fabrication process of the functional composited surface coatings.

Source: Panels (a) and (b) reproduced with permission from Ref. 161. Copyright 2021 Elsevier. Panel (c) reproduced with permission from Ref. 163. Copyright 2021 Elsevier.

groups and higher surface roughness. When treated with a 35,000-ppm saline solution, the surface-modified CHFM membranes achieved a salt rejection of >99.98% and a flux of 6.7 L/m$^2 \cdot$h [Fig. 8(b)]. This work demonstrated an effective and environmentally friendly approach to fabricating hydrophobic ceramic membranes for MD desalination.

To investigate the effects that different polysiloxane matrices have on superhydrophobic glass coatings for electrical insulators, Ribeiro *et al.* used trimethoxy (2,2,4-trimethyl pentyl)-based silane, a mixture of methyl methoxy siloxane and methylsilsesquioxane, and a mixture of dimethyl siloxane, methylsilsesquioxanes, and *n*-octyl silsesquioxanes and mixed them with fillers such as alumina trihydrate, octyl silane-treated silica, and PDMS-treated silica.[162] It was noted that the type of polysiloxane did not affect the WCA significantly but the sliding angle ranged from 5° to 90° due to the different micrometric spacing and roughness values in the coating. Furthermore, for the PDMS-treated silica coating, it was revealed that the type of polysiloxane affects the accumulated electrical charges on the surface while no difference was noted in the case of octyl-silica-treated silica coating. It was concluded that the mixture of methyl methoxy siloxane and methylsilsesquioxane loaded with PDMS-treated silica has the best performance due to its overall better chemical compatibility.

Using a combination of silane and SiO_2, Nie *et al.* fabricated a superhydrophobic surface coating on a metal surface with self-cleaning, antimicrobial, and anti-corrosive properties based on a "sandwich" structure.[163] As shown in Fig. 8(c), the surface of the bottom layer was chemically etched, with an SiO_2-hybridized silane layer in the middle and an Mxene (Ti_3C_2)-hybridized silane layer on the top. The chemically etched surface has a rough surface which facilitates the formation of the superhydrophobic layer. The silane molecules in the middle layer are crosslinked, forming a film with low surface energy and having superhydrophobicity and anti-corrosive properties, while Mxene at the top provides antimicrobial protection. With this design and structure, enhanced compactness, mechanical strength, wearing resistance, and anti-corrosion properties can be realized and it can potentially be applied to clinical and medical devices.

Similarly, Qi et al. prepared a coating with superhydrophobicity and flame-retardant properties on cotton fabric by using P, P-diphenyl-N-(3-(trithoxysilyl) propyl) phospinic amide (DPTES) and PDMS@SiO$_2$ via sol–gel method and brush-coating.[164] The coating has a WCA of 154° [Figs. 9(a) and 9(d)], exhibited good hydrophobicity with a slip angle (SA) of 8°, and self-cleaning properties [Figs. 9(b)–9(f)]. Limiting Oxygen Index (LOI) of the fabric increased from 18% to 26% with self-extinguishing abilities in the vertical combustion test. The peak heat release rate (PHRR) and total release heat (THR) decreased by 21.6% and 25.26%, respectively, and a high char residue of 20.3% was reported. In addition, the WCA of the coating remained greater than 120° after 30 friction tests.

Besides silane precursors and SiO$_2$, alternative materials have been used by researchers to create hydrophobic coatings as well.

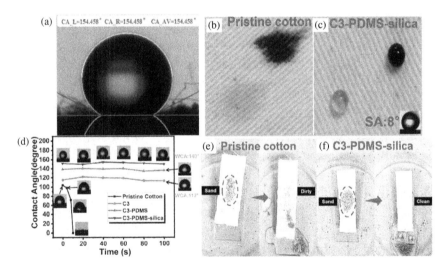

Fig. 9. (a) The WCA of the C3–PDMS–silica reached 154°. (b), (c) Hydrophobicity test of pristine cotton and C3–PDMS–silica. (d) The WCAs of the pristine cotton, C3, C3–PDMS, and C3–PDMS–silica. (e), (f) Self-cleaning performances of pristine cotton and C3–PDMS–silica.
Source: Reproduced with permission from Ref. 164. Copyright 2022 Elsevier.

Using a specially developed jet nebulizer spray pyrolysis process, Latthe et al. produced a CuO coating with superhydrophobicity and good self-cleaning capability.[165] To create a thin CuO nanolayer, $CuCl_2$ was used as a precursor. According to SEM studies, changing the precursor volume results in NPs that are between 30 nm and 100 nm in size. Due to its great roughness and extensive NP growth, which prevented air pockets, the surface was extremely hydrophobic. High-speed water jet impact did not cause any damage to the crystallite CuO coating. There are several host surface applications for this coating technique. Through electro-deposition on copper, Khorsand et al. created a nickel film with superhydrophobicity and flower-like structures to improve surface wettability.[166] The coating is perfect for hydrophobic coatings because it is anti-corrosive, chemically stable, durable, and has a self-cleaning function. Using nano-TiO_2 and water-based PU, Li et al. have created a superhydrophobic covering.[167] To improve wettability and dispersion, TiO_2 was treated with stearic acid while the coating's rolling angle and CA were found to be 4.7° and 153.5°, respectively. Asphalt may be coated with these superhydrophobic materials to alleviate the damage caused by water over time.

Using a layer-by-layer (LbL) assembly approach, Shi et al. created a multilayer coating consisting of hydrophobic modified APP (MAPP) and phosphorylated polyethyleneimine (P-PEI) on cotton fabric with desirable hydrophobicity and high fire retardancy.[168] As shown in Fig. 10(a), the WCA of APP is 25° while that of MAPP is 143°. This is attributed to the hydrophobic long alkyl chains of hexadecylamine, which was added to APP to fine-tune the hydrophobicity. It was noted that a 4BL P-PEI/MAPP coating has a WCA of 127.4° and it is almost independent of the number of BLs. Vertical burning test results [Figs. 10(b)–10(e)] showed that the coated fabrics exhibited a certain degree of flame-retardant capabilities, with LOI increasing from 19% for the pristine fabric to 34% for the 4BL coated fabric. PHRR and THR were found to decrease by 93.6% and 85.3% for the 7BL coating as well.

An interesting application of utilizing hydrophobic coatings was explored by Ghosh and Katiyar.[169] They used nano-chitosan (NCS)

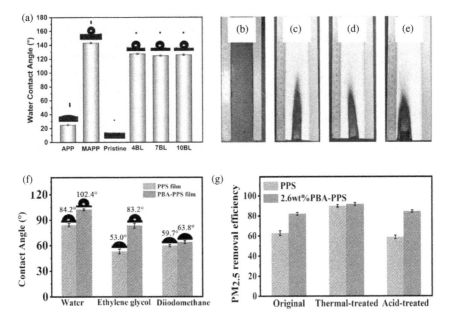

Fig. 10. (a) WCAs of different samples. Digital photos after vertical burning tests of different cotton fabrics: (b) pristine, (c) 4BL, (d) 7BL, and (e) 10BL. (f) Contact angles and photos of test liquid (water, EG, and DIM) droplets on the PPS film and PBA–PPS film. (g) $PM_{2.5}$ removal efficiencies of PPS and 2.6-wt.% PBA–PPS fabrics at high temperature (200°C).

Source: Panels (a)–(e) reproduced with permission from Ref. 168. Copyright 2022 Elsevier. Panels (f) and (g) reproduced with permission from Ref. 170. Copyright 2022 Elsevier.

to functionalize starch (ST)/guar gum (GG) biocomposite to fabricate edible food packaging for cut apples. Realizing that the shortcomings of ST and GG such as hydrophilicity and low water vapor permeability may affect their use as edible coatings, NCS was added to tailor and enhance the properties such as wettability, transparency, mechanical, etc. WCA of the ST–GG-based biocomposite increased from 46.8° to 114.3° while thermal stability improved as well. Additionally, delayed weight loss of the cut apples was reported as compared to control over a period of five days due to reduced moisture loss, thus demonstrating that overall better preservation of the cut fruit can be achieved.

Meanwhile, Bai et al. coated a thin layer of polybenzoazine (PBA) over polyphenylene sulfide (PPS) nonwoven fabric that is used in filtration bags to remove pollutants in power and waste incineration plants.[170] This additional coating served as a protective layer to offer oxidation and thermal resistance to the PPS fibers and enhance the overall structural stability. WCA test based on three different liquids indicated that the PBA–PPS film has higher hydrophobicity due to the low surface energy of PBA [Fig. 10(f)]. Tensile strength and Young's modulus of the coated fabric increased by 15% and 26%, respectively, and more importantly, the fabric with 2.6-wt.% PBA–PPS attained a higher $PM_{2.5}$ removal efficiency (89.6%) compared to the PPS fabric [Fig. 10(g)]. The introduction of hydrophobicity to the PPS fibers prevents acidic gases and moisture from affecting the filtration performance by protecting it against acid droplet corrosion.

Popular coating methods for surface modification and roughening include LbL, electrospinning, lithography, sol–gel, solution immersion, plasma, and template approaches. Superhydrophobic coatings on a variety of substrates provide the built environment with anti-fouling, anti-adhesion, self-cleaning, and improved mechanical properties. Even while there have been a lot of research works, much of them are still lab-based and not applied to the production of things for sale. Future difficulties include tying coating performance and cost together.

There are limits to superhydrophobic coating. Large surfaces that are sprayed or spun-coated may have irregular hydrophobic coats.[171] Second, each approach uses a specific base material since a superhydrophobic coating that works on one substrate could not work on another. Additionally, harmful or harsh substances, such as fluorine-based goods, may pose a risk to human health and the environment.

6. Solar Reduction Coatings

Towns are heating up and becoming less comfortable as a result of rapid urbanization and climate change. To improve energy efficiency, living comfort, and pollution emissions, many strategies utilizing cutting-edge technologies are being implemented. Santamouris *et al.* examined a variety of cooling options for buildings, such as phase-change, cool-colored, reflective, and cooling options.[172] To mitigate the urban heat island effect, Santamouris studied ways to raise the albedo of cities.[173] The enhanced albedo of reflecting roofs (cool roofs) displayed a significant moderating influence. An albedo of 0.9 for the roof resulted in decrements of 0.6 K and 0.3 K for the daily maximum and minimum urban temperatures, respectively. Buildings under 10 m may benefit from having green roofs (vegetables on the roof). In addition, he also assessed the advantages of indoor environmental quality improvements from passive cooling techniques in buildings and other structures.[174] Ventilation, evaporative cooling, and the usage of a heat sink are other examples of energy-efficient and user-friendly solutions. All of these passive cooling techniques may help to combat heat on islands, bring comfort to buildings without air conditioning, and ease the cooling load on buildings with thermostats.

Recent research on nanomaterial-based solar cool coatings (nSCCs) for passive cooling has been made by Zheng *et al.*[175] The coatings are transparent and by limiting the indoor heat gain, cooling loads can be reduced by up to 60%. Because they may be incorporated into paints with a small amount of 10 wt.% or less loading of nanomaterials and can be cast onto window planes, transparent nSCCs are more visually attractive than opaque ones. Metal oxides including ATO, ITO, and TiO_2 are used in transparent nSCCs to reflect solar radiation. Alkali metal-doped tungsten oxides are used in transparent nSCCs to absorb solar radiation. With the consideration that coatings of white color are not desirable on buildings or other objects in terms of functionality or esthetics, Chen *et al.* developed a paintable bilayer coating using different colors for efficient radiative cooling.[176] This bilayer consists of a thin top layer that can absorb visible wavelengths and an underlayer that reflects near-to-short wavelength infrared (NSWIR) light to reduce solar heating. Two different types of underlayers, porous poly(vinylidene fluoride-co-hexafluoropropene) [P(VdF–HFP)] and TiO_2/polymer composite paint, were used and the results showed that the bilayer performed substantially better than its monolayer counterparts regardless of the color used (Fig. 11). Using the same color as a comparison, the porous P(VdF–HFP) bilayer also increases the NSWIR reflectance by 0.51 and up to 0.89 for the blue color coating. On the other hand, Son et al. prepared a metal-free matte white PDVF/PU acrylate (PUA)(PDVF/PUA) coating via photo-initiated free-radical polymerization for radiative coolers.[177] The stability and durability of the coating were improved by utilizing

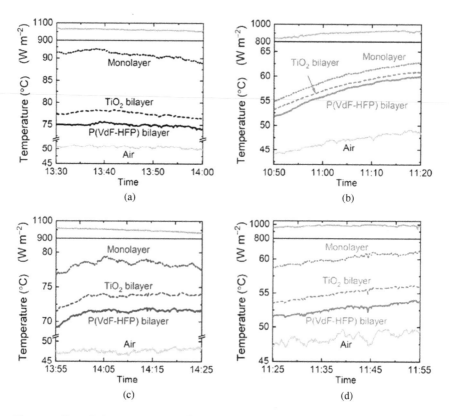

Fig. 11. Detailed solar intensity (top part of the y-axis) and temperature (lower part of the y-axis) data in the outdoor tests of the (a) black, (b) red, (c) blue, and (d) yellow cooling paints, respectively.

Source: Reproduced with permission from Ref. 176. Copyright 2020 American Association for the Advancement of Science.

PUA while the high reflectivity of PDVF and PDVF/PUA mixture reduced the solar absorption. An average emissivity of 0.9333 and a solar reflectance of 0.9336 were reported while a temperature reduction of 8.28°C was attained. The antiglare surface of the coating can reduce light pollution whilst the absence of Ag particles prevents solar absorption owing to their oxidation.

To decrease solar gains in buildings and boost energy efficiency, thermochromic (TC) coating can be used as well. Thermochromism refers to a change in the spectral properties of a material brought about by cooling or heating. Raising the temperature to above a defined temperature of a TC coating results in a change of color from a darker to a lighter tone. The molecular structure of the TC pigments will undergo a thermally reversible transformation, producing a spectral change of color. As the temperature drops below the color-changing point, the system will be reverted to a thermally stable state. TC systems can serve as energy-saving systems. During summertime, they can reflect solar energy and reduce the surface's temperature and in the wintertime, solar energy is absorbed and the surface temperature is raised. To reduce building energy consumption, Liu et al. developed TC coatings using TC microcapsule powder, Portland cement, and other additives.[178] When the ambient temperature is lower than the switching temperature, the coating has a dark color and a low solar reflectance while an ambient temperature higher than the switching temperature results in a light-colored coating with high solar reflectance. Simulation studies indicated that the TC coating can effectively reduce annual energy consumption in winter and summer seasons, in addition to having UV aging resistance and corrosion resistance properties. Its self-cleaning capability also helps to improve the long-term stability of the coating by preventing the accumulation of dust on its surface. Similarly, Zhang et al. investigated the use of TC coatings on asphalt pavement for dynamic temperature regulation and to alleviate the urban heat island effect.[179] It was revealed that in terms of visual effect and reflectance characteristics, red and black TC materials were more suitable as coatings and the optimal TiO_2 contents were 2 wt.% and 5 wt.%, respectively.

In addition, the TC coatings have a cooling effect of 2–5°C higher than those of cool coatings of the same color and can prolong the time needed for the road surface temperature to drop to 0°C considerably. Simulation results also indicated that TC pavement can lower the air temperature at 1.5 m by 1.5°C compared to conventional pavement.

A major drawback with TC materials is photodegradation, caused by prolonged exposure to the outdoor environment. Solar radiation will break or result in the crosslinking of the polymer chains, thus altering their mechanical and chemical properties and leading to the loss of the reversible TC effect. Different techniques have been used to try to decrease the degradation rate of TC coatings and to improve their outdoor performance, such as the incorporation of UV protectors, UV filters, and UV absorbers. However, none of these techniques can resolve all the issues and they prove that UV radiation is not the only factor that affects thermochromism. To investigate how thermochromism is affected by other factors besides UV radiation, Karlessi and Santamouris used a combination of optical and UV filters and applied TC coatings on concrete tiles with accelerated aging conditions.[180] Interestingly, they found that by covering the coating with a red filter which will cut off wavelengths below 600 nm, it can protect the reversible color change of the TC coating most efficiently as the solar reflectance was unaffected in the dark phase.

On hot, bright days, indoor cooling can be achieved to a certain degree by using coatings with solar reduction capabilities. These coatings can be transparent or opaque, depending on how they are applied to building exteriors. Buildings that utilize these coatings may result in lower interior temperatures by several degree centigrade, enhancing the thermal comfort. Coatings for solar reduction can be made better in the future. By improving the durability of coatings against sunlight and heat, lowering cost, adjusting the optical transparency and color hue with nano-additives while assessing and minimizing their associated risks, it may be accomplished.

7. Solar Utilization Coatings

Besides reducing the amount of solar heat gain, coatings can be used to harvest solar energy for practical applications. Solar energy is the cleanest and most abundant renewable energy source and it can be harnessed for a variety of applications. Solar energy conversion technologies such as PVs and photocatalysis have received much attention due to the need to meet global energy demand.[181–186]

PV technology is the most cost-effective method for producing electricity from renewable sources and therefore, the installation of solar PV modules has risen rapidly globally to reduce carbon emissions and accelerate renewable energy efforts.[187] Even though this technology has existed for over half a century, it is still costly due to its maintenance cost and low conversion efficiency. PV modules suffer from decreasing power output due to the high-temperature effect and reflection losses. Approximately 8% of incident light in the wavelength range of 380–1100 nm is reflected off the PV cover glass due to the difference between the refractive indexes of air and the glass or materials.[188] Nevertheless, this issue can be mitigated by employing an anti-reflection coating with a low refraction index to increase transmitted light and reduce reflectance losses. To achieve high optical and photocatalytic properties for PV modules, Biswas *et al.* fabricated a dual-functional coating (self-cleaning and anti-reflective) using highly crystalline TiO_2 and MgF_2 NPs via the solvothermal method.[189] Besides enhancing the transmittance by 5% compared to uncoated glass in the 380–1100-nm wavelength range, there is a net 5–33% increase in transmittance from 10° to 80° angle of incidence with high omnidirectional broadband anti-reflective property. The PV module with the coated glass also displayed a 3.5% enhancement in power conversion efficiency compared to its uncoated counterpart. On the other hand, to control the visibility of embedded solar cells in building-integrated PV (BIPV) windows for esthetic reasons, Yang *et al.* modified the surface of the front glass by etching and applied a thin film consisting of niobium pentoxide as a back-coating.[190] It was noted that the surface modification of the front glass resulted in enhanced light absorption due to light trapping and anti-reflective properties while the metal oxide film reflects the incident light to the mono-Si solar cells. As a result, the BIPV module has a power conversion efficiency of 15.23% and a fill factor of 65.05%, making this strategy a promising solution to integrate BIPV windows with invisibility and high efficiency.

Photoluminescence materials can emit light at long wavelengths and absorb light at short wavelengths and it has been reported that these materials can vastly improve the conversion efficiency of PV in solar cells.[191,192] In light of this, Huang et al. developed a coating with both photoluminescence and anti-reflection properties to improve solar cells' efficiency.[193] To tune the coating's refractive index, different amounts of PEG were added to the ZnO–SiO$_2$ composite sol. It was noted that the addition of 32-wt.% PEG resulted in a superhydrophilic and high anti-reflective coating, with a maximum transmittance of 98.37% at 615 nm and an average transmittance of 95.64% between 400 nm and 1100 nm [Fig. 12(a)]. To measure the total PV conversion performance, a current–voltage (I–V) test was conducted for the coatings of different thicknesses (125, 220, 320, and 505 nm) [Fig. 12(b)]. Based on the results, a coating with a thickness of 125 nm can increase the photoelectric conversion efficiency of Si cells by 6.25%.

CSP is one of the most competitive solar energy utilization technologies due to its cost-effective thermal storage and capacity for large-scale generation of electricity. The high operating temperature of the solar receiver (700–800°C) requires the solar absorber coating (SAC) to be thermally stable and possess high optical performance. This coating is applied on the surface of the receiver tubes to absorb as much solar energy and at the same time keep the re-emittance to a minimum. To increase the efficiency of the solar receiver, higher temperatures may be used but the performances of the SAC will be compromised due to mechanical and optical degradations. To boost the thermal tolerance of metal–dielectric SAC, Wang et al. prepared a multilayer coating consisting of AlCrN/AlCrON as an absorber, Cr as an IR reflector, and AlCrO

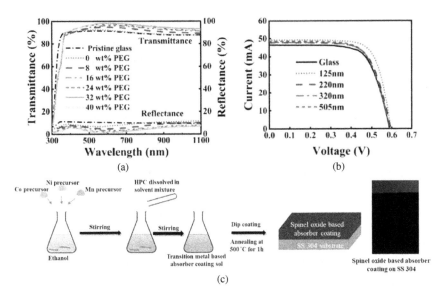

Fig. 12. (a) Transmittance and reflectance spectra of coatings prepared with different amounts of PEG. (b) The I-V curves are measured from samples of pure glass and those of glass with the $(ZnO_{40wt.\%}-SiO_2)_{68wt.\%}-PEG_{32wt.\%}$ coatings with thicknesses of 125, 220, 320, and 505 nm covered on solar cells. (c) Schematic of absorber coatings development process on SS 304 substrate with the digital image of absorber coating on SS 304.

Source: Panels (a) and (b) reproduced with permission from Ref. 193. Copyright 2023 Elsevier. Panel (c) reproduced with permission from Ref. 194. Copyright 2023 Elsevier.

as an anti-reflective layer and deposited them on an stainless steel (SS) substrate.[195] The formation of nitride NPs due to the partial recrystallization of the amorphous matrices facilitates absorption as a result of an increased optical path. This led to an increase in selectivity from 0.919 to 0.928 for the solar absorptance and a decrease from 0.128 to 0.105 for the emittance. High thermal stability was also achieved due to the formation of the amorphous matrices in the absorber and anti-reflective layer, which prevents agglomeration of NPs and interface degradation.

To improve the photothermal conversion efficiency of SAC in solar thermal systems, Kumar et al. developed a cobalt-rich transition metal oxide-based absorber coating.[194] The absorber coating sol was prepared via a wet chemical route and dip-coated on SS 304 substrate [Fig. 12(c)]. The coating has high solar emittance and absorbance of 0.13 and 0.91, respectively, while a porous structure with sub-micron-scaled surface protrusions resulted in a wide angular solar absorptance range of 0.97–0.96 over the incidence angles of 10–40°. More importantly, it displayed good thermal stability at 400°C for 100 h with a photothermal conversion of 87%. The performance of an SAC is typically investigated using heat treatments but varying the concentration of the flux conditions will also provide valuation information such as the photothermal degradation performance. Martínez-Manuel et al. studied the thermal efficiencies of different SACs [Pyromark®2500, Solkote®, Thurmalox®250, Comex®, and a new Soot from Forest Biomass (SFB)-based coating] by applying low (100 ± 3 kW/m^2) and high (415 ± 12 kW/m^2) levels of concentrated irradiance using a high-flux solar simulator.[196] For the low flux level, the results indicated that the different coatings have largely similar thermal efficiencies, ranging from 83.24% to 91.74%. This suggests that other factors such as availability, cost, and environmental friendliness are important considerations as well. At a high flux level, the efficiencies ranged from 72.69% to 88.69%, with Pyromark®2500 having the highest efficiency in both cases. Solar absorptance of an SAC is the determining factor here although resistance to thermal stress and degradation may affect the coating efficiency. Interestingly, SFB only has 3.7% lower coating efficiency than Pyromark®2500 at low flux level and can potentially be used for low-to-medium-temperature levels (≤150°C). Being a by-product of combustion waste, it may be implemented as a sustainable strategy to prepare SACs, in addition to having low cost, environmental friendliness, and ease of fabrication.

8. Insight

This review summarizes recent developments in functional coatings pertinent to the built environment using nanomaterials. They include PCM, photocatalytic, hydrophilic, hydrophobic, solar reduction, and solar utilization coatings. By absorbing outside heat rather than allowing it to enter the building, PCM coatings control interior temperature. Both hydrophobic and hydrophilic coatings can clean themselves, particularly the photocatalytic coatings that break down contaminants in the presence of light. Solar heat gain can be reduced using solar heat reduction coatings, which reflect sunlight to use less energy for cooling the inside while solar energy can be harvested for a myriad of applications. Together, these coatings on building components, particularly the exterior façade, have the potential to create "smart" self-regulating systems that use water and energy more effectively, providing and moving in the direction of a more sustainable and environmentally friendly way of life.

Technologies for functional coatings are becoming more common. The generation of an efficient coating may be hampered by the diverse properties of different materials as well as the challenges faced during the coating process. Although flexible, quick, and reasonably priced, spray coating needs careful attention to spray angle, pattern, and distance. While being flexible and long-lasting, sol–gel coating is a relatively slow process. Regardless of the technique used, the benefits ought to be enhanced and drawbacks be reduced. The cost of materials, stability, and safety must all be taken into consideration to encourage widespread use. Reduced usage of nano-additive loadings or identification of substitutes of lower cost that have comparable properties and performance to cut material prices can be explored. Numerous types of NPs still pose safety risks and are hazardous to human health. To prevent NPs from leaking into the environment, innovative engineering can be employed to encapsulate or immobilize them. The long-term stabilities of these coating materials are imperative because they will be subjected to harsh weather conditions for a very long period. R&D on cutting-edge coating materials should continue in the meantime. Organic polymers and nanomaterials offer lower costs and toxicity.

By maximizing resource efficiency and occupant comfort, health, and well-being, functional coatings play a significant role in the sustainability of the built environment such as the development of green buildings. They now perform new tasks in addition to serving esthetic and protective purposes. The broad field of sustainable living may also be relevant to and significant for other nanotechnology-based functional coatings, such as antimicrobial and anti-fouling coatings. The future of functional coatings with enhanced performance and usage for the built environment is anticipated to be facilitated by ongoing research in sustainable materials.

Conflict of Interest

The authors declare no conflict of interest.

Acknowledgments

The authors acknowledge the support under the Structural Metal Alloy Program (SMAP) (Grant No. A18b1B0061) of the Agency for Science, Technology and Research (A*STAR) in Singapore.

ORCID

Pin Jin Ong https://orcid.org/0000-0002-7284-9906
Suxi Wang https://orcid.org/0000-0003-3638-9401
Ady Suwardi https://orcid.org/0000-0002-7342-0431
Jing Cao https://orcid.org/0000-0002-7853-573X
Fuke Wang https://orcid.org/0000-0002-6980-5857
Xuesong Yin https://orcid.org/0000-0003-1335-8506
Pei Wang https://orcid.org/0000-0002-0881-0871
Fengxia Wei https://orcid.org/0000-0002-2058-5056
Dan Kai https://orcid.org/0000-0002-2330-3480
Enyi Ye https://orcid.org/0000-0002-2398-0676
Jianwei Xu https://orcid.org/0000-0003-3945-5443
Ming Hui Chua https://orcid.org/0000-0002-3789-7750
Warintorn Thitsartarn https://orcid.org/0000-0002-9855-9421
Qiang Zhu https://orcid.org/0000-0002-1184-0860
Xian Jun Loh https://orcid.org/0000-0001-8118-6502

References

1. S. Sugiarto, R. Pong, Y. Tan, Y. Leow, T. Sathasivam, Q. Zhu, X. Loh and D. Kai, *Mater. Today Chem.*, 2022, **26**, 101022.
2. F.-J. Wang, L.-Q. Wang, X.-C. Zhang, S.-F. Ma and Z.-C. Zhao, *J. Food Eng.*, 2022, **332**, 111039.
3. V. C. C. Checchin, A. Gonzalez, M. Bertuola and M. A. F. L. de Mele, *Prog. Org. Coat.*, 2022, **172**, 107068.
4. D. Tan, B. Xu, Y. Gao, Y. Tang, Y. Liu, Y. Yang and Z. Li, *Nano Energy*, 2022, **104**, 107873.
5. A. Babu, G. Perumal, H. Arora and H. Grewal, *Renew. Energy*, 2021, **180**, 1044–1055.
6. I. Rodrigues, T. M. Mata and A. A. Martins, *J. Clean. Prod.*, 2022, **367**, 133011.
7. K. Khaled and U. Berardi, *Energy Build.*, 2021, **244**, 111022.
8. V. Bui, H. Liu, Y. Low, T. Tang, Q. Zhu, K. Shah, E. Shidoji, Y. Lim and W. Koh, *Energy Build.*, 2017, **157**, 195–203.
9. Q. Zhu, M. H. Chua, P. J. Ong, J. J. C. Lee, K. L. O. Chin, S. Wang, D. Kai, R. Ji, J. Kong and Z. Dong, *Mater. Today Adv.*, 2022, **15**, 100270.
10. R. Majidi, M. Keramatinia, B. Ramezanzadeh and M. Ramezanzadeh, *Polym. Degrad. Stab.*, 2023, **207**, 110211.
11. P. J. Ong, Z. M. Png, X. Y. D. Soo, X. Wang, A. Suwardi, M. H. Chua, J. W. Xu and Q. Zhu, *Mater. Chem. Phys.*, 2022, **277**, 125438.
12. Y. Wu, J. Wei, X. Shi and W. Zhao, *J. Mater. Sci. Technol.*, 2023, **148**, 222–234.
13. M. C. Chen, P. W. Koh, V. K. Ponnusamy and S. L. Lee, *Prog. Org. Coat.*, 2022, **163**, 106660.
14. Y. Wu, J. Du, G. Liu, D. Ma, F. Jia, J. J. Klemeš and J. Wang, *Renew. Energy*, 2021, **185**, 1034–1061.
15. S. Nundy, A. Ghosh and T. K. Mallick, *ACS Omega*, 2020, **5**, 1033–1039.
16. N. K. Ilango, P. Gujar, A. K. Nagesh, A. Alex and P. Ghosh, *Cem. Concr. Compos.*, 2021, **115**, 103856.
17. A. F. Baldissera, M. R. da Silva Silveira, C. H. Beraldo, N. S. Tocchetto and C. A. Ferreira, *J. Mater. Res. Technol.*, 2019, **8**, 2832–2845.

18. Y. Chen, Y. Wu and W. Zhao, *Carbon*, 2023, **202**, 196–206.
19. X. Zhou, Q. Fu, Z. Zhang, Y. Fang, Y. Wang, F. Wang, Y. Song, C. U. Pittman, Jr. and Q. Wang, *J. Clean. Prod.*, 2021, **321**, 128949.
20. J. Machotová, A. Kalendová, M. Voleská, D. Steinerová, M. Pejchalová, P. Knotek and L. Zarybnicka, *Prog. Org. Coat.*, 2020, **147**, 105704.
21. S. Wang, P. J. Ong, S. Liu, W. Thitsartarn, M. J. B. H. Tan, A. Suwardi, Q. Zhu and X. J. Loh, *Chem. Asian J.*, 2022, **17**, e202200608.
22. M. M. El-Kady, I. Ansari, C. Arora, N. Rai, S. Soni, D. K. Verma, P. Singh and A. E. D. Mahmoud, *J. Mol. Liq.*, 2022, **370**, 121046.
23. Z. Xu, D. Zhao, J. Lu, J. Liu, G. Dao, B. Chen, B. Huang and X. Pan, *Chem. Eng. J.*, 2022, **455**, 140927.
24. M. M. Azim, A. Arifutzzaman, R. Saidur, M. Khandaker and D. Bradley, *J. Mol. Liq.*, 2022, **360**, 119443.
25. M. A. Shah, B. M. Pirzada, G. Price, A. L. Shibiru and A. Qurashi, *J. Adv. Res.*, 2022, **38**, 55–75.
26. M. Haris, T. Hussain, H. I. Mohamed, A. Khan, M. S. Ansari, A. Tauseef, A. A. Khan and N. Akhtar, *Sci. Total Environ.*, 2022, **857**, 159639.
27. L. Wang, M. P. Teles, A. Arabkoohsar, H. Yu, K. A. Ismail, O. Mahian and S. Wongwises, *Sustain. Energy Technol. Assess.*, 2022, **54**, 102864.
28. A. K. Parameswaran, J. Azadmanjiri, N. Palaniyandy, B. Pal, S. Palaniswami, L. Dekanovsky, B. Wu and Z. Sofer, *Nano Energy*, 2022, **105**, 107994.
29. S. Sikiru, O. A. Ayodele, Y. K. Sanusi, S. Y. Adebukola, H. Soleimani, N. Yekeen and A. A. Haslija, *J. Environ. Chem. Eng.*, 2022, **10**, 108065.
30. T. P. Vo, M. H. Chua, S. J. Ang, K. L. O. Chin, X. Y. D. Soo, Z. M. Png, T. L. D. Tam, Q. Zhu, D. J. Procter and J. Xu, *Polym. Chem.*, 2022, **13**, 6512–6524.
31. N. Ding, X. Wang, Y. Hou, S. Wang, X. Li, D. W. H. Fam, Y. Zong and Z. Liu, *Solid State Ion.*, 2018, **323**, 72–77.
32. T. Sreethawong, K. W. Shah, S.-Y. Zhang, E. Ye, S. H. Lim, U. Maheswaran, W. Y. Mao and M.-Y. Han, *J. Mater. Chem. A*, 2014, **2**, 3417–3423.
33. J. Zheng, C.-G. Wang, H. Zhou, E. Ye, J. Xu, Z. Li and X. J. Loh, *Research*, 2021, **2021**, 3750689.

34. X. J. Loh, S. H. Goh and J. Li, *J. Phys. Chem. B*, 2009, **113**, 11822–11830.
35. X. J. Loh, W. Guerin and S. M. Guillaume, *J. Mater. Chem.*, 2012, **22**, 21249–21256.
36. X. J. Loh, *J. Appl. Polym. Sci.*, 2013, **127**, 992–1000.
37. X. J. Loh, V. P. N. Nguyen, N. Kuo and J. Li, *J. Mater. Chem.*, 2011, **21**, 2246–2254.
38. Z. Li and X. J. Loh, *Wiley Interdiscip. Rev. Nanomed. Nanobiotechnol.*, 2017, **9**, e1429.
39. E. Ye, H. Tan, S. Li and W. Y. Fan, *Angew. Chem., Int. Ed.*, 2006, **45**, 1120–1123.
40. Z. W. Seh, S. Liu, S.-Y. Zhang, K. W. Shah and M.-Y. Han, *Chem. Commun.*, 2011, **47**, 6689–6691.
41. K. W. Shah, T. Sreethawong, S.-H. Liu, S.-Y. Zhang, L. S. Tan and M.-Y. Han, *Nanoscale*, 2014, **6**, 11273–11281.
42. O. Y. Wen, M. Z. M. Tohir, T. C. S. Yeaw, M. A. Razak, H. S. Zainuddin and M. R. A. Hamid, *Prog. Org. Coat.*, 2023, **175**, 107330.
43. M. S. Özer and S. Gaan, *Prog. Org. Coat.*, 2022, **171**, 107027.
44. Q. Zhang, L. Ma, T. Xue, J. Tian, W. Fan and T. Liu, *Compos. B, Eng.*, 2023, **248**, 110377.
45. Z. Jiang, H. Li, C. Zhang and P. Zhu, *Surf. Interfaces*, 2022, **34**, 102409.
46. R. Teijido, L. Ruiz-Rubio, A. G. Echaide, J. L. Vilas-Vilela, S. Lanceros-Mendez and Q. Zhang, *Prog. Org. Coat.*, 2022, **163**, 106684.
47. C. I. Idumah, C. M. Obele, E. O. Emmanuel and A. Hassan, *Surf. Interfaces*, 2020, **21**, 100734.
48. S. Pourhashem, A. Seif, F. Saba, E. G. Nezhad, X. Ji, Z. Zhou, X. Zhai, M. Mirzaee, J. Duan and A. Rashidi, *J. Mater. Sci. Technol*, 2022, **118**, 73–113.
49. S. S. A. Kumar, S. Bashir, K. Ramesh and S. Ramesh, *Prog. Org. Coat.*, 2021, **154**, 106215.
50. M. Cheng, Q. Fu, B. Tan, Y. Ma, L. Fang, C. Lu and Z. Xu, *Prog. Org. Coat.*, 2022, **167**, 106790.
51. S. García, H. Fischer and S. van der Zwaag, *Prog. Org. Coat.*, 2011, **72**, 211–221.
52. X. Han, S.-X. Wang, F. Zhong and Y. Lu, *Synthesis*, 2011, **2011**, 1859–1864.

53. D. Adak, R. Bhattacharyya and H. C. Barshilia, *Renew. Sustain. Energy Rev.*, 2022, **159**, 112145.
54. K. Manoharan and S. Bhattacharya, *J. Micromanuf.*, 2019, **2**, 59–78.
55. M. Bahri, S. H. Gebre, M. A. Elaguech, F. T. Dajan, M. G. Sendeku, C. Tlili and D. Wang, *Coord. Chem. Rev.*, 2023, **475**, 214910.
56. M. Bello and S. Shanmugan, *J. Alloys Compd.*, 2020, **839**, 155510.
57. L. Piperno, A. Vannozzi, G. Sotgiu and G. Celentano, *J. Eur. Ceram. Soc.*, 2021, **41**, 2193–2206.
58. H. Wang, B. Patterson, J. Yang, D. Huang, Y. Qin and H. Luo, *Appl. Mater. Today*, 2017, **9**, 402–406.
59. F. Rey-García, B. J. Sieira, C. Bao-Varela, J. R. Leis, L. A. Angurel, J. B. Quintana, R. Rodil and G. F. de la Fuente, *Sci. Total Environ.*, 2020, **742**, 140507.
60. N. Hong, Q. Zhao, D. Chen, K. M. Liechti and W. Li, *Carbon*, 2023, **201**, 712–718.
61. X. Wu, I. Wyman, G. Zhang, J. Lin, Z. Liu, Y. Wang and H. Hu, *Prog. Org. Coat.*, 2016, **90**, 463–471.
62. R. Nistico, D. Scalarone and G. Magnacca, *Microporous Mesoporous Mater.*, 2017, **248**, 18–29.
63. L. Wang, L. Wu, Y. Wang, J. Luo, H. Xue and J. Gao, *Nano Mater. Sci.*, 2022, **4**, 178–184.
64. R. K. Nishihora, P. L. Rachadel, M. G. N. Quadri and D. Hotza, *J. Eur. Ceram. Soc.*, 2018, **38**, 988–1001.
65. X. Y. D. Soo, S. Wang, C. C. J. Yeo, J. Li, X. P. Ni, L. Jiang, K. Xue, Z. Li, X. Fei and Q. Zhu, *Sci. Total Environ.*, 2022, **807**, 151084.
66. S. Wu, W. Shi, K. Li, J. Cai and L. Chen, *J. Environ. Chem. Eng.*, 2022, **10**, 108921.
67. J. K. Muiruri, J. C. C. Yeo, Q. Zhu, E. Ye, X. J. Loh and Z. Li, *ACS Sustain. Chem. Eng.*, 2022, **10**, 3387–3406.
68. Z. U. Arif, M. Y. Khalid, M. F. Sheikh, A. Zolfagharian and M. Bodaghi, *J. Environ. Chem. Eng.*, 2022, **10**, 108159.
69. S. Wang, J. K. Muiruri, X. Y. D. Soo, S. Liu, W. Thitsartarn, B. H. Tan, A. Suwardi, Z. Li, Q. Zhu and X. J. Loh, *Chem. Asian J.*, 2022, **18**, e202200972.
70. H. Acuña-Pizano, M. González-Trevizo, A. Luna-León, K. Martínez-Torres and F. Fernández-Melchor, *Constr. Build. Mater.*, 2022, **344**, 128083.

71. J. Zheng, S. F. D. Solco, C. J. E. Wong, S. A. Sia, X. Y. Tan, J. Cao, J. C. C. Yeo, W. Yan, Q. Zhu and Q. Yan, *J. Mater. Chem. A*, 2022, **10**, 19787–19796.
72. Y. Shi, C. Lee, X. Tan, L. Yang, Q. Zhu, X. Loh, J. Xu and Q. Yan, *Small Struct.*, 2022, **3**, 2100185.
73. P. Cai, B. Hu, W. R. Leow, X. Wang, X. J. Loh, Y. L. Wu and X. Chen, *Adv. Mater.*, 2018, **30**, 1800572.
74. E. Yildirim, Q. Zhu, G. Wu, T. L. Tan, J. Xu and S.-W. Yang, *J. Phys. Chem. C*, 2019, **123**, 9745–9755.
75. H. Zhou, M. H. Chua, Q. Zhu and J. Xu, *Compos. Commun.*, 2021, **27**, 100877.
76. M. H. Chua, S. H. G. Toh, P. J. Ong, Z. M. Png, Q. Zhu, S. Xiong and J. Xu, *Polym. Chem.*, 2022, **13**, 967–981.
77. Q. Zhu, E. Yildirim, X. Wang, A. K. K. Kyaw, T. Tang, X. Y. D. Soo, Z. M. Wong, G. Wu, S.-W. Yang and J. Xu, *Mol. Syst. Des. Eng.*, 2020, **5**, 976–984.
78. W. T. Neo, Q. Ye, M. H. Chua, Q. Zhu and J. Xu, *Macromol. Rapid Commun.*, 2020, **41**, 2000156.
79. X. Yong, G. Wu, W. Shi, Z. M. Wong, T. Deng, Q. Zhu, X. Yang, J.-S. Wang, J. Xu and S.-W. Yang, *J. Mater. Chem. A*, 2020, **8**, 21852–21861.
80. T. Tang, A. K. K. Kyaw, Q. Zhu and J. Xu, *Chem. Commun.*, 2020, **56**, 9388–9391.
81. K. W. Shah, S.-X. Wang, Y. Zheng and J. Xu, *Appl. Sci.*, 2019, **9**, 1511.
82. J. Cao, X. Y. Tan, N. Jia, J. Zheng, S. W. Chien, H. K. Ng, C. K. I. Tan, H. Liu, Q. Zhu and S. Wang, *Nano Energy*, 2022, **96**, 107147.
83. P. J. Ong, Z. X. J. Heng, Z. Xing, H. Y. Y. Ko, P. Wang, H. Liu, R. Ji, X. Wang, B. H. Tan and Z. Li, *Trans. Tianjin Univ.*, 2023, **29**, 225–234.
84. J. J. C. Lee, S. Sugiarto, P. J. Ong, X. Y. D. Soo, X. Ni, P. Luo, Y. Y. K. Hnin, J. S. Y. See, F. Wei, R. Zheng, P. Wang, J. Xu, X. J. Loh, D. Kai and Q. Zhu, *J. Energy Storage*, 2022, **56**, 106118.
85. J. J. C. Lee, N. J. X. Lim, P. Wang, H. Liu, S. Wang, C.-L. K. Lee, D. Kai, F. Wei, R. Ji, B. H. Tan, S. Ge, A. Suwardi, J. Xu, X. J. Loh and Q. Zhu, *World Sci. Annu. Rev. Funct. Mater.*, 2022, **1**, 2230007.
86. S. R. A. Dantas, F. Vittorino and K. Loh, *J. Build. Eng.*, 2022, **46**, 103829.

87. A. Maestri, D. L. Marinoski, R. Lamberts and S. Guths, *Energy Build.*, 2021, **245**, 111057.
88. E. Gámez-Espinosa, C. Deyá, F. Ruiz and N. Bellotti, *J. Build. Eng.*, 2022, **50**, 104148.
89. F. Farsinia, E. K. Goharshadi, N. Ramezanian, M. M. Sangatash and M. Moghayedi, *Mater. Chem. Phys.*, 2023, **297**, 127355.
90. Z. Huang and H. Ghasemi, *Adv. Colloid Interface Sci.*, 2020, **284**, 102264.
91. J. Feng, S. Liu, J.-Y. Hwang, W. Mo, X. Su, S. Ma and Z. Wei, *Constr. Build. Mater.*, 2022, **349**, 128755.
92. Z. M. Png, X. Y. D. Soo, M. H. Chua, P. J. Ong, A. Suwardi, C. K. I. Tan, J. Xu and Q. Zhu, *Sol. Energy*, 2022, **231**, 115–128.
93. K. W. Shah, P. J. Ong, M. H. Chua, S. H. G. Toh, J. J. C. Lee, X. Y. D. Soo, Z. M. Png, R. Ji, J. Xu and Q. Zhu, *Energy Build.*, 2022, **262**, 112018.
94. Z. M. Png, X. Y. D. Soo, M. H. Chua, P. J. Ong, J. Xu and Q. Zhu, *J. Mater. Chem. A*, 2022, **10**, 3633–3641.
95. S. Sikiru, T. L. Oladosu, T. I. Amosa, S. Y. Kolawole and H. Soleimani, *J. Energy Storage*, 2022, **53**, 105200.
96. J. Xu, X. Zhang and L. Zou, *J. Energy Storage*, 2022, **54**, 105341.
97. C. Li, X. Wen, W. Cai, H. Yu and D. Liu, *J. Build. Eng.*, 2022, **65**, 105763.
98. P. J. Ong, Y. Leow, X. Y. D. Soo, M. H. Chua, X. Ni, A. Suwardi, C. K. I. Tan, R. Zheng, F. Wei, J. Xu, X. J. Loh, D. Kai and Q. Zhu, *Waste Manag. (Oxford)*, 2023, **157**, 339–347.
99. Z. Ling, Y. Zhang, X. Fang and Z. Zhang, *J. Energy Storage*, 2021, **41**, 102963.
100. M. Chen, H. Liu, H. Zhang and X. Wang, *J. Energy Storage*, 2023, **57**, 106232.
101. J. Nam, J. Y. Choi, H. Yuk, Y. U. Kim, S. J. Chang and S. Kim, *Build. Environ.*, 2022, **224**, 109534.
102. B. S. Yilbas, H. Ali, A. Al-Sharafi and N. Al-Aqeeli, *RSC Adv.*, 2018, **8**, 938–947.
103. M. Shamshiri, R. Jafari and G. Momen, *Prog. Org. Coat.*, 2023, **177**, 107414.
104. W. Chalco-Sandoval, M. J. Fabra, A. López-Rubio and J. M. Lagaron, *J. Food Eng.*, 2017, **192**, 122–128.

105. F. Rodriguez-Cumplido, E. Pabon-Gelves and F. Chejne-Jana, *J. Energy Storage*, 2019, **24**, 100821.
106. X. Y. D. Soo, Z. M. Png, M. H. Chua, J. C. C. Yeo, P. J. Ong, S. Wang, X. Wang, A. Suwardi, J. Cao, Y. Chen, Q. Yan, X. J. Loh, J. Xu and Q. Zhu, *Mater. Today Adv.*, 2022, **14**, 100227.
107. S. Drissi, T.-C. Ling, K. H. Mo and A. Eddhahak, *Renew. Sustain. Energy Rev.*, 2019, **110**, 467–484.
108. G. Alva, Y. Lin, L. Liu and G. Fang, *Energy Build.*, 2017, **144**, 276–294.
109. X. Y. D. Soo, Z. M. Png, X. Wang, M. H. Chua, P. J. Ong, S. Wang, Z. Li, D. Chi, J. Xu, X. J. Loh and Q. Zhu, *ACS Appl. Polym. Mater.*, 2022, **4**, 2747–2756.
110. G. Liu, H. Xia, Y. Niu, X. Zhao, G. Zhang, L. Song and H. Chen, *Chem. Eng. J.*, 2021, **409**, 128187.
111. A. Seifi, D. Salari, A. Khataee, B. ÇoΔut, L. Ç. Arslan and A. Niaei, *Ceram. Int.*, 2023, **49**, 1678–1689.
112. C. Zhang, Y. Li, H. Shen and D. Shuai, *Chem. Eng. J.*, 2021, **403**, 126365.
113. I. Grčić, B. Erjavec, D. Vrsaljko, C. Guyon and M. Tatoulian, *Prog. Org. Coat.*, 2017, **105**, 277–285.
114. U. S. Meda, K. Vora, Y. Athreya and U. A. Mandi, *Process Saf. Environ. Prot.*, 2022, **161**, 771–787.
115. J. S. Appasamy, J. C. Kurnia and M. K. Assadi, *Sol. Energy*, 2020, **196**, 80–91.
116. L. Valenzuela, A. Iglesias, M. Faraldos, A. Bahamonde and R. Rosal, *J. Hazard. Mater.*, 2019, **369**, 665–673.
117. X. Wang, H. Ding, G. Lv, R. Zhou, R. Ma, X. Hou, J. Zhang and W. Li, *Ceram. Int.*, 2022, **48**, 20033–20040.
118. M. Luna, J. J. Delgado, I. Romero, T. Montini, M. A. Gil, J. Martínez-López, P. Fornasiero and M. J. Mosquera, *Constr. Build. Mater.*, 2022, **338**, 127349.
119. H. Wen, Y.-I. Hsu, T.-A. Asoh and H. Uyama, *J. Mater. Sci.*, 2021, **56**, 12224–12237.
120. V. A. Ganesh, H. K. Raut, A. S. Nair and S. Ramakrishna, *J. Mater. Chem.*, 2011, **21**, 16304–16322.
121. P. Ragesh, V. A. Ganesh, S. V. Nair and A. S. Nair, *J. Mater. Chem. A*, 2014, **2**, 14773–14797.

122. R. da Silva Cardoso, S. M. de Amorim, G. Scaratti, C. D. Moura-Nickel, R. P. M. Moreira, G. L. Puma and R. de Fatima Peralta Muniz Moreira, *RSC Adv.*, 2020, **10**, 17247–17254.
123. S. Altınışık, A. Kortun, A. Nazlı, U. Cengiz and S. Koyuncu, *Prog. Org. Coat.*, 2023, **175**, 107352.
124. Y. Higaki, M. Kobayashi, D. Murakami and A. Takahara, *Polym. J.*, 2016, **48**, 325–331.
125. Y. Xiang, R.-G. Xu and Y. Leng, *Langmuir*, 2018, **34**, 2245–2257.
126. C. Leng, H.-C. Hung, S. Sun, D. Wang, Y. Li, S. Jiang and Z. Chen, *ACS Appl. Mater. Interfaces*, 2015, **7**, 16881–16888.
127. A. Aguilar-Sanchez, B. Jalvo, A. Mautner, V. Rissanen, K. S. Kontturi, H. N. Abdelhamid, T. Tammelin and A. P. Mathew, *RSC Adv.*, 2021, **11**, 6859–6868.
128. J. Fang, S.-H. Ye, V. Shankarraman, Y. Huang, X. Mo and W. R. Wagner, *Acta Biomater.*, 2014, **10**, 4639–4649.
129. C. Wang, C. Ma, C. Mu and W. Lin, *RSC Adv.*, 2017, **7**, 27522–27529.
130. P. Kaner, A. V. Dudchenko, M. S. Mauter and A. Asatekin, *J. Mater. Chem. A*, 2019, **7**, 4829–4846.
131. X. Fan, B. H. Tan, Z. Li and X. J. Loh, *ACS Sustain. Chem. Eng.*, 2017, **5**, 1217–1227.
132. C. Thompson, R. Fleming and M. Zou, *Sol. Energy Mater. Sol. Cells*, 2013, **115**, 108–113.
133. N. Li, J. Kuang, Y. Ren, X. Li and C. Li, *Ceram. Int.*, 2021, **47**, 18743–18750.
134. X. Yang, L. Zhu, Y. Chen, B. Bao, J. Xu and W. Zhou, *Appl. Surf. Sci.*, 2015, **349**, 916–923.
135. W. E. Ward, *J. Aesthet. Art Crit.*, 1952, **11**, 135–146.
136. R. Xu, C. Yin, J. You, J. Zhang, Q. Mi, J. Wu and J. Zhang, *Green Energy Environ.*, 2022, doi:10.1016/j.gee.2022.10.009.
137. S. Maharjan, K.-S. Liao, A. J. Wang and S. A. Curran, *Constr. Build. Mater.*, 2020, **243**, 118189.
138. M. Li, M. Prabhakar and J.-I. Song, *Prog. Org. Coat.*, 2022, **172**, 107144.
139. R. Pirich, *Proc. SPIE*, 2013, **8876**, 88760Q.
140. S. Rabbani, E. Bakhshandeh, R. Jafari and G. Momen, *Prog. Org. Coat.*, 2022, **165**, 106715.
141. W. Li, Y. Zhan and S. Yu, *Prog. Org. Coat.*, 2021, **152**, 106117.

142. S. S. A. Kumar, S. Bashir, K. Ramesh and S. Ramesh, *FlatChem*, 2021, **31**, 100326.
143. G. Jena and J. Philip, *Prog. Org. Coat.*, 2022, **173**, 107208.
144. H. Zhu, Y. Chen, H. Li, S. X. Wang, X. Li and Q. Zhu, *Macromol. Rapid Commun.*, 2019, **40**, 1800252.
145. V. P. N. Nguyen, N. Kuo and X. J. Loh, *Soft Matter*, 2011, **7**, 2150–2159.
146. X. J. Loh, Y.-L. Wu, W. T. J. Seow, M. N. I. Norimzan, Z.-X. Zhang, F.-J. Xu, E.-T. Kang, K.-G. Neoh and J. Li, *Polymer*, 2008, **49**, 5084–5094.
147. Z. Li, P. L. Chee, C. Owh, R. Lakshminarayanan and X. J. Loh, *RSC Adv.*, 2016, **6**, 28947–28955.
148. X. J. Loh, Z.-X. Zhang, K. Y. Mya, Y.-L. Wu, C. B. He and J. Li, *J. Mater. Chem.*, 2010, **20**, 10634–10642.
149. X. J. Loh, B. J. H. Yee and F. S. Chia, *J. Biomed. Mater. Res. A*, 2012, **100**, 2686–2694.
150. E. A. Appel, M. J. Rowland, X. J. Loh, R. M. Heywood, C. Watts and O. A. Scherman, *Chem. Commun.*, 2012, **48**, 9843–9845.
151. L. Gan, G. R. Deen, X. Loh and Y. Gan, *Polymer*, 2001, **42**, 65–69.
152. D. A. Schaeffer, G. Polizos, D. B. Smith, D. F. Lee, S. R. Hunter and P. G. Datskos, *Nanotechnology*, 2015, **26**, 055602.
153. M. Mao, H. Xu, K.-Y. Guo, J.-W. Zhang, Q.-Q. Xia, G.-D. Zhang, L. Zhao, J.-F. Gao and L.-C. Tang, *Compos. A, Appl. Sci.*, 2021, **140**, 106191.
154. Q. Li, H. Liu, S. Zhang, D. Zhang, X. Liu, Y. He, L. Mi, J. Zhang, C. Liu and C. Shen, *ACS Appl. Mater. Interfaces*, 2019, **11**, 21904–21914.
155. Q. Wu, L.-X. Gong, Y. Li, C.-F. Cao, L.-C. Tang, L. Wu, L. Zhao, G.-D. Zhang, S.-N. Li and J. Gao, *ACS Nano*, 2018, **12**, 416–424.
156. G.-D. Zhang, Z.-H. Wu, Q.-Q. Xia, Y.-X. Qu, H.-T. Pan, W.-J. Hu, L. Zhao, K. Cao, E.-Y. Chen and Z. Yuan, *ACS Appl. Mater. Interfaces*, 2021, **13**, 23161–23172.
157. K.-Y. Guo, Q. Wu, M. Mao, H. Chen, G.-D. Zhang, L. Zhao, J.-F. Gao, P. Song and L.-C. Tang, *Compos. B, Eng.*, 2020, **193**, 108017.
158. C.-F. Cao, G.-D. Zhang, L. Zhao, L.-X. Gong, J.-F. Gao, J.-X. Jiang, L.-C. Tang and Y.-W. Mai, *Compos. Sci. Technol.*, 2019, **171**, 162–170.
159. F. Qiang, L.-L. Hu, L.-X. Gong, L. Zhao, S.-N. Li and L.-C. Tang, *Chem. Eng. J.*, 2018, **334**, 2154–2166.

160. C. Xie, C. Li, Y. Xie, Z. Cao, S. Li, J. Zhao and M. Wang, *Surf. Interfaces*, 2021, **22**, 100833.
161. Z. S. Tai, M. H. D. Othman, A. Mustafa, J. Ravi, K. C. Wong, K. N. Koo, S. K. Hubadillah, M. A. Azali, N. H. Alias and B. C. Ng, *J. Membr. Sci.*, 2021, **637**, 119609.
162. A. Ribeiro, B. Soares, J. Furtado, A. Silva and N. Couto, *Prog. Org. Coat.*, 2022, **168**, 106867.
163. Y. Nie, S. Ma, M. Tian, Q. Zhang, J. Huang, M. Cao, Y. Li, L. Sun, J. Pan and Y. Wang, *Surf. Coat. Technol.*, 2021, **410**, 126966.
164. L. Qi, S. Qiu, J. Xi, B. Yu, Y. Hu and W. Xing, *J. Colloid Interface Sci.*, 2022, **607**, 2019–2028.
165. S. S. Latthe, P. Sudhagar, C. Ravidhas, A. J. Christy, D. D. Kirubakaran, R. Venkatesh, A. Devadoss, C. Terashima, K. Nakata and A. Fujishima, *CrystEngComm*, 2015, **17**, 2624–2628.
166. S. Khorsand, K. Raeissi, F. Ashrafizadeh and M. Arenas, *Chem. Eng. J.*, 2015, **273**, 638–646.
167. H. Li, X. Lin and H. Wang, *Materials*, 2021, **14**, 211.
168. X.-H. Shi, Q.-Y. Liu, X.-L. Li, A.-K. Du, J.-W. Niu, Y.-M. Li, Z. Li, M. Wang and D.-Y. Wang, *Polym. Degrad. Stab.*, 2022, **197**, 109839.
169. T. Ghosh and V. Katiyar, *Int. J. Biol. Macromol.*, 2022, **211**, 116–127.
170. M. Bai, J. Wang, R. Zhou, Z. Lu, L. Wang and X. Ning, *J. Hazard. Mater.*, 2022, **432**, 128735.
171. X.-X. Zhang, B.-B. Xia, H.-P. Ye, Y.-L. Zhang, B. Xiao, L.-H. Yan, H.-B. Lv and B. Jiang, *J. Mater. Chem.*, 2012, **22**, 13132–13140.
172. M. Santamouris, A. Synnefa and T. Karlessi, *Sol. Energy*, 2011, **85**, 3085–3102.
173. M. Santamouris, *Sol. Energy*, 2014, **103**, 682–703.
174. M. Santamouris and D. Kolokotsa, *Energy Build.*, 2013, **57**, 74–94.
175. L. Zheng, T. Xiong and K. W. Shah, *Sol. Energy*, 2019, **193**, 837–858.
176. Y. Chen, J. Mandal, W. Li, A. Smith-Washington, C.-C. Tsai, W. Huang, S. Shrestha, N. Yu, R. P. Han and A. Cao, *Sci. Adv.*, 2020, **6**, eaaz5413.
177. S. Son, Y. Liu, D. Chae and H. Lee, *ACS Appl. Mater. Interfaces*, 2020, **12**, 57832–57839.
178. H. Liu, T. Jiang, F. Wang, J. Ou and W. Li, *Energy Build.*, 2021, **251**, 111374.

179. X. Zhang, H. Li, N. Xie, M. Jia, B. Yang and S. Li, *Sustain. Cities Soc.*, 2022, **83**, 103950.
180. T. Karlessi and M. Santamouris, *Int. J. Low Carbon Technol.*, 2015, **10**, 45–61.
181. M. Liu, Z. Xing, Z. Li and W. Zhou, *Coord. Chem. Rev.*, 2021, **446**, 214123.
182. S. Gorjian, H. Sharon, H. Ebadi, K. Kant, F. B. Scavo and G. M. Tina, *J. Clean. Prod.*, 2021, **278**, 124285.
183. Y.-C. Wang, X.-Y. Liu, X.-X. Wang and M.-S. Cao, *Chem. Eng. J.*, 2021, **419**, 129459.
184. J. Cao, Y. Sim, X. Y. Tan, J. Zheng, S. W. Chien, N. Jia, K. Chen, Y. B. Tay, J. F. Dong and L. Yang, *Adv. Mater.*, 2022, **34**, 2110518.
185. S. Y. Tee, E. Ye, C. P. Teng, Y. Tanaka, K. Y. Tang, K. Y. Win and M.-Y. Han, *Nanoscale*, 2021, **13**, 14268–14286.
186. K. Y. Tang, J. X. Chen, E. D. R. Legaspi, C. Owh, M. Lin, I. S. Y. Tee, D. Kai, X. J. Loh, Z. Li and M. D. Regulacio, *Chemosphere*, 2021, **265**, 129114.
187. S. Shalaby, *Renew. Sustain. Energy Rev.*, 2017, **73**, 789–797.
188. K. Ilse, C. Pfau, P.-T. Miclea, S. Krause and C. Hagendorf, Proc IEEE 46th Photovoltaic Specialists Conf (PVSC), Chicago, IL, USA, 2019, pp. 2883–2888.
189. D. Biswas, N. Chundi, S. Atchuta, K. P. Kumar, M. S. Prasad and S. Sakthivel, *Sol. Energy*, 2022, **246**, 36–44.
190. K.-Y. Yang, W. Lee, J.-Y. Jeon, T.-J. Ha and Y.-H. Kim, *Sol. Energy*, 2020, **197**, 99–104.
191. X. Huang, S. Han, W. Huang and X. Liu, *Chem. Soc. Rev.*, 2013, **42**, 173–201.
192. A. Polman, M. Knight, E. C. Garnett, B. Ehrler and W. C. Sinke, *Science*, 2016, **352**, aad4424.
193. J. Y. Huang, G. T. Fei, S. H. Xu and B. Wang, *Compos. B, Eng.*, 2023, **251**, 110486.
194. K. P. Kumar, S. Mallick and S. Sakthivel, *Renew. Energy*, 2023, **203**, 334–344.
195. X. Wang, X. Yuan, D. Gong, X. Cheng and K. Li, *J. Mater. Res. Technol.*, 2021, **15**, 6162–6174.
196. L. Martínez-Manuel, N. G. González-Canché, L. B. López-Sosa, J. G. Carrillo, W. Wang, C. A. Pineda-Arellano, F. Cervantes, J. J. A. Gil and M. I. Peña-Cruz, *Sol. Energy*, 2022, **239**, 319–336.

© 2025 World Scientific Publishing Company
https://doi.org/10.1142/9789819806409_0009

Direct Growth of Lithium Niobate Thin Films for Acoustic Resonators*

Zhen Ye,[†] Qibin Zeng, Celine Sim, Baichen Lin, Hui Kim Hui, Anna Marie Yong, Chee Kiang Ivan Tan,[‡] Seeram Ramakrishna,[§] and Huajun Liu[¶]

[†]*Institute of Materials Research and Engineering (IMRE)Agency for Science, Technology and Research (A*STAR)2 Fusionopolis Way, Innovis #08-03, Singapore 138634 Republic of Singapore Department of Mechanical Engineering, National University of Singapore 9 Engineering Drive 1 Block EA #07-08 Singapore 117575 Republic of Singapore*
[‡]*Institute of Materials Research and Engineering (IMRE) Agency for Science Technology and Research (A*STAR) 2 Fusionopolis Way, Innovis #08-03, Singapore 138634 Republic of Singapore*
[§]*Department of Mechanical Engineering, National University of Singapore 9 Engineering Drive 1 Block EA #07-08 Singapore 117575, Republic of Singapore*
[¶]*Institute of Materials Research and Engineering (IMRE) Agency for Science Technology and Research (A*STAR) 2 Fusionopolis Way Innovis #08-03, Singapore 138634 Republic of Singapore*
liu_huajun@imre.a-star.edu.sg

Lithium niobate ($LiNbO_3$, LN) thin films have been extensively studied for applications in acoustic and photonic devices, due to their outstanding piezoelectric, ferroelectric and electro-optical properties. With the

---

[†]Corresponding authors.
*To cite this article, please refer to its earlier version published in the *World Scientific Annual Review of Functional Materials*, Volume 2, 2330003 (2024), DOI: 10.1142/S2810922823300039.

increasing demand for high speed and low latency wireless communication, LN thin films with high electromechanical coupling coefficients are very attractive to improve the performance of acoustic resonators for radio frequency filters. The current bottleneck for LN-based devices is the synthesis of high-quality LN thin films, which is typically fabricated by expensive and inefficient process of ion slicing and layer transfer from bulk single crystals. This review paper focuses on the direct growth of high-quality LN thin films, which has the potential to scale up and lower the cost of LN thin films. We first introduce the crystal structure and piezoelectric properties of LN, followed by an overview of the state-of-the-art LN acoustic resonators. After a summary of the challenges in the fabrication of LN thin films, we review the direct growth of LN thin films by sputtering, pulsed laser deposition, metalorganic chemical vapor deposition and molecular beam epitaxy. With the progress in optimizing the crystallinity and surface roughness, the quality of the LN thin films synthesized by direct growth has been greatly improved. As a result of the fast-growing industrial interests, we believe that the research works in direct growth of LN thin films will increase exponentially to achieve the same quality of the LN thin films as the bulk single crystals.

Keywords: Lithium niobate; micro-electromechanical systems (MEMS); resonators; radio frequency filters; sputtering; pulsed laser deposition (PLD); molecular beam epitaxy (MBE); metalorganic chemical vapor deposition (MOCVD).

1. Introduction

High-speed wireless networks such as the fifth-generation (5G) mobile networks are widely recognized as a critical hardware foundation to enable disruptive shifts in applications such as autonomous vehicles, telemedicine, and smart factories. The International Telecommunication Union reported in 2022 that mobile-cellular telephone subscriptions per 100 inhabitants globally reached 108, with even the least developed countries showing a mobile phone ownership of 78.7%.[1] The growing high demand for mobile devices creates a huge market and requires a faster data transmission speed. The 5G communication technology is currently working toward a higher frequency range up to millimeter wave (3 GHz up to 90 GHz).[2] This higher frequency range has the potential to significantly enhance data exchange rates due to the wider bandwidth available in the frequency spectrum.

As an important component in wireless communications, radio frequency (RF) filters serve the function of selecting the desired frequency bands and filtering out unwanted interfering bands. The bandwidth of RF filter directly determines the data transmission amount of the wireless network. As a result, there has been a strong demand and growing interest from the industry in developing RF filters with a wider bandwidth. Conventionally, AlN-based acoustic resonators have been widely used as the major commercial RF filters over the past few decades. AlN is known for its good thermal conductivity,[3] piezoelectricity,[4] dielectric[5] and chemical stability.[3] Moreover, AlN thin films have high acoustic velocities, 12,350 m/s and 6,000 m/s for bulk and surface waves, respectively.[4] However, the moderate piezoelectric properties (electromechanical coupling coefficients k_t^2 of 4%–7%)[6,7] of AlN thin films limit the bandwidth of RF filters. Chemical doping has been explored to improve the piezoelectric performance of AlN thin films. In 2009, Akiyama et al.[8] found scandium (Sc) doping of AlN thin films increased d_{33} by 400%. By using Sc-doped AlN thin films, Moreira et al.[9] achieved 12% k_t^2 on bulk acoustic resonators. However, the increased film roughness and stress limit the amount of Sc that can be doped into AlN. As a result, researchers are exploring new material systems with higher electromechanical coupling coefficients to enhance the bandwidth of RF filters further. In this regard, LN is very attractive due to its high electromechanical coupling coefficients k_t^2, which can reach up to 46.4%[7], compared to 4%–7% k_t^2 for AlN thin films.[6,7] LN-based acoustic devices are being developed rapidly worldwide, such as surface acoustic wave (SAW),[10] bulk acoustic wave (BAW),[11,12] thin-film bulk acoustic resonator (FBAR),[13] laterally excited bulk-wave resonators (XBAR).[14,15] These devices present extraordinary performances at ultra-high working frequency, making LN-based devices ideal for next-generation acoustic filters. The recent development of XBAR by Plessky et al.[15] demonstrated a working frequency of up to 38 GHz, promising for the second phase 5G wireless network (24–71 GHz).[16]

The main challenge for LN-based acoustic devices is the synthesis of high-quality LN thin films. Currently, high-quality LN thin films can only be fabricated by ion-slicing from bulk single crystals and transferring to desired substrates of interest.[17] This ion-slicing and layer-transfer process is tedious, expensive, and difficult to scale up. Direct growth of LN thin films, although challenging, is the only way to drive LN devices toward mass production and commercialization. In this paper, we aim to review the recent progress in the direct growth of LN thin films by sputtering, pulsed laser deposition (PLD), molecular beam epitaxy (MBE), metalorganic chemical vapor deposition (MOCVD). Key learnings from previous works are summarized to provide a guideline for further development in the direct growth of LN thin films.

2. Crystal Structure and Piezoelectric Properties of LN

Before discussing the growth of LN thin films, we first review the crystal structure and piezoelectric properties of LN. Bulk LN single crystals have been synthesized since 1966 and crystal structures have been reported with detailed atomic arrangements.[18–20] Bulk LN belongs to the space group *R3c*. The lattice constants in hexagonal lattice at 23°C are $a = b = 5.148$ Å, $c = 13.863$ Å.[20] Face-shared oxygen octahedra are arranged along the c direction, with one-third of them filled by Li ions, one-third by Nb ions, and one-third empty [Fig. 1(a)].[21] From the top view, oxygen atoms in the octahedra are arranged in a distorted hexagonal close-packed configuration [Fig. 1(b)]. The oxygen atoms rotate in each layer of the octahedra, as shown in Fig. 1(c). Below the Curie temperature of 1210°C, a small displacement of the Li and Nb cations is involved with respect to the oxygen layers and along the c-axis,[20,22] leading to spontaneous polarizations in LN crystals.

The piezoelectric properties of bulk LN single crystals have been fully characterized. In the design of LN-based acoustic resonators, the important material properties are the elastic stiffness constants, the piezoelectric coefficients and dielectric constants, which are listed as follows.[23]

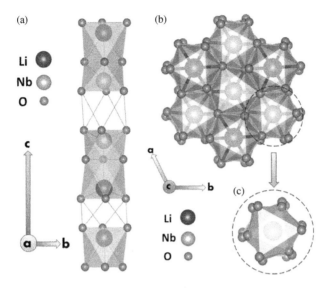

Fig. 1. (a) The unit cell of LN in hexagonal lattice. (b) A schematic drawing of LN crystal structure and (c) an enlarged view of a single column of octahedra projected along c-axis. These figures are drawn by VESTA software.[24]

Elastic stiffness constants (10^{11} N/m^2):

$$C_{ij} = \begin{pmatrix} C_{11} & C_{12} & C_{13} & C_{14} & 0 & 0 \\ C_{12} & C_{11} & C_{13} & -C_{14} & 0 & 0 \\ C_{13} & C_{13} & C_{33} & 0 & 0 & 0 \\ C_{14} & -C_{14} & 0 & C_{44} & 0 & 0 \\ 0 & 0 & 0 & 0 & C_{44} & C_{14} \\ 0 & 0 & 0 & 0 & C_{14} & \frac{1}{2}(C_{11}-C_{12}) \end{pmatrix}$$

$$= \begin{pmatrix} 2.030 & 0.573 & 0.752 & 0.085 & 0 & 0 \\ 0.573 & 2.030 & 0.752 & -0.085 & 0 & 0 \\ 0.752 & 0.752 & 2.424 & 0 & 0 & 0 \\ 0.085 & -0.085 & 0 & 0.595 & 0 & 0 \\ 0 & 0 & 0 & 0 & 0.595 & 0.085 \\ 0 & 0 & 0 & 0 & 0.085 & 0.728 \end{pmatrix} \quad (1)$$

Piezoelectric coefficients (C/m²):

$$e_{ij} = \begin{pmatrix} 0 & 0 & 0 & 0 & e_{15} & -e_{22} \\ -e_{22} & e_{22} & 0 & e_{15} & 0 & 0 \\ e_{31} & e_{31} & e_{33} & 0 & 0 & 0 \end{pmatrix}$$

$$= \begin{pmatrix} 0 & 0 & 0 & 0 & 3.76 & -2.43 \\ -2.43 & 2.43 & 0 & 3.76 & 0 & 0 \\ 0.23 & 0.23 & 1.33 & 0 & 0 & 0 \end{pmatrix} \quad (2)$$

Dielectric constants:

$$\varepsilon_{ij} = \begin{pmatrix} \varepsilon_{11} & 0 & 0 \\ 0 & \varepsilon_{11} & 0 \\ 0 & 0 & \varepsilon_{33} \end{pmatrix} = \begin{pmatrix} 43.60 & 0 & 0 \\ 0 & 43.60 & 0 \\ 0 & 0 & 29.16 \end{pmatrix} \quad (3)$$

3. State-of-the-Art LN Acoustic Resonators

LN-based acoustic resonators have been widely investigated using the thin films prepared by ion slicing and transferring process.[25–28] In this section, we summarize the recent achievements and state-of-the-art performance of LN acoustic resonators (Table 1).

In 2017, Yang et al. proposed a super-high frequency (SHF) resonator operating at 5 GHz, and the device demonstrated electromechanical coupling (k_t^2) of 29% and a quality factor (Q) of 527 simultaneously.[28] Later, they designed a class of LN-based antisymmetric resonators (from A1 to A7) toward the Ka band with a working frequency up to 40 GHz.[38] They further increased working frequency to 60 GHz by increasing the antisymmetric lamb wave mode higher order in the subsequent work.[45] Yang et al. also designed a ladder filter with bandwidth of 70 MHz at 10.8 GHz using one series and two shunt resonators.[44] Instead of an antisymmetric mode, Lu et al. designed a symmetric mode resonator that reaches resonance at 15.1 GHz.[35] This design leads to high Q and large k^2 value. Specifically, the resonator exhibited resonance at the

Table 1. The list of device performances of LN acoustic resonators and filters. f_r is resonance frequency, f_a is antiresonance frequency, k_t^2 is the electromechanical coupling coefficient, Q is the quality factor, FoM is the figure of merit, FoM = $k_t^2 Q$, R-a-R is the relative Resonance-anti-Resonance distance, $(R\text{-}a\text{-}R) = (f_a - f_r)/f_a$.

f_r (GHz)	f_a (GHz)	k_t^2 (%)	Q	FoM	f_r*Q Hz	R-a-R	LN orientation	Ref.
0.96	1.05	21.80	1316	286.88	1259.6	8.86%	X-cut	29
1.65	NA	14.00	3112	435.68	5134.8	NA	Y-cut	27
1.71	NA	6.30	5341	336.48	9133.1	NA	Y-cut	25
1.94	2.2	26.00	610	158.60	1183.4	11.82%	Y+163°-cut	30
2.28	NA	26.90	1228	330.33	2799.8	NA	X-cut	31
2.41	2.99	39.20	250	98	604	19.09%	X-cut	32
3.00	3.32	16.63	183	30.43	550.6	9.45%	Y43°-cut	12
3.02	3.21	14.06	120	16.87	363.4	5.70%	Y+43°-cut	33
3.04	3.31	20.20	92	18.58	279.6	8.21%	Y+43°-cut	34
3.05	NA	21.50	657	141.25	2003.8	NA	Z-cut	35
3.20	NA	46.40	598	277.47	1913.6	NA	128° Y-cut	7
4.11	4.4	15.40	390	60.06	1602.9	6.59%	Y+36°-cut	36
4.22	4.72	23.60	295	69.62	1244.9	10.59%	Y+163°-cut	36
4.26	4.32	3.50	295	10.325	1256.7	1.39%	Z-cut	36
4.26	4.43	9.20	260	23.92	1107.6	3.84%	Y+64°-cut	36
4.35	NA	29.00	527	152.83	2292.4	NA	Z-cut	28
4.35	4.81	21.40	255	54.57	1109.2	9.56%	Y+36°-cut	37
4.50	NA	24.00	118	28.32	531	NA	Z-cut	38
4.50	NA	28.00	420	117.60	1890	NA	Z-cut	26
4.60	NA	28.00	300	84	1380	NA	Z-cut	39
4.72	4.77	2.70	300	8.10	1416	1.05%	Z-cut	40
4.78	5.18	17.60	190	33.44	908.2	7.72%	Y+163°-cut	41
5.09	5.70	25.00	154	38.50	784.16	10.70%	ZY-cut	42
5.43	5.45	8.00	40	3.20	217.2	0.37%	Z-cut	40

(*Continued*)

Table 1. (*Continued*)

f_r (GHz)	f_a (GHz)	k_t^2 (%)	Q	FoM	f_r*Q Hz	R-a-R	LN orientation	Ref.
9	NA	6.72	376	25.26	3384	NA	ZX-cut	43
9.05	NA	3.71	636	23.59	5755.8	NA	Z-cut	35
10.80	NA	3.60	337	12.13	3639.6	NA	Z-cut	44
12.90	NA	3.70	224	8.28	2889.6	NA	Z-cut	38
13.00	NA	3.80	372	14.13	4836	NA	Z-cut	45
21.40	NA	1.50	287	4.30	6141.8	NA	Z-cut	38
21.60	NA	1.20	566	6.79	12225.6	NA	Z-cut	45
29.90	NA	0.94	328	3.08	9807.2	NA	Z-cut	38
30.20	NA	0.74	715	5.29	21593	NA	Z-cut	45
38.80	NA	0.45	539	2.42	20913.2	NA	Z-cut	45
47.40	NA	0.31	474	1.46	22467.6	NA	Z-cut	45
55.00	NA	0.22	340	0.74	18700	NA	Z-cut	45

second-order symmetric (S2) mode at 3.05 GHz with Q of 657 and k^2 of 21.5%. In addition, the device also demonstrated resonance at the sixth-order symmetric (S6) mode at 9.05 GHz with Q of 636 and k^2 of 3.71%. Similarly, Faizan *et al.*[39] demonstrated an LN-based bulk acoustic resonator (FBAR) using a Z-cut LN membrane with interdigitated metal electrodes on top. This device demonstrated k_t^2 and Q values of 28% and 300, respectively. In 2019, Yang *et al.*[46] proposed another resonator design on Y-cut LN thin film by optimizing the electrode gap. A1 mode resonator shows a k_t^2 of 14% and quality factor of 3112 in the 1–6 GHz range. This resonator is further developed into a ladder filter capable of operating at 4.5 GHz.[26] This filter has two distinct filter designs. The first design offers a fractional bandwidth (FBW) of 10% and a low insertion loss (IL) of 1.7 dB. The second design provides a slightly narrower FBW of 8.5% but still manages to achieve a low IL of 2.7 dB. Yang *et al.*[47] optimized the filter structure at the frequency of 8.4 GHz. The bandwidth reached 290 MHz, which corresponds to a fractional bandwidth (FBW) of 3.45% with a minimum insertion loss (IL) of 2.7 dB.

In addition to conventional c-oriented LN devices, recent works have explored LN resonators in other orientations. Marie et al.[48] investigated the relationship between device performance and LN orientation. They demonstrated a Y+163° cut-based FBAR with higher performance. This device exhibited k_t^2 ~26% and Q ~600, which is higher than their previous FBAR device based on X-cut LN (Q ~250).[32] Xing et al.[11] fabricated a Y43°-cut BAW device and further reduced the LN surface roughness to 3.6 nm by introducing a low-energy irradiation process. The modified BAW device exhibited a 16.3% k_t^2 and Q of 132, resulting in an 8.5% bandwidth with a 3 GHz center frequency in the final 3-stage filter. Bai et al.[12] introduced a unique Y43°-cut device by introducing Mo/Ti multilayer Bragg reflector, which can prevent the dissipation of acoustic energy effectively [Fig. 2(a)]. The improved device exhibited k_t^2 of 16.63% and Q-factor of 183 at 3.3 GHz. Lu et al.[7] and Gorisse et al.[32] also discussed the effect of LN orientation and explored the 128° Y-cut LN thin films. This orientation gives a piezoelectric constant e_{15} of 4.47 C/m^2, a high k^2 of 46.4% and Q of 598 at 3.2 GHz.

In 2019, Plessky et al.[15] demonstrated a novel type of resonator, laterally excited bulk wave resonators (XBAR), as shown in Fig. 2(b). It can overcome the limitation of both SAW and FBAR devices to reach the working frequency above 3 GHz. Figure 2(c) shows the resonance behavior of XBAR. Under A1 mode, the resonator had a high resonance frequency (f_r) of 4.5 GHz and a high k_t^2 of 24%. Moreover, XBAR can operate at 13 GHz under A3 mode and even reach up to 38 GHz at A9 mode. This enables RF filters for millimeter-wave applications. This work provides a new solution for resonator design. Inspired by the success of XBAR, Turner et al.[49] designed a ladder filter consisting of five ZY-cut LN film XBAR, as shown in Fig. 2(d). They measured a 4.7 GHz bandpass ladder-type filter with 1 dB mid-band loss and 600 MHz bandwidth [Fig. 2(e)], which meet the 5G Band n79 requirements. Yandrapalli et al.[42] optimized the structure geometry by a systematic study of design parameters. Their simulation demonstrated a 10% expanding of

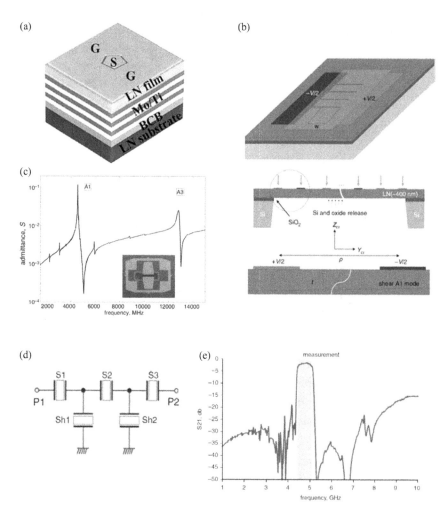

Fig. 2. (a) A schematic configuration of the LN-BAW resonator with the Mo/Ti Bragg reflector. (b) A schematic diagram of the z-cut XBAR structure in its top view and the cross-sectional view (c) The resonance spectra of XBAR showing A1 and A3 resonance modes. The inset is the photo of the device. (d) The design of improved ladder filter, composed of five resonators (three series resonators and two shunt resonators). (e) Wideband wafer-probe measurements of an improved XBAR filter from 1–10 GHz.

Source: (a) Reprinted with permission of AIP Publishing from Ref. 12, (b) and (c) Reproduced with permission of IET Publishing from Ref. 15, (d) and (e) Reproduced with permission of IET Publishing from Ref. 49.

Direct Growth of Lithium Niobate Thin Films for Acoustic Resonators 427

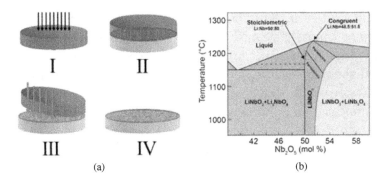

Fig. 3. (a) LN thin films fabricated by ion slicing and layer transfer process, consisting of ion implantation (I), wafer bonding process (II), laser irradiation and splitting (III), and annealing and polishing of transferred layer on Si substrate (IV). (b) Phase diagram of the Li_2O–Nb_2O_5 pseudobinary system in the vicinity of LN.[53]

Source: (a) Reproduced with permission of Wiley Publishing from Ref. 17. (b) Reproduced under open access Creative Common CC BY license from MDPI publication.

bandwidth and a low insertion loss of 1.4 dB. More recently, Yang et al.[50] achieved a high-quality factor of 692 and k_t^2 of 28% at 3 GHz by micro-machining the lithium niobate thin film to balance electrical and mechanical loadings of electrodes. Dai et al.[51] demonstrated an ultra-high Q device by adjusting the 3D Euler angle of the thin-film piezoelectric resonator body. The research of LN thin film-based acoustic resonators is still fast developing, with great promise to be the major material candidate for next-generation acoustic filters in 5G and 6G wireless networks.

4. Challenges in the Fabrication of LN Thin Films

All the LN devices reviewed in the section above used LN thin films fabricated by ion slicing and layer transfer process from bulk LN single crystal. The processes are shown in Fig. 3(a).[17] Firstly, ion implantation is performed on the LN single crystal to create defects or microcavities at the desired depth of hundreds of nanometers to a few micrometers. Secondly, the implanted LN crystal is bonded to the carrier substrate by adhesive or direct bonding. Before bonding, the surface of the LN crystal and the carrier substrate need to be carefully cleaned to avoid contamination at the interface. Thirdly, the stack of LN crystal and carrier substrate is heated to expand the microcavities so that the LN crystals can be split from the transferred LN thin layer. Lastly, the transferred LN layer needs to be annealed and polished to remove the surface damage induced in the process. Overall, this process is very challenging to obtain high quality LN thin layers in a large quantity. Therefore, the high price and low quantity of LN thin films fabricated by this method only allow small scale research and development in academic research, preventing wide adoption by industry. In addition, the stoichiometry of transferred LN thin layer is limited by the bulk single crystals. As most of the LN single crystals have congruent composition of Li/(Li+Nb) = 48.5% instead of stoichiometric composition of Li/(Li+Nb) = 50%, the intrinsic defects associated with Li deficiency in the transferred LN thin layers are detrimental for acoustic and optical devices.

Direct growth on the substrates of interest is the ideal method to scale up the fabrication of LN thin films and devices. However, the direct growth of LN thin films is still challenging after decades of research.[52] As shown in Fig. 3(b), Li–Nb–O material system has a complicated phase diagram.[53] Firstly, the congruent melting point favors lithium-deficient composition (Li: Nb = 48.5: 51.5), leading to challenges to obtain stoichiometric (Li: Nb = 50: 50) $LiNbO_3$ phase. Secondly, the phase diagram shows coexistence of multiple phases, such as $LiNb_3O_8$ and Li_3NbO_4, which do not show similar functional properties of $LiNbO_3$ phase. This means a precise control of Li/Nb composition is critical to obtain pure LN phase. In the next section, we will review the progress of the direct growth of LN thin films by sputtering, PLD, MOCVD, and MBE.

5. Direct Growth of LN Thin Films

5.1. *Sputtering*

Sputtering is a major deposition tool for thin films in both industry and academic laboratories. With the increasing demands for functional films, sputtering has already become an essential method for exploring and growing new films.[54] In past decades, researchers have been using sputtering to grow LN thin films. In 1969, Foster[55] grew LN thin films on a c-oriented sapphire by triode sputtering. Later, Meek et al.[56] investigated the effect of temperature on the growth of high-quality single crystal LN thin films on c-sapphire. At temperatures below 200°C, the layers formed a uniform film with a refractive index in the range of 2.00–2.20. At temperatures above 600°C, the refractive index reached 2.30 and 2.20, which is close to the bulk crystal. This study also found that ensuring the initial clean surface of the substrate can improve both the adhesion and transparency of the film. Additionally, using source material with excess lithium can result in stoichiometric LN thin films. This provides a useful guideline for LN thin films grown by the sputtering process. In 1991, Fujimura et al.[57] further explored the temperature effect. They found that variation in temperature led to the formation of the secondary phase. Figure 4(a) illustrates that with an increase in temperature, the dominant phase transitions from $LiNb_3O_8$ to $LiNbO_3$. $LiNb_3O_8$ phases can also be converted to $LiNbO_3$ by post-annealing at 1000°C. Later on, Lansiaux et al.[58] also confirmed the change of crystalline phase with temperatures in Fig. 4(b) and indicated only a narrow temperature window to obtain stoichiometric LN thin films. In 1994, Fujimura et al.[59] found Li concentration in the film can lead to the change of LN crystal orientation. In their research work, the orientation of the LN thin film on R-cut (01-12) sapphire changed from (01-12) to (10-10) via (11-20) by increasing the Li concentration in the film.

In addition to sapphire substrates, more research works on the sputtering growth of LN thin films focused on using silicon wafers with the potential towards large-scale manufacturing. In 1992, when Rost et al.[60] deposited LN on silicon by RF magnetron sputtering. They performed a systematic study of the effects of flow rate, pressure, gas ratio, temperature, RF power and Li_2O concentration. Under optimized conditions, polycrystalline LN thin films with preferred c orientation are grown that exhibit a columnar grain structure with the polar axis normal to the substrate surface.

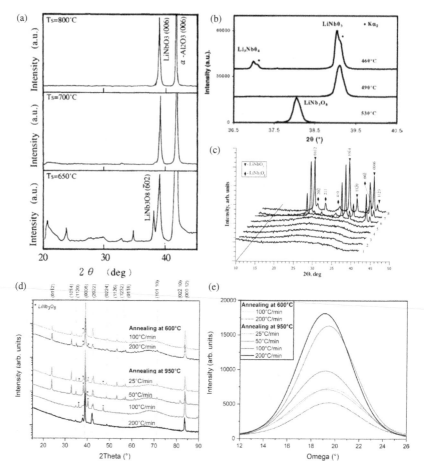

Fig. 4. (a) X-ray diffraction patterns of LN thin films deposited at 650°C, 700°C, and 800°C. (b) X-ray diffraction patterns of LN thin films deposited on c-sapphire at 460°C, 490°C, and 530°C. (c) X-ray diffraction patterns of LN thin films after post-annealing for 1 h, from curves 1 to 8 were 350, 400, 425, 450, 475, 500, 525°C, respectively. (d) X-ray diffraction patterns of the LN thin films annealed at 600°C for temperature ramping rates of 100°C and 200°C/min and annealed at 950°C for temperature ramping rates of 25, 50, 100 and 200°C/min. (e) Rocking curves of the samples annealed at 600°C for temperature ramping rates of 100°C/min and 200°C/min and of the samples annealed at 950°C for temperature ramping rates of 25, 50, 100 and 200°C/min.

Source: (a) Reproduced with permission of Elsevier Publishing from Ref. 57. (b) Reprinted with permission of AIP Publishing from Ref. 58. (c) Reproduced with permission of Springer Publishing from Ref. 61. (d) and (e) Reproduced with permission of Elsevier Publishing from Ref. 62.

The effect of post-annealing temperature is discussed by Sumet et al.[61] They grew LN thin films on Si at room temperature and annealed from 350°C to 525°C [Fig. 4(c)]. Films remained amorphous below 400°C. Crystallization of the films begins at 450°C, and crystallinity increases with the annealing temperature. But the $LiNb_3O_8$ phase appeared at 500°C. In 2021, Sauze et al.[62] investigated the effect of temperature and ramping rate on LN thin film quality [Fig. 4(d)]. The amount of the secondary phase is minimized (reaching 99.35% of preferential c-orientation) after annealing at 950°C with a temperature ramping rate of 100°C/min. Full width at half maximum (FWHM) of (0006) X-ray diffraction peak of c-oriented LN thin films decreased with increasing ramping rate at both 600°C and 950°C [Fig. 4(e)]. In addition to the temperature effect, Sumets et al.[63] found that low gas pressure can suppress the formation of the secondary phase. In a pure Ar environment, with decreasing pressure, the secondary phase was suppressed and formed (001) textured polycrystalline LN thin films. The defects in LN thin films were found to be reduced by introducing oxygen gas during sputtering.

In addition to deposition conditions, multi-step growth and buffer layers are also investigated to obtain high quality LN thin films by sputtering. Bornand et al.[65] developed a 2-step growth process that combines sputtering and pyrosol techniques to separate nucleation and growth stages. This method successfully grew LN thin films on (111) silicon and (001) sapphire substrates. The FWHM of (006) diffraction peaks are 0.2° and 0.14° for LN thin films grown on Si and sapphire, respectively. Akazawa et al.[64] proposed the use of ZnO as a buffer layer on SiO_2/Si substrate before the growth of LN thin films. c-oriented ZnO layer was deposited on amorphous SiO_2 as a template to promote the crystallinity of subsequent LN thin films. The introduction of ZnO reduced the surface roughness of LN film, resulting in a smooth surface with lower roughness under different deposition temperatures and thicknesses as shown by the AFM images in Fig. 5. Although ZnO provided a lattice matching interface, the results showed that LN films were not epitaxial. The FWHM of ZnO and LN is 6° and 4°, respectively.

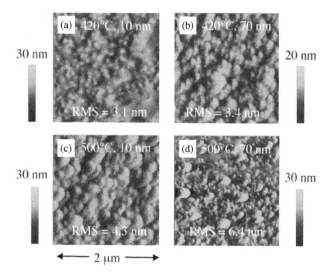

Fig. 5. AFM images of LN thin films on ZnO/SiO$_2$/Si substrates. Deposition temperatures of LN thin films and the thicknesses of ZnO buffer layers are (a) 420°C and 10 nm, (b) 420°C and 70 nm, (c) 500°C and 10 nm, and (d) 500°C and 70 nm. Root Mean Square (RMS) roughness is 3.1, 3.4, 4.3 and 6.4 nm, respectively.

Source: Reproduced with permission of Elsevier Publishing from Ref. 64.

5.2. *PLD*

Pulsed Laser Deposition (PLD) is a highly effective method for depositing high-quality films with complex compositions. It uses a pulsed laser beam to ablate the target, unlike using Ar ions to bombard the target in the sputtering process. The first LN that grew on sapphire was demonstrated by Marsh *et al.*[66] in 1992. They grew on α-sapphire substrate at 730°C and oxygen pressure of 1 Torr. This film exhibited strong texture in the direction perpendicular to the substrate plane and indicated an index of refraction close to the ideal value of bulk LN. Inspired by this work, Shibata *et al.*[67] optimized the orientation of the substrate (*c*-sapphire) to improve film quality. The effect of orientation is further investigated by Pershukov *et al.*[68] Under the same condition (0.2 mbar, 600°C, stoichiometric LiNbO$_3$ target), the roughness of c-sapphire and α-sapphire is 1.8 nm and 9.46 nm, respectively.

The quality of LN thin film is affected by deposition conditions, including substrate, temperature and deposition rate. Temperature is a crucial parameter during PLD processing. LN films are usually crystallized at substrate temperatures of 500–800°C.[68] The effect of temperature was first discussed and optimized by Marsh et al.[66] In the temperature range from 650°C to 800°C, they grew epitaxial LN thin films at the optimal temperature of 730°C on α-sapphire. In the parametric study of Pershukov et al.[68] they highlighted that excessively high temperatures can lead to the formation of secondary phases, such as Li_3NbO_4 and $LiNb_3O_8$, which can affect the quality of the films.

Deposition rate can be controlled by gas pressure and laser frequency in PLD. A high O_2 pressure leads to more scattering between atoms ablated from the target and O_2 gas, resulting in a lower deposition rate. Son et al.[69] investigated the impact of O_2 pressure from 100–500 mTorr. According to Fig. 6(a), low pressure introduces the Li-deficient phase, while high pressure introduces the Li-excess phase. The deposition flux can also be controlled by laser frequency in the PLD system. Lee et al.[70] explored the effect of deposition flux on the surface roughness of LN thin films on sapphire substrates. Figure 6(b) shows that lower roughness can be obtained by using higher laser frequency, thus a higher deposition rate. The higher deposition rate introduces a high density of nucleation sites, therefore reducing the roughness. However, it could lead to lower crystallinity due to the large density of grain boundaries.

In addition to sapphire substrates, Si substrates have also been used for LN thin film growth by PLD. In 1998, Guo et al.[71] optimized $LN/SiO_2/Si$ heterostructure by introducing an in-situ electrical field during deposition by PLD. c-direction electric fields along out-of-plane direction help to form a c-domain LN crystal and suppress other orientations. Wang et al.[72] grew LN thin films on (001) silicon substrate at 600°C with a low FWHM of 0.21°for the rocking curves of (0006) diffraction peak. This work proved that highly oriented LN film could be grown without introducing any electric field or buffer layer. Wang et al.[73] grew LN thin films on SiO_2/Si substrates and explored the effect of substrate temperature on film

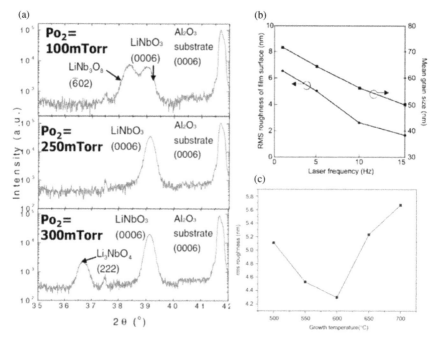

Fig. 6. (a) X-ray diffraction patterns of LN thin films grown at different oxygen partial pressures. (b) Variation of RMS surface roughness and mean grain size of LN thin films as a function of laser frequency in PLD. (c) Variation of RMS roughness of LN thin films as a function of substrate temperatures in PLD.

Source: (a) Reproduced with permission of Elsevier Publishing from Ref. 69. (b) Reproduced with permission of Elsevier Publishing from Ref. 70. (c) Reproduced with permission of Elsevier Publishing from Ref. 73.

roughness. Figure 6(c) shows the change of roughness at different substrate temperatures. The lowest roughness can be obtained at the optimal temperature of 600°C. Below this temperature, the low mobility of adatoms results in the growth of islands and, thus, higher roughness. Above this temperature, the grain size increases with temperature, leading to higher roughness.

5.3. MOCVD

MOCVD is a widely used thin-film growth technique. It uses metalorganic sources as precursors for chemical reactions on the substrates in the process chamber. Compared with other growth methods, MOCVD has several advantages, such as easier compositional control by using separate metal sources, scalability, and superior step coverage.[74]

In 1975, Position et al.[75] reported LN thin films grew on $LiTaO_3$ substrates by MOCVD at 450°C. By post-heat treatment at 1000°C, they obtained single crystal LN thin films. In 1993, Wernberg et al.[76] showed that the crystallinity of LN thin films improved with increasing substrate temperature. The best sample was deposited at 590°C with an FWHM of 0.48° for (006) diffraction peak. In 1998, Lee et al.[77] investigated the effect of temperature on roughness and proposed a two-step growth method to achieve both low roughness and high crystal quality. They demonstrated that the grain size of LN thin films increases with substrate temperature, leading to a rougher surface [Figs. 7(a) and 7(b)]. A smooth surface can be obtained at a lower temperature but with lower crystallinity. By using two-step growth, both low roughness and high crystallinity can be achieved, as shown in Fig. 7(c).

In addition to temperature, deposition pressure is another factor that affects the quality of LN thin films in MOCVD. Margueron et al.[74] found that lower deposition pressure can lead to the formation of the secondary phase. The $LiNb_3O_8$ was observed under a deposition pressure of 2 Torr. LN thin films deposited at 5–10 Torr consisted of a single stoichiometric phase but with different orientations, while purely (001)-oriented LN thin films were obtained at a high pressure of 20 Torr. In MOCVD, the source composition of the Li/Nb ratio is an important factor in forming single-phase LN thin films. In 1994, Lu et al.[78] investigated the effect of source composition and found that a single phase only exists at a 65Li:35Nb source composition. A two-phase combination region of $LiNbO_3 + Li_3NbO_4$ and $LiNbO_3 + LiNb_3O_8$ exists on the Li-rich side and Li-poor side, respectively. This result is further confirmed by Akiyama et al.,[79] who obtained a phase diagram in Fig. 7(d) that presents the narrow phase window of epitaxial LN thin films. At any temperature, under Li-rich conditions outside the window, a mixed oxide film consisting of Li_3NbO_3 and $LiNbO_3$ grows, while under Li-poor

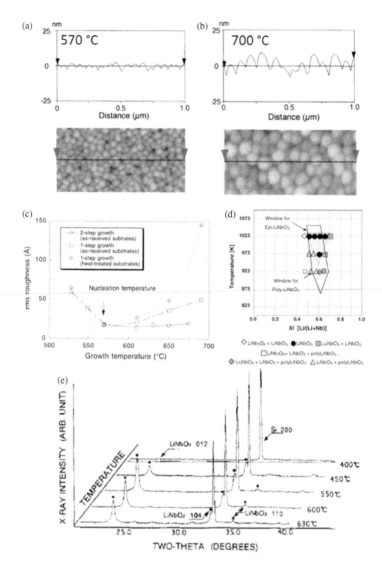

Fig. 7. The cross-section of 150 nm thick LN thin films grown on sapphire substrates at (a) 570°C (RMS = 1.4 nm, peak-to-valley value = 4.2 nm) and (b) 700°C (RMS = 4.5 nm, peak-to-valley value = 12 nm).[77] (c) The effect of 2-step growth on the RMS roughness of LN thin films.[77] (d) Phase diagram of Li–Nb–O thin films on sapphire. (e) X-ray diffraction patterns of LN thin films deposited at different temperatures on Si substrates.

Source: (b) and (c) Reproduced with open access Creative Commons CC-BY-NC-ND license from Elsevier publication. (d) Reproduced with permission of Elsevier Publishing from Ref. 79. (e) Reproduced with permission of Elsevier Publishing from Ref. 80.

conditions outside the window, the deposited film contains $LiNbO_3$ and $LiNb_3O_8$.

Besides deposit on sapphire, silicon has also been used to grow LN thin films by MOCVD. Tanaka et al.[80] investigated the growth of LN thin films on SiO_2/Si by optimizing temperature. Figure 7(e) shows that LN thin film started to crystallize from 450°C, and crystal quality increased with temperature. Due to the large lattice mismatch, the LN thin films were polycrystalline at all temperatures. In 1999, Lee et al.[81] studied the effect of temperature to optimize the thin film quality on the silicon substrate. They found that LN thin films presented poor crystal quality under low temperatures as shown by weak and board XRD peaks. As the growth temperature increased, the orientation of LN thin films gradually changed from (006) to (012). They explained this change by surface energy, as the (012) plane has the lowest surface energy. As the growth temperature increases, atoms have enough mobility to arrange themselves to minimize total surface energy and form (012) textured LN thin films.

5.4. MBE

MBE is a well-established tool to grow high quality epitaxial thin films with precisely controlled flux for semiconductors.[82] Early oxide film grown by MBE was reported in 1979 by Ploog et al.[83] In 1985, Betts et al.[84] demonstrated the first LN thin films grown by MBE using Nb and Li as sources and studied the effect of substrate temperature. The LN thin films grown at ambient temperature is amorphous. At 450°C, the LN thin films are polycrystalline with preferred c orientation. Single crystal LN thin films were formed at 550°C but with significant density of twins. These twins are greatly reduced at 650°C.

In 2006, Carver et al.[85] utilized chlorine chemistries by introducing $NbCl_5$ source to replace Nb metal source. They obtained epitaxial $LiNbO_2$ thin films on SiC substrates, which can be converted to LN thin films by annealing in oxygen environment. This method is further developed by Henderson et al.[86] and Greenlee et al.[87] to achieve the FWHM of (006) diffraction peak of 150 arcsec for LN thin films.

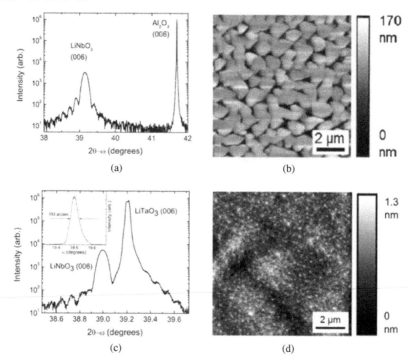

Fig. 8. (a) X-ray diffraction patterns and (b) AFM image of LN thin films grown on c-sapphire. (c) X-ray diffraction patterns and (d) AFM image of LN thin films grown on LiTaO$_3$ substrate.

Source: Reproduced with permission of Elsevier Publishing from Ref. 88.

With optimization of growth conditions, LN thin films reached high crystallinity close to bulk single crystal with FWHM as low as 8.6 arcsec.[88] Figure 8(a) presents the XRD pattern of LN thin films grown on sapphire substrates. Due to the 8.7% lattice mismatch between sapphire and LN, the AFM image in Fig. 8(b) shows an island structure and rough surface (maximum height reaching 170 nm). To reduce lattice mismatch, the authors used LiTaO$_3$ (LTO) as the substrate with only 0.1% lattice mismatch. This leads to good crystallinity [Fig. 8(c)] and very smooth LN thin films [Fig. 8(d)].

Tellekamp *et al.*[89] performed a systematic study of the effect of Li/Nb ratio and temperature on the growth of thin films with chemical composition in Li–Nb–O family. Figure 9(a) shows the

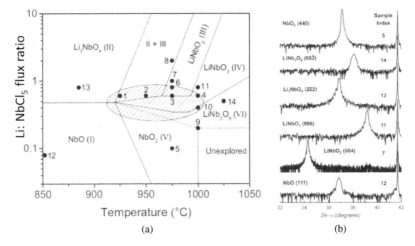

Fig. 9. (a) Empirical phase diagram of Li–Nb–O material systems with Li: $NbCl_5$ flux ratio and substrate temperature as tuning parameters. (b) X-ray diffraction patterns of Li–Nb–O thin films grown on c-oriented Al_2O_3 substrates.

Source: Reproduced with permission of Elsevier Publishing from Ref. 89.

empirical phase diagram for MBE growth of multiple phases including NbO, Li_3NbO_4, $LiNbO_2$, $LiNbO_3$, NbO_2 and $LiNb_3O_8$. It suggests that a high Li/Nb ratio and temperature above 1000°C is necessary to obtain single-phase LN thin films. The XRD diffraction patterns of LN thin films under growth condition 11 show a strong LN (006) diffraction peak. This provides a useful guideline for further development of LN films grown by MBE.

6. Conclusion

Based on the direct growth methods discussed above, each technique has its strengths and limitations when it comes to the growth of LN thin films. In Table 2, we provide a summary of these methods, focusing on their current status, challenges and potential solutions. By understanding the advantages and limitations of each growth method and considering the challenges associated with lattice mismatch and surface roughness, we can explore strategies to enhance the growth and quality of LN thin films for improved

Table 2. Summary of the direct growth methods of LN thin films.

Method	Current status	Challenges	Possible solutions
Sputtering	Most LN thin films grown by sputtering are polycrystalline with high surface roughness.	• Control of stoichiometry of LN thin films. • Surface roughness control by reducing island growth.	• Use lithium-rich targets to control the stoichiometry. • Develop separate source of lithium and niobium, potentially hybrid with MBE or MOCVD source.
PLD	Epitaxial LN thin films can be grown by PLD with moderate surface roughness.	• Control of stoichiometry of LN thin films. • Small sample size	• Use lithium-rich targets to control the stoichiometry. • Develop separate source of lithium and niobium, potentially hybrid with MBE or MOCVD source.
MOCVD	Stoichiometric LN thin films with low surface roughness can be achieved by two-step growth using MOCVD.	• Precise control of vapor pressure of Li and Nb precursors. • Narrow growth window of LN thin films with low growth rate.	• Develop stable Li and Nb precursors with wide growth window and high growth rate.
MBE	The LN thin films grown by MBE achieved the lowest FWHM of (006) diffraction peak (8.6 arcsec) but with high roughness [Fig. 8(b)].	• Nb is a refractory metal and it is difficult to achieve high and stable vapor pressure required for MBE growth. • $NbCl_5$ is proposed as the Nb source. However, the corrosive chlorine damages the metal parts in the chamber.	• Develop an efficient Nb source. Could potentially hybrid with metalorganic-based Nb source. • Optimize the surface roughness and crystal quality of LN thin films by fine control of Li to Nb flux ratio and temperature.

device performance. We summarize the key learnings below for further developments in the direct growth of LN thin films.

(1) On the choice of substrates. Although Al_2O_3 has large lattice mismatch with LN, it is the most popular choice for epitaxial growth of LN thin films due to the availability of high-quality substrates and lower refractive index than LN. Si substrates have also been widely investigated for direct growth of LN thin films. However, most LN thin films grown on Si are polycrystalline. SiC, $LiTaO_3$ and $LiNbO_3$ substrates have also been used for direct growth of LN thin films.

(2) On the chemical composition. Precise control of stoichiometry is a great challenge in direct growth of LN thin films. Multiple phases exist for non-stoichiometric thin films including Li_3NbO_4 and $LiNb_3O_8$. The first step to grow LN thin films by any method should focus on optimizing the chemical composition, which is largely determined by source composition, temperature and deposition rate. MOCVD and MBE have higher flexibility in stoichiometry control as they typically use separate Li and Nb source, instead of a single source for both Li and Nb in PLD and sputtering.

(3) On the balance between crystallinity and roughness. It is generally observed that the LN thin films show better crystallinity at higher temperatures but at the price of increasing surface roughness. A compromised condition may be obtained at moderate temperature. Other methods have also shown to be effective to reach both high crystallinity and low roughness, such as using buffer layers or two-step growth.

The outstanding piezoelectric and electro-optical properties of LN thin films are very attractive for next-generation acoustic resonators and photonic devices. This will drive more research work to overcome the difficulties in the direct growth of LN thin films. We believe that the quality of LN thin films, including crystallinity and roughness, will soon reach the level of bulk single crystal to enable wide adoption of LN-based acoustic resonators in telecommunication industry.

Acknowledgement

The authors acknowledge the Central Research Fund from the Agency for Science, Technology and Research (A*STAR) for the funding support. Z. Ye acknowledges the support from A*STAR Graduate Scholarship (AGS).

References

1. Geneva, *Facts and Figures: Focus on Least Developed Countries*, Switzerland, 2022.
2. N. Al-Falahy and O. Y. Alani, *IT Prof*, 2017, **19**, 12–20.
3. R. Rounds, B. Sarkar, A. Klump, C. Hartmann, T. Nagashima, R. Kirste, A. Franke, M. Bickermann, Y. Kumagai, Z. Sitar and R. Collazo, *Appl. Phys. Exp.*, 2018, **11**, 071001.
4. G. Bu, D. Čiplys, M. Shur, L. J. Schowalter, S. Schujman and R. Gaska, *IEEE Trans. Ultrason. Ferroelectr. Freq. Control*, 2006, **53**, 251–254.
5. S. Kume, M. Yasuoka, S. K. Lee, A. Kan, H. Ogawa and K. Watari, *J. Eur. Ceram. Soc.*, 2007, **27**, 2967–2971.
6. G. Wingqvist, F. Tasnádi, A. Zukauskaite, J. Birch, H. Arwin and L. Hultman, *Appl. Phys. Lett.*, 2010, **97**, 112902.
7. R. Lu, Y. Yang, S. Link and S. Gong, *J. Microelectromech. Syst.*, 2020, **29**, 313–319.
8. M. Akiyama, T. Kamohara, K. Kano, A. Teshigahara, Y. Takeuchi and N. Kawahara, *Adv. Mater.*, 2009, **21**, 593–596.
9. M. Moreira, J. Bjurström, I. Katardjev and V. Yantchev, *Vacuum*, 2011, **86**, 23–26.
10. M. Hu and F. Li Duan, *Solid State Electron.*, 2018, **150**, 28–34.
11. Y. Xing, Y. Shuai, X. Wang, L. Lv, X. Bai, S. Huang, K. Jian, T. Yang, W. Luo, C. Wu and W. Zhang, *Mater. Exp.*, 2020, **10**, 1504–1510.
12. X. Bai, Y. Shuai, L. Lv, S. Huang, Y. Xing, J. Zhao, W. Luo, C. Wu, T. Yang and W. Zhang, *J. Appl. Phys.*, 2020, **128**, 094503.
13. A. Bartasyte, S. Margueron, T. Baron, S. Oliveri and P. Boulet, *Adv. Mater. Interf.*, 2017, **4**, 1600998.
14. S. Yandrapalli, V. Plessky, J. Koskela, V. Yantchev, P. Turner and L. G. Villanueva, *IEEE Int Ultrasonics Symp*, IEEE, 2019, pp. 185–188.

15. V. Plessky, S. Yandrapalli, P. J. Turner, L. G. Villanueva, J. Koskela and R. B. Hammond, *Electron. Lett.*, 2019, **55**, 98–100.
16. A. Narayanan, E. Ramadan, J. Carpenter, Q. Liu, Y. Liu, F. Qian and Z.-L. Zhang, *Proc Web Conference*, ACM, New York, NY, USA, 2020, pp. 894–905.
17. Y. B. Park, B. Min, K. J. Vahala and H. A. Atwater, *Adv. Mater.*, 2006, **18**, 1533–1536.
18. S. C. Abrahams, W. C. Hamilton and J. M. Reddy, *J. Phys. Chem. Solids*, 1966, **27**, 1013–1018.
19. S. C. Abrahams, J. M. Reddy and J. L. Bernstein, *J. Phys. Chem. Solids*, 1966, **27**, 997–1012.
20. S. C. Abrahams, H. J. Levinstein and J. M. Reddy, *J. Phys.Chem. Solids*, 1966, **27**, 1019–1026.
21. R. S. Weis and T. K. Gaylord, *Lithium Niobate: Summary of Physical Properties and Crystal Structure*, Vol. 37, 1985.
22. T. Volk and M. Wöhlecke, *Lithium Niobate: Defects, Photorefraction and Ferroelectric Switching*, Springer Science & Business Media, 2008, vol. 115.
23. H. Bhugra and G. Piazza, *Piezoelectric MEMS Resonators*, Springer, 2017, pp. 103–105.
24. K. Momma and F. Izumi, *J. Appl. Crystallogr.*, 2011, **44**, 1272–1276.
25. Y. Yang, R. Lu, T. Manzaneque and S. Gong, *IEEE MTT-S Int Microwave Symp Digest*, Institute of Electrical and Electronics Engineers Inc., 2018, vol. 2018 pp. 563–566.
26. Y. Yang, R. Lu, L. Gao and S. Gong, *J. Microelectromech. Syst.* 2019, **28**, 575–577.
27. Y. Yang, R. Lu and S. Gong, *J. Microelectromech. Syst.* 2020, **29**, 135–143.
28. Y. Yang, A. Gao, R. Lu and S. Gong, *IEEE 30th Int Conf on Micro Electro Mechanical Systems*, IEEE, 2017, pp. 942–945.
29. A. Kochhar, A. Mahmoud, Y. Shen, N. Turumella and G. Piazza, *J. Microelectromech. Syst.*, 2020, **29**, 1464–1472.
30. M. Bousquet, M. Bertucchi, P. Perreau, G. Castellan, C. Maeder-Pachurka, D. Mercier, J. Delprato, A. Borzi, S. Sejil, G. Enyedi, J. Dechamp, M. Zussy, P. Sylvia, P. Kuisseu, F. Mazen, C. Billard and A. Reinhardt, *Single-mode high frequency $LiNbO_3$ Film Bulk Acoustic Resonator*, 2019.

31. S. Zhang, R. Lu, H. Zhou, S. Link, Y. Yang, Z. Li, K. Huang, X. Ou and S. Gong, *IEEE Trans. Microw. Theory Tech.*, 2020, **68**, 3653–3666.
32. M. Gorisse, R. Bauder, H.-J. Timme, H.-P. Friedrich, L. Dours, P. Perreau, A. Ravix, R. Lefebvre, G. Castellan, C. Maeder-Pachurka, M. Bousquet and A. Reinhardt, *Joint Conf IEEE Int Frequency Control Symp and European Frequency and Time Forum*, IEEE, 2019, pp. 1–2.
33. S. Huang, W. Luo, X. Bai, L. Lv, X. Pan, Y. Shuai, C. Wu and W. Zhang, *IEEE Trans. Ultrason. Ferroelectr. Freq. Control*, 2021, **68**, 2585–2589.
34. X. Bai, Y. Shuai, L. Lv, Y. Xing, J. Zhao, W. Luo, C. Wu, T. Yang and W. Zhang, *AIP Adv*, 2020, **10**, 075002.
35. R. Lu, Y. Yang, S. Link and S. Gong, *J. Microelectromech. Syst.*, 2020, **29**, 1332–1346.
36. M. Bousquet, P. Perreau, A. Joulie, F. Delaguillaumie, C. Maeder-Pachurka, G. Castellan, G. Envedi, J. Delprato, F. Mazen, A. Reinhardt, I. Huyet, T. Laroche, A. Clairet and S. Ballandras, *IEEE Int Ultrasonics Symp*, IEEE, 2021, pp. 1–4.
37. A. Reinhardt, M. Bousquet, A. Joulie, C.-L. Hsu, F. Delaguillaumie, C. Maeder-Pachurka, G. Enyedi, P. Perreau, G. Castellan and J. Lugo, in *Joint Conf European Frequency and Time Forum and IEEE Int Frequency Control Symp*, IEEE, 2021, pp. 1–4.
38. Y. Yang, R. Lu, T. Manzaneque and S. Gong, in *IEEE Int Frequency Control Symp*, IEEE, 2018, pp. 1–5.
39. M. Faizan, A. De Pastina, S. Yandrapalli, P. J. Turner, R. B. Hammond, V. Plessky and L. G. Villanueva, *20th Int Conf on Solid-State Sensors, Actuators and Microsystems & Eurosensors XXXIII (TRANSDUCERS & EUROSENSORS XXXIII)*, IEEE, 2019, pp. 1752–1755.
40. A. Almirall, S. Oliveri, W. Daniau, S. Margueron, T. Baron, P. Boulet, S. Ballandras, S. Chamaly and A. Bartasyte, *Appl. Phys. Lett.*, DOI:10.1063/1.5086757.
41. M. Bousquet, P. Perreau, C. Maeder-Pachurka, A. Joulie, F. Delaguillaumie, J. Delprato, G. Enyedi, G. Castellan, C. Eleouet, T. Farjot, C. Billard and A. Reinhardt, *IEEE Int Ultrasonics Symp*, IEEE Computer Society, 2020, vol. 2020.
42. S. Yandrapalli, S. E. K. Eroglu, V. Plessky, H. B. Atakan and L. G. Villanueva, *J. Microelectromech. Syst.*, 2022, **31**, 217–225.

43. Y. Zhang, L. Wang, Y. Zou, Q. Xu, J. Liu, Q. Wang, A. Tovstopyat, W. Liu, C. Sun and H. Yu, *Proc IEEE Int Conf Micro Electro Mechanical Systems*, Institute of Electrical and Electronics Engineers Inc., 2021, vol. 2021, pp. 466–469.
44. Y. Yang, R. Lu and S. Gong, *IEEE Int Electron Devices Meeting*, IEEE, 2018, pp. 39.6.1-39.6.4.
45. Y. Yang, R. Lu, L. Gao and S. Gong, *IEEE Trans. Microw. Theory Tech.*, 2020, **68**, 5211–5220.
46. R. L. and S. G. Yansong Yang, *IEEE 32nd Int Conf Micro Electro Mechanical Systems*, 2019, pp. 875–878.
47. Y. Yang, L. Gao and S. Gong, *IEEE Trans. Microw. Theory Tech.*, 2021, **69**, 1602–1610.
48. M. Bousquet, G. Enyedi, J. Dechamp, M. Zussy, P. S. Pokam Kuisseu, F. Mazen, C. Billard, A. Reinhardt, M. Bertucchi, P. Perreau, G. Castellan, C. Maeder-Pachurka, D. Mercier, J. Delprato, A. Borzi and S. Sejil, *IEEE Int Ultrasonics Symp*, IEEE, 2019, pp. 84–87.
49. P. J. Turner, B. Garcia, V. Yantchev, G. Dyer, S. Yandrapalli, L. G. Villanueva, R. B. Hammond and V. Plessky, *Electron. Lett.*, 2019, **55**, 942–944.
50. Y. Yang, L. Gao, R. Lu and S. Gong, *IEEE Trans. Ultrason. Ferroelectr. Freq. Control*, 2021, **68**, 1930–1937.
51. Z. Dai, X. Liu, H. Cheng, S. Xiao, H. Sun and C. Zuo, *IEEE Electron. Dev. Lett.*, 2022, **43**, 1105–1108.
52. B. Zivasatienraj, M. B. Tellekamp and W. A. Doolittle, *Crystals (Basel)*, 2021, **11**, 397.
53. O. Sánchez-Dena, C. D. Fierro-Ruiz, S. D. Villalobos-Mendoza, D. M. Carrillo Flores, J. T. Elizalde-Galindo and R. Farías, *Crystals (Basel)*, 2020, **10**, 973.
54. P. J. Kelly and R. D. Arnell, *Vacuum*, 2000, **56**, 159–172.
55. N. F. Foster, *J. Appl. Phys.*, 1969, **40**, 420–421.
56. P. R. Meek, L. Holland and P. D. Townsend, *Thin Solid Films*, 1986, **141**, 251–259.
57. N. Fujimura, T. Ito and M. Kakinoki, *J. Cryst. Growth*, 1991, **115**, 821–825.
58. X. Lansiaux, E. Dogheche, D. Remiens, M. Guilloux-Viry, A. Perrin and P. Ruterana, *J. Appl. Phys.*, 2001, **90**, 5274–5277.
59. N. Fujimura, M. Kakinoki, H. Tsuboi and T. Ito, *J. Appl. Phys.*, 1994, **75**, 2169–2176.

60. T. A. Rost, H. Lin, T. A. Rabson, R. C. Baumann and D. L. Callahan, *J. Appl. Phys.*, 1992, **72**, 4336–4343.
61. M. Sumets, V. Ievlev, V. Dybov, A. Kostyuchenko, D. Serikov, S. Kannykin, G. Kotov and E. Belonogov, *J. Mater. Sci.: Mater. Electron.*, 2019, **30**, 15662–15669.
62. L. C. Sauze, N. Vaxelaire, D. Rouchon, R. Templier, D. Remiens, G. Rodriguez and F. Dupont, *Thin Solid Films*, 2021, **726**, 138660.
63. M. Sumets, V. Ievlev, A. Kostyuchenko and V. Kuzmina, *Molecul. Cryst. Liq. Cryst.*, 2014, **603**, 202–215.
64. H. Akazawa, *Thin Solid Films*, 2022, **748**, 139148.
65. V. Bornand, I. Huet, J. F. Bardeau, D. Chateigner and P. Papet, *Integr. Ferroelect.*, 2002, **43**, 51–64.
66. A. M. Marsh, S. D. Harkness, F. Qian and R. K. Singh, *Appl. Phys. Lett.*, 1993, **62**, 952–954.
67. Y. Shibata, K. Kaya, K. Akashi, M. Kanai, T. Kawai and S. Kawai, *Appl. Phys. Lett.*, 1992, **61**, 1000–1002.
68. I. Pershukov, J. Richy, E. Borel, E. Soulat, M. Bousquet, F. Dupont and B. Vilquin, *IEEE Int Symp Applications of Ferroelectrics*, IEEE, 2022, pp. 1–4.
69. J.-W. Son, S. S. Orlov, B. Phillips and L. Hesselink, *MRS Proc.*, 2003, **784**, C11.24.
70. G. H. Lee, B. C. Shin and B. H. Min, *Mater. Sci. Eng.: B*, 2002, **95**, 137–140.
71. X. L. Guo, W. S. Hu, Z. G. Liu, S. N. Zhu, T. Yu, S. B. Xiong and C. Y. Lin, *Mater. Sci. Eng.: B*, 1998, **53**, 278–283.
72. X. Wang, Z. Ye, J. He, J. Hung and B. Zhao, *Int. J. Mod. Phys. B*, 2002, **16**, 4343–4346.
73. X. Wang, Z. Ye, G. Li and B. Zhao, *J. Cryst. Growth*, 2007, **306**, 62–67.
74. S. Margueron, A. Bartasyte, V. Plausinaitiene, A. Abrutis, P. Boulet, V. Kubilius and Z. Saltyte, *Oxide-based Materials and Devices IV*, SPIE, 2013, vol. 8626, p. 862612.
75. B. J. Curtis and H. R. Brunner, *Mater. Res. Bull.*, 1975, **10**, 515–520.
76. A. A. Wernberg, H. J. Gysling, A. J. Filo and T. N. Blanton, *Appl. Phys. Lett.*, 1993, **62**, 946–948.
77. S. Y. Lee and R. S. Feigelson, *Reduced Optical Losses in MOCVD Grown Lithium Niobate Thin Films on Sapphire by Controlling Nucleation Density*, 1998, vol. 186.

78. Z. Lu, R. Hiskes, S. A. DiCarolis, R. K. Route, R. S. Feigelson, F. Leplingard and J. E. Fouquet, *J. Mater. Res.*, 1994, **9**, 2258–2263.
79. Y. Akiyama, K. Shitanaka, H. Murakami, Y. S. Shin, M. Yoshida and N. Imaishi, *Thin Solid Films*, 2007, **515**, 4975–4979.
80. A. Tanaka, K. Miyashita, T. Tashiro, M. Kimura and T. Sukegawa, *J. Cryst. Growth*, 1995, **148**, 324–326.
81. S. Y. Lee and R. S. Feigelson, *J. Mater. Res.*, 1999, **14**, 2662–2667.
82. W. A. Doolittle, A. G. Carver and W. Henderson, *J. Vacuum Sci. Technol. B: Microelectron. Nanometer Struct.*, 2005, **23**, 1272.
83. K. Ploog, A. Fischer, R. Trommer and M. Hirose, *J. Vac. Sci. Technol.*, 1979, **16**, 290–294.
84. R. A. Betts and C. W. Pitt, *Electron. Lett.*, 1985, **21**, 960.
85. W. A. Doolittle, A. Carver, W. Henderson and W. L. Calley, *ECS Trans.*, 2006, **2**, 103–114.
86. W. E. Henderson, W. Laws Calley, A. G. Carver, H. Chen and W. Alan Doolittle, *J. Cryst. Growth*, 2011, **324**, 134–141.
87. J. D. Greenlee, W. L. Calley, W. Henderson and W. A. Doolittle, *Phys. Status Solidi (C) Curr. Top. Solid State Phys.*, 2012, **9**, 155–160.
88. M. B. Tellekamp, J. C. Shank, M. S. Goorsky and W. A. Doolittle, *J. Electron. Mater.*, 2016, **45**, 6292–6299.
89. M. B. Tellekamp, J. C. Shank and W. A. Doolittle, *J. Cryst. Growth*, 2017, **463**, 156–161.

© 2025 World Scientific Publishing Company
https://doi.org/10.1142/9789819806409_0010

Recent Advances in Bio-Based Concrete Materials: A Critical In-Depth Review*

Abirami Manoharan[†], Rajapriya Raja[‡], Himanshu Sharma[§] and Sanjeev Kumar[¶,‖]

[†]Remainenergy, India
[‡]Department of Civil Engineering
Jerusalem College of Engineering
Pallikaranai, Chennai 600100, India
[§]Department of Civil Engineering
Maharaja Agrasen Institute of Technology
Maharaja Agrasen University, Baddi 174103, India
[¶]Norfolk State University, Engineering Technology, USA
[‖]kssanjeev@nsu.edu

With the growth of the construction industry, the role and importance of sustainable construction practices are also increasing. This study reviews the various advancements in the field of bio-based construction materials, including bio-aggregates, bio-binders, and bio-bricks. The origins of our centuries-old construction methods can be found in bio-based concrete materials, which have been revived and modernized to meet the needs of present-day construction. The integration of recovered construction and agricultural wastes plays a significant role in bringing these materials into the competitive building sector, aligning with sustainability goals. This study provides an in-depth examination of the processes and compositions required to achieve the desired strength and durability of bio-based materials. By comprehensively analyzing global research data, this study offers insights into the successful

---

[‖]Corresponding authors.
*To cite this article, please refer to its earlier version published in the *World Scientific Annual Review of Functional Materials*, Volume 2, 2430004 (2024), DOI: 10.1142/S2810922824300046.

incorporation of biomaterials, emphasizing their potential to reduce carbon emissions and promote the responsible utilization of natural raw materials. This pathway moving toward more sustainable construction practices underscores the environmental benefits of bio-based concrete materials.

Keywords: Biomaterials; biomineralization; polymer impregnation; strength; durability.

1. Introduction

Biomaterials have been used in construction since the most ancient construction era. The construction industry's evolution from bamboo and timber constructions to hardened concrete materials has been a long journey. Though the journey towards attaining this goal has succeeded, the hierarchy of achieving sustainable construction practices is descending.[1] Cement holds a very strong and unavoidable position in the construction industry as a binder material despite its high carbon footprint for its strength and rapid construction capability. The increased pollution caused by massive industrialization, and increasing population pressures the building sector for newer construction which in turn consumes about 40% of the global energy releasing a maximum of 30% of carbon dioxide and harmful greenhouse gases like methane and chlorofluorocarbons into the atmosphere.[2,3] This is the status as of 2007 and this trend has been increasing as per the reports in 2022 which on accounting implies a 6% increase in global energy and a 7% increase in carbon dioxide emissions.[3] The 2022 Global Status Report for Buildings and Construction states that 100 billion tons of waste are produced by the construction industry and a total of 9% of the carbon dioxide emissions are by construction materials.[4] The increasing carbon dioxide emissions into the atmosphere can cause global warming and climate change which has been gaining awareness recently and the impacts of these two major issues areas are a threat globally.[5] The increasing global warming can melt the glaciers of the Arctic and Antarctic regions and lead to an increase in sea levels by submerging the coastal areas.[6] Climate change on the other hand paves the way for unpredictable seasons and extreme climatic conditions.

The huge harmful gas emission into the atmosphere during the industry's production processes also leads to the formation of "Acid rain" which contains elevated acidic precipitates in them. Acid rain is capable of polluting the entire authentic environment by its acidic nature. The evolution of genetic problems and diseases that affect the skin and respiratory system of humans are the major threats caused by this rain and that also disturbs the ecological balance in Flora and fauna.[7,8]

For many years, concrete has been the material of preference for civil engineers all around the world. It is chosen for its superior performance, increased lifespan, and minimal maintenance requirements. Every year, small-scale structures are demolished, and newer, larger ones are built to accomplish fast urbanization. Majority of these demolished materials, which are often concrete, are frequently landfilled and are never used again. This practice has an impact on soil fertility. Scientists and engineers throughout the world are looking for sustainable and reusable construction materials since the wave of sustainability is also affecting the construction industry. Hence the term "Biomaterials" has been coined for materials that are derived from available biomass or materials that are reused from biomass. Biomaterials in construction come in various forms like aggregates, binders, and admixtures using plant/animal products as construction materials in a sustainable way. Concrete with recycled aggregate is one such substance. The hindrances caused by the construction industry can be coped with by these alternate construction materials that can minimize this global issue to an extent.[1] Conventional constructions with other alternate material construction can be compared with construction materials' Life cycle assessment (LCA). This can be a vital parameter in making a product sustainable or unsustainable by predicting the amount of carbon emissions of a construction material.[9,10] The use of biomaterials is capable of producing minimum carbon emissions compared to the conventional construction methods through various studies.[11]

The need for sustainable construction materials and an alternative to cement is increasing with the awareness of achieving sustainable growth in the construction industry. The alternate materials

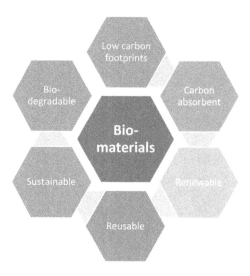

Fig. 1. Characteristics of a biomaterial.

have been researched in all the ingredients of the basic matrix needed for construction, which includes binding material and fine and coarse aggregates. The goal of achieving sustainable construction materials mainly looks back into the evolution of the construction industry that started from the most basic ingredients that were locally available for use in construction.

Figure 1 represents the basic properties exhibited by biomaterials which fulfil all the eligibility of being a sustainable construction material. Some of the most primarily used biomaterials in the construction industry are timber, bamboo (forms of wood), hemp, corn cobs, sunflower piths, coconut shells, and rice husk. Agricultural products occupy the major part of biomaterials, which assures agricultural waste is available as long as the basic need for food from agriculture is required. The promising results of agricultural products in the construction industry tend to be great alternates such as aggregates, binders, and admixtures.[12]

Figure 2 shows the visualization of biomaterials from the collected literatures. The industrial by-products are the second major abundantly available materials that are rich in alumina,

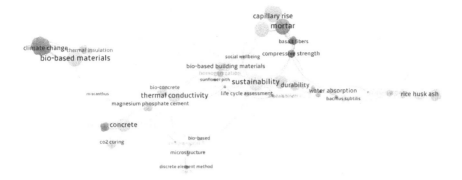

Fig. 2. Overall visualization of biomaterials from the collected literatures.

silica contents that are being researched as a feasible replacement to cement partially and the attainment of complete replacement is not very far away from today's intensive research works on bringing biomaterials into the construction industry. The cement kiln dust, silica sand, waste foundry sand, fly ash, palm oil fuel ash, GGBS, Silica fume, and sawdust have a major role as a construction material.[13] On the overall observation, the use of bio-based materials not only improves strength but also they satisfy thermal performance through their insulating properties, minimizing carbon emissions, and in providing healthy indoor air quality.[14]

The novelty of this study lies in its comprehensive evaluation of the successful utilization of construction and agricultural waste as bio-construction materials for sustainable and eco-friendly construction practices. Unlike previous studies, this research particularly focuses on bio-aggregates, bio-binders, and bio-bricks, examining their effective utilization, enhanced properties, and potential applications. By evaluating global research data on these biomaterials, this study offers valuable insights into their successful incorporation, highlighting their ability to reduce carbon emissions, and promote the responsible use of natural resources. This study not only aims to align the use of biomaterials with current sustainability goals but also paves the way for future advancements in environmentally friendly building materials.

2. Need for Biomaterials in the Construction Industry

The use of biomaterials can be seen from the evolution of the construction era of ancient times, the utilization of abundant agricultural waste materials in the mortar mix for building constructions was commonly termed as vernacular constructions. These vernacular constructions made the living environment healthy, and these constructions encouraged sustainable construction practices that hold the construction methodology environmentally friendly.[15] It has been proved in many studies that the thermal comfort of the region is adaptable by using vernacular construction practices[16] which is the need for artificial comfort modes. Researchers have been keenly studying on bringing these vernacular construction methods to the modern construction era in the following ways.

The recent advances in the use of these materials in an updated form with cement have been a trending one with improved sustainability compared to the use of modern construction materials completely. The use of modern construction materials as stabilizers in the vernacular binder has been giving positive results that have been implemented successfully in many parts of the world.[15] The basis of achieving sustainability through green buildings by utilizing bio-based materials is rooted in the vernacular construction practices and materials from the past.[17] Some of the most commonly

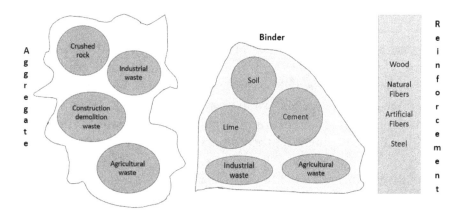

Fig. 3. Vernacular and present-day construction materials.

Fig. 4. Bio-aggregates: Recycled aggregate,[20] Hemp shiv,[21] corn stalk,[22] sunflower pith, maize pith,[23] Peach and Apricot shell,[24] and corn stalk.[22]

used bio-materials from a vernacular background are timber/bamboo, natural fibers to enhance the properties of the mortar/concrete matrix, Ash and raw agricultural wastes in the matrix for thermal performance and strength properties, extracts from plants and animals as admixtures that are accelerators/retarders and superplasticizers in an eco-friendly way.[16,18,19]

Figure 4 depicts some of the commonly used bio-aggregates in the concrete by processing them with microorganisms and chemical solutions into utilizable forms. The processing of recycled aggregates is highly mandatory due to the porous nature of the aggregates, the reactive transition zone between the recycled aggregates and the new concrete material, higher water absorption rate which will be the main reasons for lowering the mechanical and durable properties of a normal concrete.[25–27]

3. Bio-based Concrete Materials

The overuse of present construction materials has created the need for new eco-friendly construction alternatives that satisfy the conventional materials in a sustainable way.[28] The intensive research on utilizing waste and abundantly available byproducts of agricultural and industrial products has been trending in the last few years but there is a long way to achieve a methodology that can fit into the use of these alternate materials as a partial and complete replacement in the concrete. The need for a design mix that can be utilized standardly for bio-based concrete materials can be the initial step in making the construction material reach a safely guaranteed form of construction material in the market and satisfy the mindset of people to make use of it by a commoner. Reference 30 has proposed a mixed design model for the utilization of plant aggregates in cement and lime using a numerical optimization module that specifically concentrates on improving the porosity of the matrix using packing density of the aggregates in the matrix material.

4. Bio-Aggregates

60–70% of the concrete's volume is held by the fine and coarse particles. Massive mineral over-extraction and mountain demolition will affect the natural system as a result of the enormous aggregate extraction from the rock layers. This restriction creates a need to open the door for the use of bio-aggregates in concrete in line with sustainability principles. One of the most promising alternatives to recycled aggregates in concrete is crushed aggregate made from construction site detritus or demolition waste. According to the place of the aggregate previously used, they are classified into the following types of recycled aggregates.

The only obstacle to using these crushed aggregates is the old mortar that has been applied to them, which causes the matrix to be reactive to chemicals, to have more microcracks, and to absorb more water.[32-37] Various chemical treatments and the reaction of microorganisms heal or close the pores present in the aggregates, increasing their density and improving or decreasing their water absorption property. This helps to overcome these limitations and make efficient use of these readily available construction demolition wastes.

The table and Fig. 5 summarizes the research of various recycled aggregates, aggregates from agricultural wastes being processed

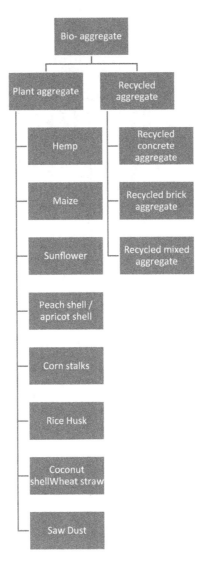

Fig. 5. Classification of bio-aggregate (see Refs. 30, 31, PMW Quarries, Specify concrete, and concrete construction).

and used in cement/lime as binders with successful results of aggregates that have been self-healed by the process of carbonation filing the excess pores.[62] The presence of microcracks plays a major role in lowering the quality of the concrete, which is healed by calcium

Table 1. Various research studies for the efficient use of bio-aggregates using microorganisms.

S. no	Aggregate composition	Term given for aggregate	Properties of aggregate					Microorganisms utilized	Binder	Properties studied	Property enhanced	Reference
			Specific gravity	Fineness modulus	Water absorption (%)	Crushing strength value (%)						
1	Crushed concrete	Recycled aggregate	2.12	7.81	6.2	26.96	Fungi–Fusarium oxysporum and Trichoderma longibrachiatum	OPC	Compression and Tension, Acid resistance.	Crack healing performance.	38
2	Crushed concrete	Recycled aggregate	2.4	5.85	—	—	Bacteria– Bacillus subtilis	OPC	Compressive strength, split tensile strength, Crack healing.	Crack healing performance.	39
3	Crushed concrete	Recycled aggregate	2.66	—	6.1	25	Bacillus Subtilis and Bacillus Sphaericus	—	Specific gravity, density, abrasion, and impact test.	Bio-deposition improves the strength of aggregates.	20
4	Crushed concrete	Recycled aggregate	2.40	—	5.40	23.74	Bacillus pasteurii	OPC	Compressive strength, Tensile strength, Specific gravity, and water absorption of aggregates.	Microbial carbonate precipitation of recycled aggregates are efficient in concrete.	26
5	Hemp	Bio-aggregate	—	—	—	—	—	Lime	Compression and flexural strength.	Increasing binder content improved flexural strength and thermal conductivity.	21, 40

6 Crushed concrete	Recycled aggregate	—	Bacteria–*Staphylococcus Pasteurii*	PC	Water absorption, compressive strength, thermal stability, and freez-thaw.	Bio-mineralization reduced water absorption of recycled aggregates.	27
7 Maize and Sunflower bark chips	Bio-aggregate	—	—	Metakaoline and lime	Pore structure, thermal stability, compressive strength, and sorption.	Composition of sunflower and metakaolin yielded optimum results.	41
8 Corn stalks	Plant aggregate	—	—	MPC, PC	Hygrothermal, mechanical and microstructural properties.	Magnesium phosphate cement (MPC) performed well with plant aggregates compared to PC.	22
9 Hemp, Sunflower pith, and Maize pith	Bio-based composite	—	—	Lime	Specific gravity, density, abrasion, impact test, water absorption, and Energy absorbing capacity.	10% basalt fibers improved the strength of geopolymer with rice husk ash.	23

(*Continued*)

Table 1. (*Continued*)

S. no	Aggregate composition	Term given for aggregate	Properties of aggregate					Microorganisms utilized	Binder	Properties studied	Property enhanced	Reference
			Specific gravity	Fineness modulus	Water absorption (%)	Crushing strength value (%)					
10	Peach shell and apricot shell	Bio-aggregate	—	—	—	—	—	PC	Compressive strength, flexural strength, drying shrinkage, freeze thaw.	Heat treated bio-aggregates had improved properties.	24
11	Hemp	Light weight aggregate	—	—	—	—	—	Geopolymer	Compressive strength, bulk density, and flexural strength.	Mineralizer treated hemp improved the properties of geopolymer.	42
12	Corn separates, crushed straw, and green biomass	Bio-based aggregate	—	—	—	—	—	PPC	Bulk density, open porosity, flexural and compressive strength, and thermal and acoustical property.	Carbonized lightweight aggregates improved the overall properties.	43
13	Crushed concrete	Recycled aggregate	—	—	—	—	*Bacillus pasteurii*	OPC	Water absorption, specific gravity, Density, compressive strength, and tensile strength.	Bacterial treatment improved the properties compared to cement slurry coating.	25

14	Crushed concrete	Recycled aggregate	1.47	—	—	B. pseudofirmus	PC	Water absorption, crushing value, compressive strength, and density.	Bio deposited aggregates performed well compared to others.	44
15	Crushed concrete	Recycled aggregate	—	—	—	H4 bacillus	PC	Porosity, compressive strength, and flexural strength.	Bio-calcium carbonate improved hydration and reduced formation of cracks	45
16	Cornstalk	Bio-aggregate	—	—	—	—	OPC	Water absorption, setting time, compressive strength, and mineralogical characterization.	Hydrophobic treatment improved hydration process.	46
17	Cornstalk	Plant aggregate	—	—	—	—	Magnesium phosphate cement	Sorption isotherm, capillary water uptake, drying kinetics, moisture diffusivity, compressive strength, microstructural analysis, saturations, and thermal conductivity.	The magnesium phosphate cement gave higher compressive strength with corn stalk aggregates compared to OPC as a binder. Thermal performance improved with	47

(Continued)

Table 1. (*Continued*)

S. no	Aggregate composition	Term given for aggregate	Properties of aggregate					Microorganisms utilized	Binder	Properties studied	Property enhanced	Reference
			Specific gravity	Fineness modulus	Water absorption (%)	Crushing strength value (%)						
18	Steel slag and Miscanthus powder	Bio-based lightweight aggregate	—	—	179.1	—	—	OPC	Compressive strength, bulk density, thermal study, and mineralogical characterization study.	Bulk density is minimized and the thermal property is enhanced by adding bio-based aggregate.	48
19	Crushed concrete	Recycled aggregate	—	—	6.4	16.7	*Bacillus mucilaginous*	PC	Surface morphology, adhesion, Elastic modulus, water absorption, and carbon dioxide capture.	Calcium carbonate covers the aggregate and improves its properties by the process of carbonation and microorganisms.	49
20	Dura oil palm shell (DOPS) and tenera oil palm shell	Bio-based lightweight aggregate	1.46	5.5	14.6	—	—	OPC	Slump, densities, compressive strength (air curing, water curing), water absorption, and elastic modulus.	The grouting process on the aggregate improved the slump value with the enhancement in the decrease in the aggregate size.	50

							interfacial bonding between the aggregates.	
21	Crushed concrete	Recycled aggregate	2.73	0.21	24.1	*Bacillus megaterium*	—	Water absorption, weight increase, and Mineralogical study.	The calcium carbonate precipitation reduced the porous nature of the aggregates and enhanced the water absorption properties.	51
22	Crushed concrete	Fly ash aggregate	—	—	—	*Bacillus species* bacteria	PC	Water absorption, compressive strength, split tensile strength, and flexural strength.	The review shows enhancement in the properties of aggregates by the bacterial inclusion of fibers.	52
23	Rice husk (RH)	Bio-aggregate	—	—	—	—	PC + Rice husk ash	Compressive strength, shrinkage in concretes, thermal conductivity, and mineralogical study.	The thermal performance is enhanced by adding RH and it was compatible with the concrete matrix.	34

(*Continued*)

Table 1. (*Continued*)

S. no	Aggregate composition	Term given for aggregate	Properties of aggregate				Microorganisms utilized	Binder	Properties studied	Property enhanced	Reference
			Specific gravity	Fineness modulus	Water absorption (%)	Crushing strength value (%)					
24	Crushed concrete	Recycled aggregate	—	—	—	—	*Bacillus sphaericus* bacteria	OPC	Weight increase, water absorption, ultrasonic absorbance, fragmentation resistance, workability, compressive strength, permeability, and sulphate attack.	The achievement in strength was 40% more using the microorganisms' infused aggregates than the conventional concrete mix.	53
25	Crushed bricks	Recycled aggregate	—	—	—	—	*Lysinibacillus boronitolerans*	PC	Compressive strength, split tensile strength, flexural strength test, sorptivity test, crack width analysis, self-healing potential, and mineralogical study.	The microorganisms infused mix improved the self-healing potential. The mineralogical study confirms the crystalline formation in the crack healing.	54

26	Sunflower stem/pith/ bark and hemp wood	Vegetable aggregates	—	—	PC+Wood ash	Bulk density, water absorption, thermal and chemical property, mineralogical study, compressive strength, flexural strength, and thermal conductivity.	Sunflower pith minimized the strength achievement compared to the conventional mix, increasing the cement content increased the mechanical strength properties.	55
27	Coconut shell	High strength lightweight aggregate	—	—	OPC	Compressive strength, flexural strength, split tensile strength, water absorption, and sorptivity test.	Superior quality concrete achievement is possible with coconut shell aggregate with a higher cement content.	56
28	Wheat straw	Vegetable aggregates	—	—	Lime	Bulk density, thermal conductivity, compressive strength, mineralogical study, water absorption, porosity, and particle size density.	The treated wheat aggregates enhanced the strength and thermal property with increase in the percentage of the mix.	57

(*Continued*)

Table 1. (Continued)

S. no	Aggregate composition	Term given for aggregate	Properties of aggregate				Microorganisms utilized	Binder	Properties studied	Property enhanced	Reference
			Specific gravity	Fineness modulus	Water absorption (%)	Crushing strength value (%)					
29	Spent tea	Plant aggregates	—	—	—	—	—	OPC	The mineralogical study, water absorption, thermal study (hot disk method, flash method), and compressive strength.	The overall strength and thermal performance of spent tea inclusion in the concrete improved the properties drastically.	58
30	Hemp	Bio-aggregate	—	—	—	—	—	Lime	Heat conduction problem, Iterative approach, and thermal conductivity.	Anisotropic bio-based building materials compared with the model approach and experimental data matched highly.	59

| 31 | Saw dust | Bio-aggregate | — | — | — | Magnesium oxysulfate cement + OPC | Thermal study, strength, water absorption, water resistance, and microstructural study. | Thermal insulation of panels. | 60 |
| 32 | — | Bio-admixture | — | — | — | Euphorbia Tortilis cactus | OPC | Review content | OPC sustainable construction materials. | 61 |

carbonate precipitation by the microorganism's activity in the aggregates. The exploration of utilizing agricultural waste as aggregates has been a challenging one to compete with conventional aggregates for their strength attainment, whereas agricultural products as aggregates have the minimum carbon footprint that is the need of the hour in the construction industry.[63,64] The concretes that use agriculture byproducts and a stage are also termed vegetal concrete and have resulted in very good thermal performance and hygrothermal properties.[41] The lifecycle assessment of agricultural waste materials in construction has been proven as a positive alternative in concrete for their compatible nature in lime/cement and for exhibiting a very minimum disturbance to nature compared to the conventional cement and aggregates obtained from rocks.[65]

4.1. Bio-mineralization

Bio-mineralization is the use of microorganisms to deposit minerals in a substance. Bio-mineralization can be utilized to fortify recycled aggregates by depositing calcium carbonate on their surface. This method can assist in improving the qualities of recycled aggregates, such as water absorption and mechanical strength, making them more acceptable for use in concrete production.

The process is initiated using microorganisms that are decomposers of urea present in the recycled aggregates.[36] This process is classified into various types according to the type of exposure of microorganisms given to the aggregates; MICP is one of the common process where microbially induced calcium carbonate precipitate process happens, bio-deposition is a process of depositing microorganisms in the recycled aggregates, and at times encapsulated microorganisms are used in these bio-aggregate concrete mix to selfheal during the occurrence of microcracks.

4.2. Polymer impregnation

The process of treating polymer ingression into recycled aggregates for the welfare of the mechanical and durability properties has

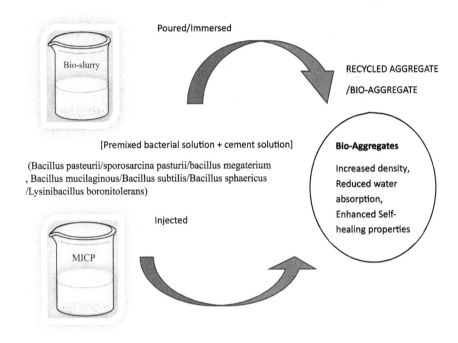

Fig. 6. Treatment of Construction demolition waste to recycled aggregates.[66,67]

been a successful one. A process schematic is shown in Fig. 6.[68] Some of the most commonly used polymers in this process are polyvinyl chloride, polyester resin & epoxy resin,[69] and silicon-based polymers.[70]

As shown in Fig. 7, polymer impregnation is one of the efficient processes that minimize the water absorption levels of the aggregate by filling all the pores with a polymer that doesn't allow water ingress in them.[66] The recycled aggregates are immersed in the polymer and then they are placed in the desiccator to settle inside the pores and to remove the moisture content in them. The most commonly used polymers for the process of polymer impregnation are polyvinyl alcohol, paraffin wax, silica fume solution, siloxane emulsion, bitumen emulsion, and residual cooking oil and bio-oil are used.[71,72]

Fig. 7. Polymer impregnation process.[37,68–70]

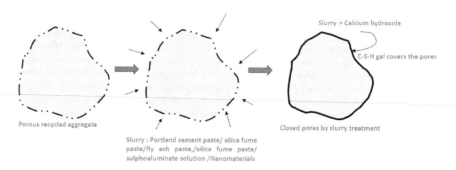

Fig. 8. Pozzalonic slurry treatment.[66,67]

4.3. *Pozzolanic slurry immersion*

The process consists of immersing recycled aggregates into pozzolan-rich binder content slurry that reacts with the mortar content in the recycled aggregate to form a stable layer that minimizes the porosity of the aggregates as shown in Fig. 8. This is one of the simpler forms of processing the recycled aggregates with slurry immersion and drying them to form stable recycled aggregates with better performance.

5. Bio-Binders

Binder material in the concrete is responsible for the chemical reaction taking place between them and water that bonds the concrete to form a rigid and stable matrix. The production of cement involves both a chemical process and the use of energy, both of which result in the emission of carbon dioxide (CO_2). When limestone, or calcium carbonate ($CaCO_3$), is burned in a cement kiln to produce lime, carbon dioxide (CO_2) is released as a byproduct. About half of the pollution caused by making cement can be attributed to this.

The results of the references compiled in Table 2 are depicted in the graph in Fig. 9. The composition of different bio-aggregates has been compiled to illustrate the compatible material content in these bio-aggregates, which can serve as substitutes for the conventional binder employed in concrete as shown in Fig. 10. Overall, the fly ash content indicates a greater concentration of calcium oxide in comparison to conventional cement, which is presently an exceptional blended binder for strength attainment. Adjacent to fly ash, the powdered cashew nut shells constitute a cost-effective byproduct that can be incorporated into the binder material. Investigating the feasibility of substituting natural and organic materials in the construction sector may prove to be a fruitful endeavor in the pursuit of sustainability. By incorporating silica oxide-rich materials (e.g., recycled brick powder and zeolite) with other constituent-rich replacement materials, the mechanical properties can be significantly enhanced.

6. Bio-Bricks

Bio-bricks are manufactured in place of conventional bricks composed of discarded natural or industrial materials. The substitution of alternative materials for soil in the production of bricks conserves a substantial amount of resources that are vital to the ecological survival of other ways of life. Table 3 presents a compilation of frequently investigated materials that are currently being employed as substitutes and alternatives in the fabrication process of bio-bricks.

Mass urbanization led to the over-extraction of natural resources to fulfill the needs of the construction industry to build more structures. Hence, experimentation and research in making bio-bricks using agro-based and industrial waste play a major role in replacing

Table 2. Bio-binders in bio-concrete materials.

S. no.	Binder composition	Bio-fibers/ microorganisms	Role	Enhanced properties	Reference
1	Magnesium sulfate cement	Recycled wood and basalt fiber.	Composite reinforcement	Flexural strength, Carbon sequestration (Paste and mortar)	73
2	Calcium sulfoaluminate cement	B. subtilis bacteria.	Self-healing	Strength and durability (Paste and mortar)	74
3	Rosin esters from pine trees	Waste cooking oil.	Energy-efficient roof	Thermal performance	75
4	Coal fly ash	Paenibacillus mucilaginosus bacteria.	Concrete	Low carbon high strength	76
5	Sugarcane bagasse ash	Alumina silicates for alkali activation.	Geopolymer	Reduce greenhouse gas emissions.	77
6	Waste concrete powder	—	Utilization of hardened cement and concrete powder.	The porous nature was reduced by utilizing waste concrete powder.	78

7	A composition of calcined magnesium oxide, fly ash, GGBS, OPC, and NH$_4$H$_2$PO$_4$	Alkalization and hydrophobic pretreatment.	Improved mechanical property.	Compatibility of binder and aggregate was improved; water absorption was minimized.	79
8	OPC + Sodium-based bentonite	—	Improved strength and stable thermal performance.	Compatibility of coke and the improved strength version of coke briquette.	80
9	Biohydrometallurgy waste + OPC	—	Minimizing the use of cement as a binder.	Strength attainment and suitability to be used as a binder.	81
10	Bitumen PG 64-22	Bio-oils	Oxidating agent in the binder, minimizing the carbon dioxide emission.	Bio-oils delayed the aging process, enhancing the durability of the material.	82
12	Magnesium oxysulfate (MOS) cement	Wheat husk and hemp hurd as bio-aggregate.	Alternate materials to improve sustainable construction practices.	Attainment of high early strength.	83

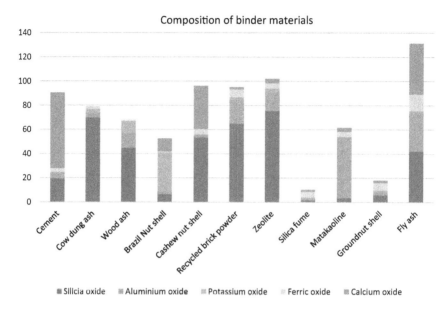

Fig. 9. Composition of binder replacement materials.

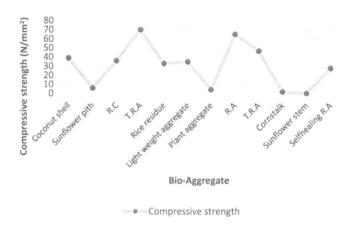

Fig. 10. Compressive strength exhibited by various bio-based aggregates [References from Table 1].

Table 3. Bio-Brick compositions.

S. no	Binder composition	Organic content/ Waste material	Role	Enhanced properties	Reference
1	Raw soil, Cement/ hydrated lime.	Egg ovalbumin	Stabilization of unfired earth bricks.	Improved compressive strength.	15
2	Fly ash	Cow dung ash and wood ash.	Partial replacement.	The durability of the fly ash brick is increased.	84
3	Nutshell waste	Ground nut, cashew nut, pistachio, hazelnut, shea nut, walnut, and Brazilnut.	As binder/fine aggregate.	Potential replacement material that matches with the strength and durability of conventional construction materials.	85
4	Soil	Potato peel powder and orange leaf powder.	Insulation and thermal performance.	The porosity, water absorption, and shrinkage are improved.	86
5	Soil	Microplastics	Replacement and waste utilization.	Effects of bio-plastic brick usage in conventional brick making and the beneficial aspects are summarized.	87
6	Soil	Landfill waste	Cost-effective and waste materials utilization.	20% of landfill waste can be used in the brick-making process.	88

(Continued)

Table 3. (Continued)

S. no	Binder composition	Organic content/ Waste material	Role	Enhanced properties	Reference
7	Brick powder	Sporosarcina pasteurii, Ottawasand, and polypropylene fiber and nature coconut fiber.	Calcium carbonate precipitation.	Improved compressive strength and cost-efficient bricks.	89
8	Brick aggregates	Lysinibacillus boronitolerans	Impact of the microstructure.	Enhanced self-healing capacity of the concrete material.	54
9	Dune sand + Cementation media	Human urine	Calcium carbonate precipitation.	Urine in calcium carbonate precipitation process enhances the properties of bio-bricks.	90
10	Clay	Juncus Acutus-fibers	Waste utilization	Increase in fiber content reduced the strength of the brick.	91
11	Soil, cementation media	Sporosarcina paseurii	To make efficient Sandstone bricks.	Low embodied energy construction materials in brick making.	92
12	Sludge	Agro waste, plastic waste, and industrial waste.	Sustainable bricks	The constraints in utilizing waste in the making process and the research gap are summarized.	93
13	Clay, fly ash	Calcium carbonate precipitation-Bacillus bacteria	Bio-Bricks	Better properties were inhibited by the bio-brick-making process compared to conventional bricks on strength, economic, and sustainability aspects.	94

the conventional use of natural resources.[95] The high demand for the natural resource soil used in brickmaking has led to over-extracting of soil and sand from the river beds which in turn has led to the change in the ecological balance of the landform and ruining the percolation of water underground, which leads to the scarcity of underground water table.

7. Durability Studies

7.1. Influence of bio aggregates

In case of bio-based aggregates, the studies showed a negative impact on durability characteristics due to their porous nature which makes them highly susceptible to environmental conditions.[96–98] It has been demonstrated that using sunflower and corn seeds as lightweight bio aggregates in concrete increases water absorption.[99] In lightweight aggregate concrete, peach, and apricot shells are utilized as bio aggregates and the research[100] showed that the absorption of water was reduced to less than 10%. The research[101] on the oil palm shell-based concrete reported that the drying shrinkage values are in the range of 405–614 $\mu\varepsilon$ at 234 days. Another study[102] found that pretreatment wood can greatly minimize shrinkage of up to 43.6% at 180 days, whereas the drying shrinkage of wood sand concrete after 90 days is approximately 1200–1900 $\mu\varepsilon$. Bio-based concrete has a much-reduced resistance to freeze-thaw cycles than normal-weight concrete because of its increased microstructure, porosity, and permeability. According to the study,[103] because calcium carbonate in seashells dissociates, the freeze-thaw performance of seashell pervious concrete is much worse than that of ordinary pervious concrete. In addition, it is important to draw attention that the utilization of the aggregate of biological origin in the construction materials necessitates the protection against fungi and bacteria.

7.2. Influence of bio binders

Biomaterials are primarily utilized as environmentally friendly materials in cementitious composites due to their exceptional impermeability and capacity to repair microcracks that occur during the hydration process.[104] The study[105] utilized *Bacillus sp.* CT-5 in cement mortar and observed a notable decrease in water absorption compared to the control specimens. This reduction can be attributed to the calcite deposition caused by *Bacillus sp.* The ureolytic bacteria were used in surface treatments of cementitious bricks and concrete in Ref. 107 and the results showed the decremental effect on capillary suction and gas permeability due to calcite deposition. Chahal *et al.*[107] conducted a study on the effect of three distinct concentrations of bacteria on concrete mixes comprising 5% and 10% silica fume relative to the weight of the cement and found that there is an improved chloride resistance due to the calcite production by the bacteria. The bacterial spores were encapsulated into hydrogels and silica gels and added in the concrete. The bacterial spores were encapsulated into hydrogels and silica gels that were added in the concrete and thus showed excellent healing characteristics. Increasing the concentration of *Bacillus sp.* in the concrete led to a decrease in durability characteristics, likely caused by smaller particle size and larger surface area.[108]

This study[109] concludes that by using agricultural biomass ashes in concrete resulted in positive aspects such as pore refinement and reduced water penetration depth due to improved fineness and ITZ, as well as negative aspects such as high porosity and increased carbonation depth due to the reduction in the formation of $Ca(OH)_2$. The improved acid resistance was found in the studies on rice husk ash,[110] rice straw ash,[111] and barely straw ash[112] due to the presence of amorphous silica. Another study[113] investigated the effect of magnesium sulphate on concrete with groundnut husk ash as a supplementary cementitious material and discovered that there is an increase in strength at 10% replacement while curing in the $MgSO_4$ solution due to the formation of ettringite and gypsum since reacting with cement.

8. Conclusion

In the current construction sector, biomaterials have great promise as an alternative or replacement material. They can help the industry move towards sustainability, reduce pollution caused by its operations thus far, and safeguard the surrounding environment and resources. This study of biomaterials shows that using bio-based materials as aggregates, binders, and admixtures has proven successful in many ways, according to researchers from all over the world.

Some of the key conclusions drawn from this research are as follows:

- Bio-based materials are highly porous and hence the permeability is high in the material which gives way to improved thermal performance of the product. Thermal comfort is the need of the hour to minimize the electricity used in providing thermal comfort indoors. Specifically, bio-based construction materials are excellent in hygrothermal properties providing a healthy indoor environment.
- The strength exhibited by bio-based materials has been highly competitive with conventional construction materials. The treated recycled aggregates have reached about 70 N/mm^2 compressive strength comparison of various bio-aggregate strength properties.
- Bio aggregates have negative impact on durability characteristics due to their porous nature which makes them highly susceptible to environmental conditions. It is important to draw attention to the utilization of the aggregate of biological origin in the construction materials that necessitates the protection against fungi and bacteria.
- Bio binders are primarily utilized as environmentally friendly materials in cementitious composites due to their exceptional impermeability and capacity to repair microcracks that occur during the hydration process. The process of self-curing is highly enhanced by the use of bio-based materials in the construction materials making them more durable.
- Since the acquired strength and durability values are in compliance with standards, it has been demonstrated that the overall performance of biomaterials as aggregates and binders has improved such that they can be used as green construction materials.

9. Research Gap

Using environmentally friendly materials helps a building earn the designation of "green building," which has become popular as a means of paving the way for the use of biomaterials to achieve sustainability. Current developments in biomaterials, such as biophase-changing materials, demonstrate how biomaterials are evolving and entering the building sector. Furthermore, as compared to conventional building materials, the high-strength achievement of biomaterials has boosted the benefits in a cost-efficient way.

The fundamental limitations of this study are that to successfully integrate these bio-materials into the building, the lower levels of the process must be carefully taught and optimized. When utilized as building materials, many of the materials experience the creation of leachate, which can reduce the material's strength and longevity. Hence, intense research on treating the bio-based materials that produce leachate without affecting the constituent of the materials has to be experimentally studied which can increase the percentage of positive results in the properties of the construction material.

Although biomaterials have been studied for a long time, only a few projects have been able to successfully use these materials. To educate the general public about the benefits of using eco-friendly waste and by-products in the building sector, more materials that have been the subject of successful research must be used in practical projects.

10. Future Recommendations

- Prior to utilizing bio-based materials in construction applications, it is important to consider the availability and quantity of such resources. Otherwise, the research will not result in commercialization.
- Majority of studies have focused on the direct use of bio-based materials in cementitious composites. However, it would be more advantageous to employ treatment methods to integrate them into performance-based mixtures.

- Previous research has primarily examined the strength properties of cement composites made from bio-based materials, with limited attention given to their longevity. Thorough investigation is required to explore the practical use of bio-based materials in structural applications. Hence, extensive investigation into the various elements affecting durability is necessary.
- It is necessary for conducting long-term investigations on various curing times. Additionally, there is a need for research on the microstructural analysis of bio-based concrete mixes in order to gain scientific understanding.

References

1. N. Aghamohammadi and M. Shahmohammadi, Towards sustainable development goals and role of bio-based building materials, *Bio-Based Materials and Waste for Energy Generation and Resource Management: Present and Emerging Waste Management Practices: Volume 5 in Advanced Zero Waste Tools*, Elsevier, January 2023, pp. 243–279, doi:10.1016/B978-0-323-91149-8.00004-1.
2. United Nations Environment Programme, Buildings and Climate Change: Current Status, Challenges and Opportunities 2007, *European commission DG ENV*, 2007.
3. C. Chandrakumar, S. J. McLaren, D. Dowdell and R. Jaques, *IOP Conf. Ser. Earth Environ. Sci.* 2019, **323**, 012183, doi:10.1088/1755-1315/323/1/012183.
4. UN Environment programme, Global Status Report for Buildings and Construction, 2022, https://unep.org/resources/publication/2022-global-status-report-buildings-and-construction.
5. R. Rehan and M. Nehdi, *Environ. Sci. Policy* 2005, **8**, 105–114, doi:10.1016/j.envsci.2004.12.006.
6. Y. Ding, S. Zhang, L. Zhao, Z. Li and S. Kang, *Sci Bull (Beijing)*, 2019, **64**, 245–253, doi:10.1016/J.SCIB.2018.12.028.
7. Y. Zhang et al., *Sci. Total Environ.* 2023, **873**, 162388, doi:10.1016/J.SCITOTENV.2023.162388.
8. Z. Liu et al., *Sci. Total Environ.* 2023, **873**, 162030, doi:10.1016/j.scitotenv.2023.162030.
9. T. Malmqvist et al., *Energy*, 2011, **36**, 1900–1907, doi:10.1016/j.energy.2010.03.026.

10. F. Fu, H. Luo, H. Zhong and A. Hill, *Math. Probl. Eng.* 2014, **2014**, 653849, doi:10.1155/2014/653849.
11. V. Cascione *et al.*, *J. Clean. Prod.* 2022, **344**, 130938, doi:10.1016/J.JCLEPRO.2022.130938.
12. C. Magniont, G. Escadeillas, M. Coutand and C. Oms-Multon, *Eur. J. Environ. Civil Eng.* 2012, **16**, s17–s33, doi:10.1080/19648189.2012.682452.
13. R. Siddique, Kunal and A. Mehta, Utilization of industrial by-products and natural ashes in mortar and concrete development of sustainable construction materials, *Nonconventional and Vernacular Construction Materials: Characterisation, Properties and Applications*, Elsevier, 2019, pp. 247–303, doi:10.1016/B978-0-08-102704-2.00011-1.
14. S. Bourbia, H. Kazeoui and R. Belarbi, A review on recent research on bio-based building materials and their applications, *Materials for Renewable and Sustainable Energy*, Springer Science and Business Media Deutschland GmbH, 2023. doi:10.1007/s40243-023-00234-7.
15. K. A. J. Ouedraogo, J. E. Aubert, C. Tribout, Y. Millogo and G. Escadeillas, *J. Cult. Herit.* 2021, **50**, 139–149, doi:10.1016/j.culher.2021.05.004.
16. S. M. Hosseinian, A. G. A. Sabouri and D. G. Carmichael, *Build. Environ.* 2023, **242**, 110605, doi:10.1016/j.buildenv.2023.110605.
17. K. Agyekum, E. Kissi and J. C. Danku, *Sci. Afr.* 2020, **8**, e00424, doi:10.1016/j.sciaf.2020.e00424.
18. B. Salgın, Ö. Bayram, A. Akgün and K. Agyekum, *Arts*, 2017, **6**, 11, doi:10.3390/arts6030011.
19. C. E. Barroso, D. V. Oliveira and L. F. Ramos, *J. Build. Eng.* 2018, **15**, 334–352, doi:10.1016/j.jobe.2017.12.001.
20. S. Jagan, T. R. Neelakantan and M. P. Lakshmi, *Mater. Today. Proc.* 2022, **49**, 1141–1147, doi:10.1016/j.matpr.2021.06.009.
21. J. Williams, M. Lawrence and P. Walker, *Constr. Build. Mater.* 2018, **159**, 9–17, doi:10.1016/j.conbuildmat.2017.10.109.
22. M. R. Ahmad and B. Chen, *Constr. Build Mater.* 2020, **251**, 118981, doi:10.1016/j.conbuildmat.2020.118981.
23. M. S. Abbas, F. McGregor, A. Fabbri and M. Y. Ferroukhi, *Constr Build Mater.* 2020, **259**, 120573, doi:10.1016/j.conbuildmat.2020.120573.
24. F. Wu, Q. Yu and C. Liu, *Constr. Build Mater.* 2021, **269**, 121800, doi:10.1016/j.conbuildmat.2020.121800.

25. A. Mistri, N. Dhami, S. K. Bhattacharyya, S. V. Barai and A. Mukherjee, *Constr. Build Mater* 2023, **369**, 130509, doi:10.1016/j.conbuildmat.2023.130509.
26. A. Mistri, N. Dhami, S. K. Bhattacharyya, S. V. Barai, A. Mukherjee and W. K. Biswas, *Resour. Conserv. Recycl.* 2021, **167**, 105436, doi:10.1016/j.resconrec.2021.105436.
27. A. Saeedi Javadi, H. Badiee and M. Sabermahani, *Constr. Build Mater.* 2018, **165**, 859–865, doi:10.1016/j.conbuildmat.2018.01.079.
28. P. K. Mehta and P. Monteiro, *Concrete: Microstructure, Properties, and Materials*, 4th edn., McGraw-Hill Education, New York, 2005.
29. E. Chabi, A. Lecomte, E. C. Adjovi, A. Dieye and A. Merlin, *Constr. Build. Mater.* 2018, **174**, 233–243, doi:10.1016/j.conbuildmat.2018.04.097.
30. R. Paiva, L. Caldas, R. Dias and T. Filho, *Role of Bio-based Building Materials in Climate Change Mitigation: Special Report of the Brazilian Panel on Climate Change. Self-healing View Project Textile Reinforced Concrete View Project*, 2018, https://www.researchgate.net/publication/327578512.
31. R. V. Silva, J. De Brito and R. K. Dhir, *Constr. Build Mater.* 2014, **65**, 201–217, doi:10.1016/j.conbuildmat.2014.04.117.
32. R. Balamuralikrishnan and J. Saravanan, *Emerg. Sci. J.* 2021, **5**, 155–170, doi:10.28991/esj-2021-01265.
33. R. Zhang, D. Xie, K. Wu and J. Wang, *Cem. Concr. Compos.* 2023, **139**, 105031, doi:10.1016/j.cemconcomp.2023.105031.
34. G. M. Amantino, N. P. Hasparyk, F. Tiecher and R. D. Toledo Filho, *J. Build. Eng.* 2022, **52**, 104348, doi:10.1016/j.jobe.2022.104348.
35. S. Rosa Latapie, M. Lagouin, V. Sabathier and A. Abou-Chakra, *J. Build. Eng.* 2023, **78**, 107664, doi:10.1016/j.jobe.2023.107664.
36. R. Zhang and J. Wang, *Constr. Build. Mater.* 2023, **403**, 133119, doi:10.1016/j.conbuildmat.2023.133119.
37. S. Yehia and M. M. Fawzy, *Int. J. Curr. Eng. Technol.* 2017, **7**, 634–640.
38. N. Khan, H. A. Khan, R. A. Khushnood, M. F. Bhatti and D. I. Baig, *Constr. Build. Mater.* 2023, **392**, 131910, doi:10.1016/j.conbuildmat.2023.131910.
39. R. A. Khushnood, Z. A. Qureshi, N. Shaheen and S. Ali, *Sci. Total Environ.* 2020, **703**, 135007, doi:10.1016/j.scitotenv.2019.135007.
40. J. Williams, M. Lawrence and P. Walker, *Constr. Build Mater.* 2016, **116**, 45–51, doi:10.1016/j.conbuildmat.2016.04.088.

41. M. Lagouin, C. Magniont, P. Sénéchal, P. Moonen, J. E. Aubert and A. Laborel-préneron, *Constr. Build Mater.* 2019, **222**, 852–871, doi:10.1016/j.conbuildmat.2019.06.004.
42. C. Narattha, S. Wattanasiriwech and D. Wattanasiriwech, *Constr. Build Mater.* 2022, **331**, 127206, doi:10.1016/j.conbuildmat.2022.127206.
43. J. Pokorný, R. Šev∆ík, J. Šál, L. Fiala, L. Zárybnická and L. Podolka, *Constr. Build Mater.* 2022, **358**, 129436, doi:10.1016/j.conbuildmat.2022.129436.
44. C. R. Wu, Y. G. Zhu, X. T. Zhang and S. C. Kou, *Cem. Concr. Compos.* 2018, **94**, 248–254, doi:10.1016/j.cemconcomp.2018.09.012.
45. C. R. Wu, Z. Q. Hong, J. L. Zhang and S. C. Kou, *Cem. Concr. Compos.* 2020, **111**, 103631, doi:10.1016/j.cemconcomp.2020.103631.
46. M. R. Ahmad, B. Chen, M. A. Haque, S. M. Saleem Kazmi and M. J. Munir, *Cem. Concr. Compos.* 2021, **121**, 104054, doi:10.1016/j.cemconcomp.2021.104054.
47. M. R. Ahmad, B. Chen, M. A. Haque and S. F. A. Shah, *J. Clean Prod.* 2020, **250**, 119469, doi:10.1016/j.jclepro.2019.119469.
48. Y. X. Chen, G. Liu, K. Schollbach and H. J. H. Brouwers, *J. Clean. Prod.* 2021, **322**, 129105, doi:10.1016/j.jclepro.2021.129105.
49. R. Wang, P. Jin, Z. Ding and W. Zhang, *J. Clean Prod.* 2021, **328**, 129537, doi:10.1016/j.jclepro.2021.129537.
50. M. K. Yew, J. H. Beh, M. C. Yew, F. W. Lee, L. H. Saw and S. K. Lim, *Case Stud. Constr. Mater.* 2022, **16**, e00910, doi:10.1016/j.cscm.2022.e00910.
51. M. Deva Kumar, K. B. Anand, V. Poornima, M. Gopal and A. Gupta, *Mater. Today. Proc.* 2021, **46**, 5138–5144, doi:10.1016/j.matpr.2020.10.662.
52. M. Abu Bakr, B. K. Singh, A. Hussain, Sustain. *Resilient Infrastruct.* 2024, 1–28, 10.1080/23789689.2024.2329390.
53. H. Sharma, S. K. Sharma, D. K. Ashish, S. K. Adhikary and G. Singh, *J. Build. Eng.* 2023, **66**, 105868, doi:10.1016/j.jobe.2023.105868.
54. H. Amjad, M. Shah Zeb, R. A. Khushnood and N. Khan, *J. Build. Eng.* 2023, **76**, 107327, doi:10.1016/j.jobe.2023.107327.
55. H. Affan, W. Arairo and J. Arayro, *Case Stud. Constr. Mater.* 2023, **18**, e01939, doi:10.1016/j.cscm.2023.e01939.
56. A. Sujatha and S. Deepa Balakrishnan, *Innov. Infrastruct. Solut.* 2024, 9, 191.

57. Z. Alshndah, F. Becquart and N. Belayachi, *J. Build. Eng.* 2023, **76**, 107199, doi:10.1016/j.jobe.2023.107199.
58. O. Horma, M. Charai, S. El Hassani, A. El Hammouti and A. Mezrhab, *J. Build. Eng.* 2022, **52**, 104392, doi:10.1016/j.jobe.2022.104392.
59. G. Huang, A. Abou-Chakra, J. Absi and S. Geoffroy, *J. Build. Eng.* 2022, **57**, 104890, doi:10.1016/j.jobe.2022.104890.
60. Z. Wei, K. Gu, B. Chen and C. Wang, Comparison of sawdust biocomposites based on magnesium oxysulfate cement and ordinary Portland cement, *J. Build. Eng.* 2023, **63**, 105514, doi:10.1016/j.jobe.2022.105514.
61. R. Mohanraj, S. Senthilkumar, P. Goel and R. Bharti, *Mater Today Proc.* 2023, https://doi.org/10.1016/j.matpr.2023.03.762.
62. J. N. Akhtar, R. A. Khan, R. A. Khan, M. N. Akhtar and J. K. Nejem, *Civ. Eng. J. (Iran)* 2022, **8**, 1069–1085, doi:10.28991/CEJ-2022-08-05-015.
63. S. Pretot, F. Collet and C. Garnier, *Build Environ.* 2014, **72**, 223–231, doi:10.1016/j.buildenv.2013.11.010.
64. K. Ip and A. Miller, *Resour. Conserv. Recycl.* 2012, **69**, 1–9, doi:10.1016/j.resconrec.2012.09.001.
65. D. Peñaloza, M. Erlandsson and A. Falk, *Constr. Build Mater.* 2016, **125**, 219–226, doi:10.1016/j.conbuildmat.2016.08.041.
66. K. Ouyang *et al.*, *J. Clean. Prod.* 2020, **263**, 121264, doi:10.1016/j.jclepro.2020.121264.
67. W. M. Shaban *et al.*, *Constr. Build Mater.* 2021, **276**, 121940, doi:10.1016/j.conbuildmat.2020.121940.
68. S. C. Kou and C. S. Poon, *Cement Concrete Compos.* 2010, **32**, 649–654, doi:10.1016/j.cemconcomp.2010.05.003.
69. J. M. L. Reis and M. A. G. Jurumenha, *Mater. Struct.* 2013, **46**, 1383–1388, doi:10.1617/s11527-012-9980-5.
70. V. Spaeth and A. Djerbi Tegguer, *Adv. Mater. Res.* 2013, **687**, 514–519, doi:10.4028/www.scientific.net/AMR.687.514.
71. Y. Zhu, Z. Pei, J. Yi, W. Zhou, X. Ai and D. Feng, *Constr. Build. Mater.* 2021, **302**, 124162, doi:10.1016/j.conbuildmat.2021.124162.
72. S. C. Kou, C. S. Poon and M. Etxeberria, *Cement Concrete Compos.* 2011, **33**, 286–291, doi:10.1016/j.cemconcomp.2010.10.003.
73. J. J. You, Q. Y. Song, D. Tan, C. Yang and Y. F. Liu, *Cement Concrete Compos.* 2023, **137**, 104934, doi:10.1016/j.cemconcomp.2023.104934.

74. B. C. Acarturk, I. Sandalci, N. M. Hull, Z. Basaran Bundur and L. E. Burris, *Cement Concrete Compos.* 2023, **141**, 105115, doi:10.1016/j.cemconcomp.2023.105115.
75. R. Álvarez-Barajas, A. A. Cuadri, C. Delgado-Sánchez, F. J. Navarro and P. Partal, *J. Clean Prod.* 2023, **393**, 136350, doi:10.1016/j.jclepro.2023.136350.
76. W. Wang, S. Guo, X. Gu, X. Li, W. Huang and A. Li, *Case Stud. Constr. Mater.* 2022, **17**, e01584, doi:10.1016/j.cscm.2022.e01584.
77. P. P. Patil and V. D. Katare, *Mater Today Proc.* 2023, doi:10.1016/j.matpr.2023.04.097.
78. R. Hu, C. Wang, J. Shen and Z. Ma, *J. Mater. Res. Technol.* 2023, **23**, 4012–4031, doi:10.1016/j.jmrt.2023.02.048.
79. M. R. Ahmad, B. Chen and S. F. Ali Shah, *Constr. Build Mater.* 2021, **281**, 122533, doi:10.1016/j.conbuildmat.2021.122533.
80. S. Kumar Patra, S. Ghorai, N. Sahu, G. U. Kapure and S. Kumar Tripathy, *Fuel*, 2022, **311**, 122502, doi:10.1016/j.fuel.2021.122502.
81. Y. Zhao *et al.*, *Powder Technol.* 2022, **398**, 117155, doi:10.1016/j.powtec.2022.117155.
82. F. Pahlavan, A. Lamanna, K. B. Park, S. F. Kabir, J. S. Kim and E. H. Fini, *Resour. Conserv. Recycl.* 2022, **187**, 106601, doi:10.1016/j.resconrec.2022.106601.
83. V. Barbieri, M. Lassinantti Gualtieri, T. Manfredini and C. Siligardi, *Constr. Build. Mater.* 2021, **284**, 122751, doi:10.1016/j.conbuildmat.2021.122751.
84. P. Indhiradevi, P. Manikandan, K. Rajkumar and S. Logeswaran, *Mater. Today: Proc.* 2021, **37**, 1190–1194, doi:10.1016/j.matpr.2020.06.355.
85. N. Jannat, R. Latif Al-Mufti, A. Hussien, B. Abdullah and A. Cotgrave, *Constr. Build. Mater.* 2021, **293**, 123546, doi:10.1016/j.conbuildmat.2021.123546.
86. M. Ghorbani, B. Dahrazma, S. Fazlolah Saghravani and G. Yousofizinsaz, *Constr. Build Mater.* 2021, **276**, 121937, doi:10.1016/j.conbuildmat.2020.121937.
87. A. Mohajerani and B. Karabatak, *Waste Manage.* 2020, **107**, 252–265, doi:10.1016/j.wasman.2020.04.021.
88. C. Dhanjode and A. Nag, *Mater. Today Proc.* 2022, **62**, 6628–6633, doi:10.1016/j.matpr.2022.04.616.
89. S. Li *et al.*, *Cem. Concr. Compos.* 2023, **139**, 105042, doi:10.1016/j.cemconcomp.2023.105042.

90. S. E. Lambert and D. G. Randall, *Water Res.* 2019, **160**, 158–166, doi:10.1016/j.watres.2019.05.069.
91. A. El-yahyaoui, I. Manssouri and H. Sahbi, *Mater. Today Proc.*, 2023, doi:10.1016/j.matpr.2023.06.064.
92. D. Bernardi, J. T. Dejong, B. M. Montoya and B. C. Martinez, *Constr. Build Mater.* 2014, **55**, 462–469, doi:10.1016/j.conbuildmat.2014.01.019.
93. J. Saravanan and P. V. Rao, *Sustain. Chem. Environ.* 2023, **3**, 100030, doi:10.1016/j.scenv.2023.100030.
94. V. Poornima, R. Venkatasubramani, V. Sreevidya and P. Chandrasekar, *Mater. Today: Proc.* 2021, **49**, 2103, doi:10.1016/j.matpr.2021.08.315.
95. P. Rautray, A. Roy, D. J. Mathew and B. Eisenbart, Bio-brick – Development of sustainable and cost effective building material, in *Proc. Int. Conf. Engineering Design, ICED* (Cambridge University Press, 2019), pp. 3171–3180, doi:10.1017/dsi.2019.324.
96. S. K. Adhikary, D. K. Ashish, H. Sharma, J. Patel, Ž. Rudžionis, M. Al-Ajamee, B. S. Thomas and J. M. Khatib, J *Resour. Conserv. Recycl. Adv.* 2022, **15**, 200107, doi:10.1016/j.rcradv.2022.200107.
97. D. K. Ashish and S. K. Verma, *Constr. Build. Mater.* 2019, **217**, 664–678, doi:10.1016/j.conbuildmat.2019.05.034.
98. D. K. Ashish and S. K. Verma, *J. Hazard. Mater.* 2021, **401**, 123329, doi:10.1016/j.jhazmat.2020.123329.
99. C. Sisman and E. Gezer, *Int. J. Environ. Waste Manage.* 2013, **12**, 203–212, doi:10.1504/IJEWM.2013.055594.
100. F. Wu, Q. Yu and C. Liu, *Constr. Build. Mater.* 2021, **269**, 121800, doi:10.1016/j.conbuildmat.2020.121800.
101. M. Aslam, P. Shafigh and M. Z. Jumaat, *J. Clean. Prod.* 2016, **127**, 183–194, doi:10.1016/j.jclepro.2016.03.165.
102. M. Bederina, M. Gotteicha, B. Belhadj, R. Dheily, M. Khenfer and M. Queneudec, *Constr. Build. Mater.* 2012, **36**, 1066–1075, doi:10.1016/j.conbuildmat.2012.06.010.
103. D. H. Nguyen, M. Boutouil, N. Sebaibi, F. Baraud and L. Leleyter, *Constr. Build. Mater.* 2017, **135**, 137–150, doi:10.1016/j.conbuildmat.2016.12.219.
104. T. Shanmuga Priya, N. Ramesh, A. Agarwal, S. Bhusnur and K. Chaudhary, *Constr. Build. Mater.* 2019, **226**, 827–838, doi:10.1016/j.conbuildmat.2019.07.172.

105. V. Achal, A. Mukherjee and M. S. Reddy, *J. Mater. Civil Eng.* 2011, **23**(6), 730–734.
106. W. De Muynck, K. Cox, N. D. Belie and W. Verstraete, *Constr. Build. Mater.* 2008, **22**(5), 875–885, doi:10.1016/j.conbuildmat.2006.12.011.
107. N. Chahal, R. Siddique and A. Rajor, *Constr. Build. Mater.* 2012, **28**(1), 351–356, doi:10.1016/j.conbuildmat.2011.07.042.
108. J. Wang, H. Soens, W. Verstraete and N. De Belie, *Cement Concrete Res.* 2014, **56**, 139–152, doi:10.1016/j.cemconres.2013.11.009.
109. S. K. Adhikary, D. K. Ashish and Ž. Rudžionis, *Sci. Total Environ.* 2022, **838**, 156407, doi:10.1016/j.scitotenv.2022.156407.
110. B. S. Thomas, *Renew. Sustain. Energy Rev.* 2018, **82**, 3913–3923, doi:10.1016/j.rser.2017.10.081.
111. A. Pandey and B. Kumar, *Heliyon* 2019, **5**, e02256, doi:10.1016/j.heliyon.2019.e02256.
112. F. Cao, H. Qiao, P. Wang and W. Li, 2021, preprint, doi:10.21203/rs.3.rs-462267/v1.
113. V. Charitha, V. S. Athira, V. Jittin, A. Bahurudeen and P. Nanthagopalan, *Constr. Build. Mater.* 2021, **285**, 122851, doi:10.1016/j.conbuildmat.2021.122851.

Design and Gene Delivery Application of Polymeric Materials in Cancer Immunotherapy*

Ying Chen[†,‡], Lingjie Ke[†,‡], Xian Jun Loh[§,¶,**] and Yun-Long Wu[†,‡,∥,**]

[†]Fujian Provincial Key Laboratory of Innovative Drug Target Research,
School of Pharmaceutical Sciences Xiamen University, Xiamen,
Fujian 361102, P. R. China
[‡]State Key Laboratory of Cellular Stress Biology, School of
Pharmaceutical Sciences,
Xiamen University, Xiamen, Fujian 361102, P. R. China
[§]Institute of Materials Research and Engineering, Agency for Science,
Technology and Research, 2 Fusionopolis Way, Singapore 138634, Singapore
[¶]lohxj@imre.a-star.edu.sg
[∥]wuyl@xmu.edu.cn

Immunotherapy has offered an alternative therapy method for cancer patients with metastatic tumors or who are not suitable for surgical resection. Different from traditional surgery, radiotherapy and chemotherapy, immunotherapy mainly restores the activity of the body's own immune cells silenced in the tumor microenvironment to achieve anti-cancer therapy. Gene therapy which corrects abnormal expression of immune cells in tumor microenvironment by delivering exogenous genes to specific immune cells, is the most widely studied immunotherapy. Although most available gene delivery vectors are still viral vectors, the further application of viral vectors is still limited by the

---

**Corresponding authors.
*To cite this article, please refer to its earlier version published in the *World Scientific Annual Review of Functional Materials*, Volume 1, 2230003 (2023), DOI: 10.1142/S2810922822300033.

immunogenicity and mutagenesis. Based on this, cationic polymeric gene vectors with high flexibility, high feasibility, low cost and high safety have been widely used in gene delivery. The structural variability of polymers allows specific chemical modifications to be incorporated into polymer scaffolds to improve their physicochemical properties for more stable loading of genes or more targeted delivery to specific cells. In this review, we have summarized the structural characteristics and application potential in cancer immunotherapy of these polymeric gene vectors based on poly(L-lysine), poly(lactic-co-glycolic acid), polyethyleneimine, poly(amidoamine) and hydrogel system.

Keywords: Gene delivery; nonviral vector; polymer; cancer immunotherapy.

1. Introduction

In recent decades, surgical resection, chemotherapy and radiation therapy have remained as the mainstay of cancer treatment. Compared with surgical resection with poor tolerance, chemotherapy and radiotherapy with large toxicity and side effects, immunotherapy using human body's own immune cells to immunosuppress tumors is gradually emerging, which changes the previous cancer treatment methods and aims to effectively improve the antitumor immune response, offers new feasibility plan for cancer patients who are not eligible for surgical resection, metastases or do wish to reduce tumor recurrence.[1,2] Gene therapy, which delivers exogenous genes (i.e. plasmid DNA and small interfering RNA) into specific cells and is transcribed to correct abnormal functional expression to achieve therapeutic effects, is the most extensive study of immunotherapy.[3] In order to avoid irreversible degradation of exogenous genes in the delivery process and achieve more controllable, extensive and safe effective gene transfection, delivery vectors with gene protection can be designed to load genes to achieve more effective targeting of the corresponding immune cells or tumor cells and reduce the probability of off-target.[4,5]

The delivery vector helps to achieve higher transfection efficiency, and the process of delivering genes mainly involves concentrated genes, specific cell uptake, endogenous degradation release genes, metastases to cell nucleus and exogenous genes to be condensed into transcribable form in the nucleus. Unnecessary nontarget toxicity and the off-target effect should be avoided in the process of gene therapy. Currently known delivery vectors for cancer immunotherapy are divided into viral vectors and nonviral vectors, and viral vectors (i.e. adenovirus, adeno-associated virus and lentivirus) are still the main clinically available gene therapy systems.[6,7] The design of viral vectors includes replacing disease-causing genomes with exogenous ones that are needed, and the protein shell of viral vectors protects the internal genome from degradation enzymes in the cells; however, severe inflammation caused by immunogenicity of viral vectors, cell death reports and insertion mutagenesis have limited the wider applications of viral vectors, and expensive production methods limit their mass production.[8,9] In contrast, nonviral vectors have attracted the attention of scientists owing to their internal security and maneuverability. Nonviral systems for gene therapy are divided into physical and chemical methods, and the physical methods include electroporation, fluid pressure, ultrasonic, genetic gun, magnetic transfection, etc.[10–12] Although the amounts of genes that can be delivered will not be restricted and lead to less immune induction by the manner of physical/mechanical relation to the plasmid vector, the physical/mechanical design is relatively complex and requires special equipment and operating experience, so physical methods in gene delivery do not have clinical applicability and demonstrate inferior cell transfection and protein expression ability to viral vectors.[3,13–15] It is worth mentioning that the nonviral system based on chemical methods can effectively overcome these problems.

Natural nonviral vectors (i.e. chitosan, agarose) do not cause rejection effect *in vivo* and can be degraded into nontoxic extracellular matrix, but the gene transfection ability of them is poor because only free genes can be released. As shown in Fig. 1, chemical methods of cationic polymer gene vectors with large molecular weight have been widely used for gene delivery to achieve cancer immunotherapy compared to natural nonviral vectors, which are developing into the most flexible, feasible and cost-effective gene vectors.[16,17] The polymer system is structurally stable, and its

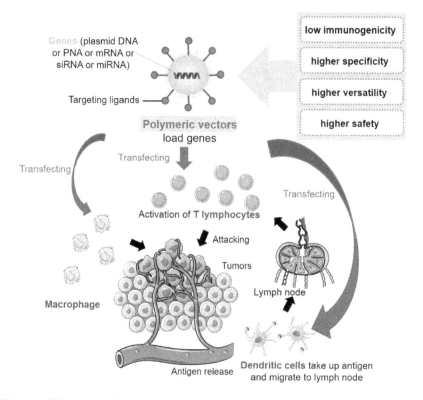

Fig. 1. Schematic illustration of the therapeutic approach of gene-delivered nonviral cationic polymeric vectors in cancer immunotherapy.

Source: Reproduced with permission from Ref. 17. Copyright 2020, Wiley VCH.

flexibility allows complex chemical modifications to be incorporated into the scaffold to improve the physicochemical properties, while increasing the possibility of repeated injections during gene therapy to maximize the most appropriate treatment.[18] The supramolecular complex system formed by cationic polymer and gene can interact with the electronegative cell membrane during gene delivery; further, it can enter the cell through adsorption endocytosis. How to increase the *in-vivo* stability of supramolecular complex system and targeting ability to target cells is the key to enhance the gene delivery efficiency of cationic polymer system *in vivo*. It is worth mentioning that the incorporation of natural molecular fragments with cationic polymers can improve the gene loading efficiency and stability.[17,19] In this paper, the structural characteristics and functional applications of various cationic polymer systems in cancer immunotherapy will be summarized and compared, and the special regulatory applications of cationic polymer systems in tumor microenvironment will be revealed from the perspective of gene delivery.

2. Structural Design of Polymers for Gene Delivery

The design of gene vectors largely determines the potential of clinical application of gene editing in cancer immunotherapy, and the first step of successful gene therapy is to transfect genes into cells to achieve effective expression.[20] Three barriers (including cell membrane, lysosomal system and nuclear membrane) need to be overcome to transfer genes into cells. First, the vector should effectively bind to the cell and cross the cell membrane in order to improve the transfection efficiency. Second, the gene should be protected from lysosomal acidic environment before release. Finally, the gene must penetrate the nuclear membrane into the nucleus before successfully expressing the nucleolar target protein.[21,22] The idea of gene therapy seems simple, but the easy degradation of genes in blood transport, low cell uptake rate, poor pharmacokinetic properties and other reasons limit the application of naked genes *in vivo*.[23] Excellent gene vectors must encapsulate genes, prolong blood circulation, protect them from degradation by circulating pathways and mediate genes entry into the nucleus for cellular transcriptional mechanisms.[24,25] At present, the most commonly used gene therapy means are virus-mediated and electroporation transduction methods. Although the transfection of T-cells by *in-vitro* electrical transfer and injection of modified CAR-T-cells into the tumor site has been proven to be effective and safe, physical electroporation is relatively expensive and has high cell mortality. The virus transfection method is also widely regarded to be with low safety and as prone to high immunogenicity. In contrast, polymeric nonviral vectors show the advantages of noninfectivity, wide sources, low immunogenicity and easy functional modification.[26] At the same time, through the surface conjugation of cationic polymers, these polymer molecules (i.e. polyvinylpyrrolidone, polyvinyl alcohol and polyethylene glycol) can form hydrophilic layer on the surface of gene vector to avoid polymer molecules aggregation, thus increasing the colloid stability, helping to realize spatial stability and prolong the *in-vivo* circulation time.[23,27–29] In this section, we summarize the structural design and applications of various cationic polymeric nonviral vectors in cancer immunotherapy, and the structural types of polymeric vector.

2.1. Poly(L-lysine)-based polymeric vector

Poly(*L*-lysine) (PLL) cationic polymers were first proposed in 1977, and often used as conjugated groups to construct conjugated systems owing to containing a large number of hydrophilic amino acid groups. Although a large number of linear or dendritic polymers have been synthesized to date, PLL-based polymer vectors still account for the largest proportion of studies, ranging from the polymerization of 19 amino acid residues up to that of 1,116 amino acid residues.[30] PLL's water-soluble side chain is often used to avoid aggregation of conjugated systems, and PLL-based conjugated systems show higher fluorescence quantum efficiency and better photostability, so PLL is often used in biosensors, cell imaging reagents and gene delivery applications.[31,32] Studies showed that with the increase of PLL amino acid residues, the positive charge also increased, and the improvement of gene binding ability and transfection efficiency was accompanied by increased cytotoxicity. So how to coordinate the molecular weight of PLL polymer, gene transfection efficiency and cytotoxicity is particularly important.[33,34] Zhang *et al.* designed a safe PLL-based replication particle for loading plasmid DNA and peptide hormones, realizing simultaneous transfection and induction of cell function without cytotoxicity [Fig. 2(a)].[35] It is worth noting that the gene transfection efficiency shown by using PLL cationic polymer alone as gene vector is unsatisfactory, but combination with a reagent can effectively overcome such shortcoming, e.g. fibroblast growth factor (EGF) combination can improve the *in-vitro* cell uptake efficiency and the fusion peptide combination can realize endosome escape in the delivery

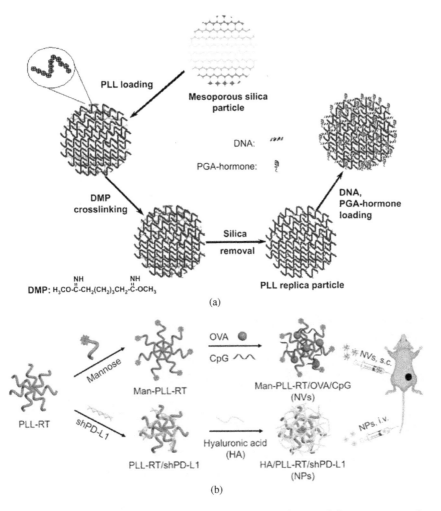

Fig. 2. Different applications of PLL-based nonviral gene delivery systems in cancer immunotherapy: (a) Schematic illustration of the synthesis and cargo loading of PLL replica particle. (b) Schematic illustration of the synthesis and delivery of Man–PLL-RT/OVA/CpG nanovaccines and HA/PLL-RT/shPD-L1 nanoparticles. (c) Schematic illustration of the regulation of tumor microenvironment by CD8-T–LP-CpG and CD8-T–LP-BMS-202.

Source: Panel (a) reproduced with permission from Ref. 35. Copyright 2010, Elsevier BV. Panel (b) reproduced with permission from Ref. 40. Copyright 2022, Elsevier BV. Panel (c) reproduced with permission from Ref. 41. Copyright 2021, The Royal Society of Chemistry.

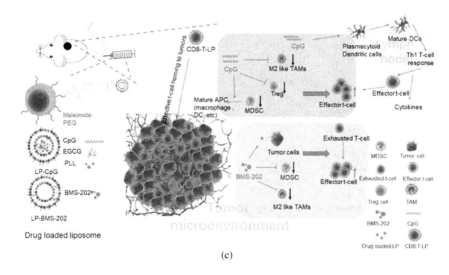

(c)

Fig. 2. (*Continued*)

process. By the way, conjugated poly(*L*-lysine) system with transferrin and polyethylene glycol has been approved for human clinical trials.[36] In addition, the high biocompatibility and biodegradability of PLL also demonstrate its advantages in photoimmunotherapy. The antitumor efficiency and photoimmunotherapy effect can be simultaneously improved by loading gene vaccine adjuvants and photoquantum-dot drugs.[37] As a typical example, Wu *et al.* designed a novel nanocomposite based on PLL cationic polymer that can simultaneously load CpG ODNs vaccine adjuvant, light-responsive quantum-dot GQD and NMR/fluorescence dual modal imaging probe, which could realize the controlled release of immune adjuvant to stimulate the immune response of tumor microenvironment and real-time monitor tumor internal conditions. And a series of photoactivated cancer immunotherapy systems have been developed as a novel treatment form.[38,39] Nanovaccines (NVs) can induce dendritic cell (DC) maturation, achieve effective cross-antigen presentation and activate cellular immune function of the body. However, the immunosuppressive microenvironment of tumors limits the anticancer efficacy of NVs. Chen *et al.* prepared Man–PLL-RT/OVA/CpG (NVs) and HA/PLL-RT/shPD-L1 (NPs) by utilizing the electrostatic adsorption properties of PLL. Among them, gene therapy of NPs adjusts the tumor immune tolerance microenvironment, alleviates the inhibition of NVs by tumor microenvironment and effectively inhibits tumor growth [Fig. 2(b)].[40] The complexity of TME is also a major obstacle to adoptive T-cell therapy (ACT) for tumor treatment. Liu et al. designed and synthesized PLL-based LP-CpG and LP-BMS-202 which contain the PD-L1 inhibitor, and combined them with CD8⁺ T-cells for the treatment of mouse melanoma to overcome the inhibition of ACT by TME and enhance T-cell viability [Fig. 2(c)].[41]

2.2. Poly(lactic-co-glycolic acid)-based polymeric vector

Poly(lactic-co-glycolic acid) (PLGA) was approved by FDA in 1969 as a copolymer of glycolic acid and lactic acid. The hydrolysates of PLGA polymer can be processed and utilized by the body after the hydrolysis process *in vivo*, so *in-vivo* cytotoxicity of this polymer is relatively low. It is often used as a biodegradable material for gene-targeted delivery and imaging.[42,43] Immune checkpoint blockers (ICBs) have been developed as one of the therapeutic drugs in cancer immunotherapy.[44] Although ICBs have a promising application prospect, they are still limited by low clinical response efficiency. For example, stimulator of interferon genes (STING) agonists need repeated administration for several times or even two years to achieve the desired effect, which is easy to cause high infection risk and poor patient compliance.[45–47] In this context, PLGA-based cubic particles with the ability to continuously release drugs reduce the adverse effects of these immunotherapy drugs. PLGA cubic particles are mainly prepared by double-emulsion solvent evaporation method; PLGA cubic particles loaded with immunoblockers slowly release the loaded blockers in the form of pulses for a long time while reducing needless endocytosis; studies have shown that this method can inhibit tumor growth and prolong patient survival with good tolerance.[48–51] Cytokines, as another cancer immunotherapy agent, can increase the immune monitoring ability of immune cells to achieve the purpose of tumor therapy, among which interleukin-1α (IL-1α) is a typical cytokine inducer, which has the potential to activate the immune response of tumor microenvironment.[52] However, in order to maintain the necessary levels of immune-activating cells in the body, patients need to be repeatedly injected with large doses of cytokines over a long period of time, which can cause serious side effects (i.e. diarrhea, fever).[53] For this, biocompatible PLGA microspheres have been designed to load cytokines to replace the system of one-time high-dose injection, which can be implemented to release low levels of cytokines within a certain interval of time, and the results also showed that the PLGA/IL-1α microspheres extended the life span of mice by 60%. Therefore, the biodegradability and high biocompatibility of PLGA are expected to improve the feasibility of cytokine reagent products for clinical immunotherapy in humans.[54] In addition to loading cytokines and immune checkpoint blockers, large molecular drugs including proteins, genes and vaccines have also been successfully integrated into PLGA or PLGA-based improved polymers.[55,56] Macrophages play an important role in cancer immunotherapy. However, since tumor cells express "Self" signals (such as CD47 molecules) to escape phagocytosis, Zhang *et al.* used PLGA-based nanoparticles to simultaneously deliver siRNA silencing the CD47 gene expression of tumor cells and mitoxantrone hydrochloride (MTO·2HCl) inducing CRT exposure on tumor surface to promote phagocytosis of tumor cells by macrophages [Fig. 3(a)].[57] While Heo and Lim designed two PLGA-based programmed nanoparticles, PLGA (R837/STAT3 siRNA) and PLGA (OVA/ICG), to transform DC cells which are the most effective antigen-presenting cells *in vitro*, improve the antigen presentation ability and achieve in-vivo tracing [Fig. 3(b)].[58]

2.3. *Polyethyleneimine-based polymeric vector*

Polyethyleneimine (PEI) is the most studied cationic gene delivery polymer after PLL. Compared with other nonviral gene vectors, it has the advantage of endosomal dissolution and has been proved by many researchers to have high *in-vitro* transfection activity.[59] Linear PEI molecules have lower cytotoxicity and higher gene transfection efficiency than dendritic PEI; the linear PEI (22 kDa)/

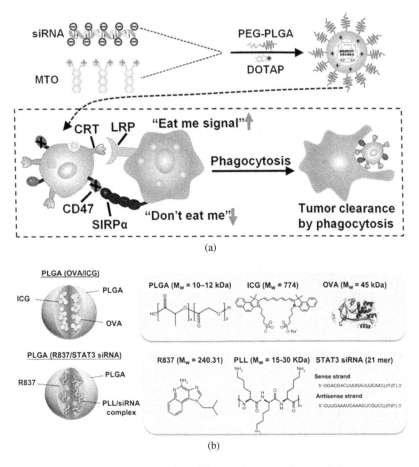

Fig. 3. Different applications of PLGA-based nonviral gene delivery systems in cancer immunotherapy: (a) Schematic illustration of VNPSiCD47/MTO synthesis and enhanced macrophage phagocytosis. (b) Schematic illustration of PLGA (OVA/ICG) and PLGA (R837/STAT3 siRNA) syntheses and the structure of each component.

Source: Panel (a) reproduced with permission from Ref. 57. Copyright 2021, American Chemical Society. Panel (b) reproduced with permission from Ref. 58. Copyright 2014, Elsevier BV.

gene complex system with particle size less than 100 nm can fully diffuse into tissues and produce high gene expression levels when injected into the ventricle.[60] On the one hand, the large amounts of primary ammonia, secondary ammonia and tertiary amino content in PEI chain endowed PEI with strong physiological buffering capacity. On the other hand, PEI can easily form circular complex system in physiological buffer environment to maintain stable aggregation. Nonviral vector system can effectively simulate the characteristics of viral vector gene delivery, such as gene concentration and enucleation into cells. On this basis, nonviral vector also introduced a new mechanism, mainly represented by the proton sponge effect of PEI.[61] However, molecular heterogeneity, *in-vivo* high toxicity and nonbiodegradability have slowed PEI's clinical research pace.[62] Similar to PLL, the transfection activity and *in-vivo* cytotoxicity of PEI were correlated with the amino content of PEI. Generally, with the increase of PEI molecular weight, the transfection activity and *in-vivo* cytotoxicity also gradually increased.[63] In order to solve the problems of limited delivery and lysosomal escape efficiency of low-molecular-weight PEI, Yin *et al.* mixed GO with PEI to produce GLP Gel for the delivery and slow release of R848 and mOVA *in vivo*. The study proved that GLP-RO Gel could not only continuously release GLP-GO NPs for 30 days, but also protect mRNA from degradation, providing a new reference for mRNA vaccine to provide a long-term and efficient cancer immunotherapy effect [Fig. 4(a)].[64] How to reduce the cytotoxicity of PEI while also retaining the high transfection efficiency is the key to the clinical restriction of PEI. In this case, the biodegradable polyethylene glycol (PEG) can be used to prevent binding by plasma proteins, thereby extending the circulation time and reducing body toxicity.[65] Nam *et al.* simplified the synthesis of PEI-based NP vaccines. In short, the cytotoxicity of PEI was reduced by PEG modification, PEI was bound to the neoantigen through simple chemical coupling, then electrostatic assembly with CpG adjuvant was carried out and finally, the self-assembly into PEI NPs with low toxicity was made, realizing the rapid and convenient production of personalized vaccine [Fig. 4(b)].[66] Similarly, Prijic *et al.* used polyacrylic acid (PAA) to reduce the cytotoxicity of PEI–DNA complex, and combined magnetic nanoparticles viz. SPIONs to construct a SPIONs–PAA–PEI gene delivery system to deliver

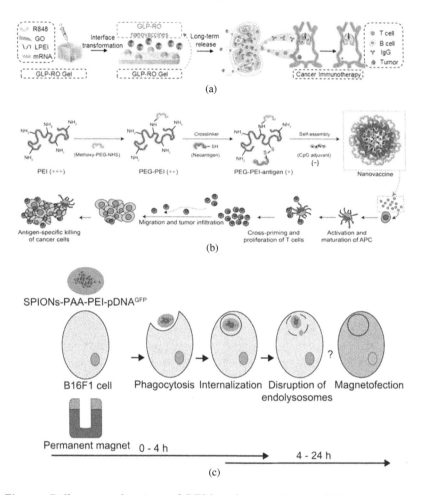

Fig. 4. Different applications of PEI-based nonviral gene delivery systems in cancer immunotherapy: (a) Schematic illustration of GLP-RO Gel releasing nanovaccines for long-term cancer immunotherapy. (b) Schematic illustration of PEI-based nanovaccine synthesis and enhanced cancer immunotherapy. (c) Schematic illustration of SPIONs–PAA–PEI–pDNAGFP cell internalization.

Source: Panel (a) reproduced with permission from Ref. 64. Copyright 2021, American Chemical Society. Panel (b) reproduced with permission from Ref. 66. Copyright 2021, Wiley-VCH Verlag. Panel (c) reproduced with permission from Ref. 67. Copyright 2012, Elsevier BV.

plasmid DNA. Studies have shown that this delivery system can achieve efficient and safe magnetic transfection of mouse cells and tumors [Fig. 4(c)].[67]

2.4. *Poly(amidoamine)-based polymeric vector*

As a typical dendritic macromolecule polymer, the core of poly(amidoamine) (PAMAM)-based polymer molecule is dominated by tertiary amine, and through the dispersion method or convergence method primary amine, acid amide or special functional group skeleton can be added to the branch chain to form a variety of dendritic polymers. The surface and internal pKa values of PAMAM dendritic macromolecules are affected by the group type. The presence of primary amine groups on the surface makes the pKa values range from 7 to 9, and the presence of tertiary amine groups in the interior makes the pKa values range from 3 to 6.[68] The conformation structure of polymer molecules with low degree of branching is circular plane or elliptic plane, while the conformation of polymer molecules with high degree of branching is spherical, in which the hydrophobic cavity inside the spherical dendritic molecule can effectively load drugs and other loads.[69] PAMAM dendritic macromolecules have been extensively studied by scientists in the past few years, mainly because of their special structural properties: controllable structure, high functional density, easy modification and high chemical homogeneity. By optimizing the chemical structure of PAMAM, scientists have demonstrated that PAMAM dendrimers can achieve gene delivery effects comparable to or even better than PLL.[70] The primary amine groups on the PAMAM polymer molecules participate in binding genes and promote cellular uptake, and the tertiary amine groups can improve the polymer's inner body buffering capacity and increase the amount of genes released into cytoplasm, but the cell transfection efficiency of PAMAM polymer is affected by polymer diameter and N/P, and presents a certain function relation. Cho *et al.* designed PAMAM-based acid-degradable core–shell nanoparticles (siRNA/PAMAM)–PK NPs. First, the primary amine of PAMAM is combined with siRNA to form a core. Then acid-degradable PK shells were formed on the surface by photoinduced polymerization to protect and control the stable release of siRNA/PAMAM dendriplexes *in vivo*. The results showed that the nanoparticles could effectively silence MnSOD gene expression and reverse TAM resistance of breast cancer cells *in vivo* and *in vitro*, thus enhancing the efficacy of chemotherapy [Fig. 5(a)].[71] Moreover, the transfection properties of PAMAM polymers with different molecular weights are also different in different cells. As a typical example, the transfection efficiency of 40-nm-diameter PAMAM dendrimer (G4) was higher than that of smaller-diameter PAMAM dendrimer. Song *et al.* prepared the G5-BGG/pDNA complex using the fifth-generation PAMAM (G5) and loaded the complex into PLGA–PEG–PLGA thermosensitive hydrogel. Therefore, G5-BGG/pDNA-loaded PLGA–PEG–PLGA hydrogel capable of achieving efficient transfection and sustained release at the post-operative site was obtained, which showed a good inhibitory effect on tumor recurrence in the *in-situ* U87MG post-operative tumor model [Fig. 5(b)].[72] Apart from delivering large nucleotides such as plasmid DNA, PAMAM can also be used to deliver small nucleotides such as CpG. Liu *et al.* obtained pH-sensitive PC_{pH} by connecting PAMAM dendrimers to self-assembled nanocarriers via amide bonds, and then combining CpG-SH with disulfide bonds to form dual-sensitive $PC_{pH\&RE}$/CpG, which can be programmed to deliver CpG to TIDCs, effectively activate T-cells and enhance antitumor immunotherapy [Fig. 5(c)].[73] In addition to increasing the grafting degree of PAMAM macromolecule, modification of PAMAM polymer molecule is another effective method to improve its gene transfection efficiency.[74] Some scientists have found that inserting quaternary ammonium salts into the surface and interior of PAMAM dendrimers by methylation can effectively improve gene delivery efficiency. Quaternary ammonium salts provide a neutral surface that reduces polymer self-aggregation problem and greatly reduces polymeric toxicity. Using quaternary ammonium salt as the core increases the electrification of PAMAM macromolecular core, then increases the gene binding sites and improves the loading capacity. Moreover, the polymer/

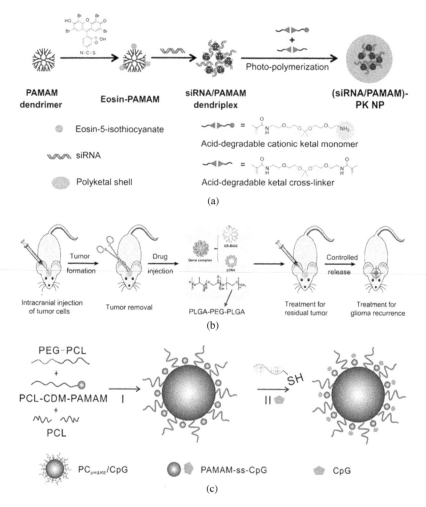

Fig. 5. Different applications of PAMAM-based nonviral gene delivery systems in cancer immunotherapy: (a) Synthetic schematic illustration of (siRNA/PAMAM)–PK NPs. (b) Schematic illustration of G5-BGG/pDNA-loaded PLGA–PEG–PLGA hydrogel for the treatment of post-operative glioma recurrence. (c) Schematic illustration of PC$_{pH\&RE}$/CpG assembly.

Source: Panel (a) reproduced with permission from Ref. 71. Copyright 2013, Elsevier BV. Panel (b) reproduced with permission from Ref. 72. Copyright 2021, Elsevier BV. Panel (c) reproduced with permission from Ref. 73. Copyright 2020, American Chemical Society.

gene complex formed by quaternary ammonium PAMAM polymer has smaller particle size and is easier to be endocytosed into the cell membrane.[75,76]

2.5. Hydrogel gene vector

Electroporation mRNA therapy is a relatively safe and economical method for T-cell transduction at present, but the short-term efficacy of the treatment requires multiple infusion of T-cells during the whole treatment process, and can induce severe allergic reactions.[77] Most of the antibodies used for immunomodulatory administration are systemic now, which can easily cause severe liver dysfunction and systemic cytokine storm syndrome.[78] Cationic polymers with high gene-loading capacity and the ability to form *in-situ* injectable hydrogels have recently been reported to induce stronger and more lasting immune cell responses than the other nonviral vectors and have been extensively designed to achieve locally sustained cancer immunotherapy.[79–81] Cyclodextrin, a cyclic oligosaccharide molecule composed of multiple glucose units, is the most typical supramolecular entity with high biocompatibility, whose internal hydrophobic cavity structure can encapsulate various molecules of appropriate size through molecular recognition to form supramolecular inclusion bodies. Supramolecular polyrotaxane and poly(pseudo)rotaxane structures formed by inserting cyclic cyclodextrin (CD) molecules into polymer chains have wide application in drug and gene delivery fields.[82,83] The formation of supramolecular hydrogels based on α-CD polymer long chains mainly includes two important stages: the hydrophilic polymer long chains infiltrate into α-CD cavity to form poly(pseudo)rotaxanes by host–guest interaction, and the aggregation between adjacent α-CD molecules happens in the form of head (secondary hydroxyl) to head and tail (primary hydroxyl) to tail by hydrogen bonding.[84] It is worth mentioning that the CD-based polyrotaxane supramolecular system has been proved to have good biocompatibility, stability and

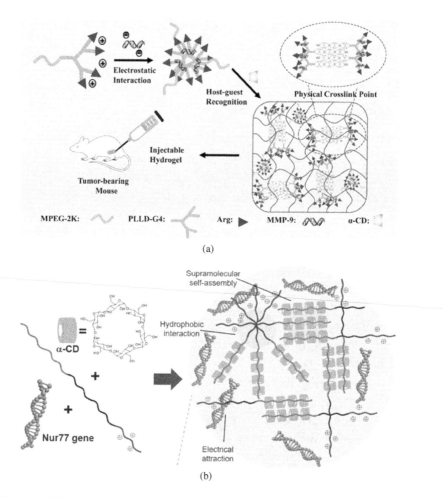

Fig. 6. Different applications of α-CD-based nonviral gene delivery systems in cancer immunotherapy: (a) Synthetic schematic illustration of MPEG–PLLD–Arg/α-CD. (b) Schematic illustration of supramolecular hydrogels co-assembled by cationic copolymer and α-CD for Nur77 gene delivery.

Source: Panel (a) reproduced with permission from Ref. 85. Copyright 2017, Elsevier BV. Panel (b) reproduced with permission from Ref. 80. Copyright 2017, John Wiley and Sons, Inc.

infiltration, which can be used for sustained local gene release, effectively prolong the transfection time and promote in-depth infiltration of solid tumors.[17] Lin *et al.* developed the injectable and highly efficient gene delivery hydrogel-PEGylated arginine-functionalized polylysine dendrites (MPEG–PLLD–Arg) self-assembled with α-CD to form a supramolecular hydrogel for sustained release and efficient delivery of MMP-9 shRNA plasmid. The results showed that the rheological properties of hydrogels, such as gel-forming time, mechanical strength and shear viscosity, could be changed by regulating the content of α-CD in the system, and the gene delivery system could make pMMP-9 release stably and continuously in the form of MPEG–PLLD–Arg/pMMP-9 complex which has higher transfection efficiency and lower cytotoxicity [Fig. 6(a)].[85] Liu et al. designed a triblock cationic copolymer MPEG–PCL–PEI for loading pDNA, which showed higher transfection efficiency and lower cytotoxicity compared with PEI (25 kDa). In this study, MPEG–PCL–PEI/Nur77 complex was mixed with α-CD complex to form supramolecular hydrogels with sustained release properties, which showed better efficacy in the treatment of drug-resistant tumors *in vivo* [Fig. 6(b)].[80] In addition, it is easy for the injectable hydrogel system based on polyvinyl alcohol (PVA) to react in the presence of reactive oxygen species, achieving specific release at the tumor site.[86]

3. Conclusion and Prospects

This review summarizes the structural characteristics of multiple polymers and their respective advantages in the field of gene delivery. In a recent report on the genetic delivery system, the rational design of the supramolecular nanoparticles of the gene is found to be of great concern in the field of nanomaterials or pharmaceuticals. Cationic polymeric delivery systems with hydrophobic segments or amphiphilic designs have been shown to facilitate intracellular transport of target genes and prevent lysosomal degradation of genetic material in immune cells. The structural characteristics of polymeric nonviral vectors such as charge density, hydrophobicity and chain compliance may have great influence on their gene-loading capacity and transfection rate.[87] Typically, supramolecular systems are formed based on physical interactions, including hydrophobic interactions, van der Waals forces and hydrogen bonds, and noncovalent bonding allows supramolecular systems to be assembled and disassembled freely.[88,89] This dynamic interaction gives it thixotropic and self-assembly properties, which can be applied to many novel supramolecular hydrogels.[90] Although nonviral polymeric vectors have demonstrated unique advantages in gene delivery, i.e. low immunogenicity and high safety *in vivo*, their high cytotoxicity is still not conducive to use *in vivo*. Therefore, in the future, solving the *in-vivo* toxicity of cationic polymer and further improving gene delivery efficiency *in vivo* are the key to promote the application of cationic polymer in human clinical immunotherapy.

Acknowledgments

This work was supported by the Natural Science Foundation of China (81971724, 82173750 and 81773661).

References

1. J. R. Quesada, E. M. Hersh, J. Manning, J. Reuben, M. Keating, E. Schnipper, L. Itri and J. U. Gutterman, *Blood*, 1986, **68**, 493–497.
2. S. A. Rosenberg, *J. Immunol.*, 2014, **192**, 5451–5458.
3. M. E. Davis, *Curr. Opin. Biotechnol.*, 2002, **13**, 128–131.
4. L. Milling, Y. Zhang and D. J. Irvine, *Adv. Drug Deliv. Rev.*, 2017, **114**, 79–101.
5. C. H. June, J. T. Warshauer and J. A. Bluestone, *Nat. Med.*, 2017, **23**, 540–547.
6. K. Lundstrom, *Trends Biotechnol.*, 2003, **21**, 117–122.
7. S. L. Ginn, A. K. Amaya, I. E. Alexander, M. Edelstein and M. R. Abedi, *J. Gene Med.*, 2018, **20**, e3015.
8. R. Gardlík, R. Pálffy, J. Hodosy, J. Lukács, J. Turna and P. Celec, *Med. Sci. Monit.*, 2005, **11**, RA110–RA121.
9. C. E. Thomas, A. Ehrhardt and M. A. Kay, *Nat. Rev. Genet.*, 2003, **4**, 346–358.
10. J. G. Fewell, F. MacLaughlin, V. Mehta, M. Gondo, F. Nicol, E. Wilson and L. C. Smith, *Mol. Ther.*, 2001, **3**, 574–583.
11. G. Zhang, V. Budker, P. Williams, V. Subbotin and J. A. Wolff, *Hum. Gene Ther.*, 2001, **12**, 427–438.

12. H. Herweijer and J. Wolff, *Gene Ther.*, 2003, **10**, 453–458.
13. H. Hirai, E. Satoh, M. Osawa, T. Inaba, C. Shimazaki, S. Kinoshita, M. Nakagawa, O. Mazda and J. Imanishi, *Biochem. Biophys. Res. Commun.*, 1997, **241**, 112–118.
14. E. S. Robertson, T. Ooka and E. D. Kieff, *Proc. Natl. Acad. Sci. USA*, 1996, **93**, 11334–11340.
15. K. Anwer, G. Kao, A. Rolland, W. Driessen and S. Sullivan, *J. Drug Target.*, 2004, **12**, 215–221.
16. S. C. De Smedt, J. Demeester and W. E. Hennink, *Pharm. Res.*, 2000, **17**, 113–126.
17. L. Ke, P. Cai, Y. L. Wu and X. Chen, *Adv. Ther.*, 2020, **3**, 1900213.
18. R. S. Riley, C. H. June, R. Langer and M. J. Mitchell, *Nat. Rev. Drug Discov.*, 2019, **18**, 175–196.
19. D. W. Pack, A. S. Hoffman, S. Pun and P. S. Stayton, *Nat. Rev. Drug Discov.*, 2005, **4**, 581–593.
20. A. Dehshahri, S. H. Alhashemi, A. Jamshidzadeh, Z. Sabahi, S. M. Samani, H. Sadeghpour, E. Mohazabieh and M. Fadaei, *Macromol. Res.*, 2013, **21**, 1322–1330.
21. Y.-L. Wu, N. Putcha, K. W. Ng, D. T. Leong, C. T. Lim, S. C. J. Loo and X. Chen, *Acc. Chem. Res.*, 2013, **46**, 782–791.
22. L. Wasungu and D. Hoekstra, *J. Control. Release*, 2006, **116**, 255–264.
23. C. Jiang, J. Chen, Z. Li, Z. Wang, W. Zhang and J. Liu, *Expert Opin. Drug Deliv.*, 2019, **16**, 363–376.
24. X. Tang, S. Zhang, R. Fu, L. Zhang, K. Huang, H. Peng, L. Dai and Q. Chen, *Front. Oncol.*, 2019, **9**, 1208.
25. A. W. Tong, C. M. Jay, N. Senzer, P. B. Maples and J. Nemunaitis, *Curr. Gene Ther.*, 2009, **9**, 45–60.
26. K. L. Kozielski, Y. Rui and J. J. Green, *Expert Opin. Drug Deliv.*, 2016, **13**, 1475–1487.
27. M. C. Woodle, C. M. Engbers and S. Zalipsky, *Bioconjug. Chem.*, 1994, **5**, 493–496.
28. H. Takeuchi, H. Kojima, H. Yamamoto and Y. Kawashima, *J. Control. Release*, 2001, **75**, 83–91.
29. R. Mohammadinejad, A. Dehshahri, V. S. Madamsetty, M. Zahmatkeshan, S. Tavakol, P. Makvandi, D. Khorsandi, A. Pardakhty, M. Ashrafizadeh and E. G. Afshar, *J. Control. Release*, 2020, **325**, 249–275.

30. E. S. Kim, C. Lu, F. R. Khuri, M. Tonda, B. S. Glisson, D. Liu, M. Jung, W. K. Hong and R. S. Herbst, *Lung Cancer*, 2001, **34**, 427–432.
31. Y. Zhang, Z. Zhang, C. Liu, W. Chen, C. Li, W. Wu and X. Jiang, *Polym. Chem.*, 2017, **8**, 1672–1679.
32. X. Zhen, C. Xie and K. Pu, *Angew. Chem.*, 2018, **130**, 4002–4006.
33. P. Goyal, K. Goyal, S. V. Kumar, A. Singh, O. P. Katare and D. N. Mishra, *Acta Pharm.*, 2005, **55**, 1–25.
34. P. Pradhan, J. Giri, F. Rieken, C. Koch, O. Mykhaylyk, M. Döblinger, R. Banerjee, D. Bahadur and C. Plank, *J. Control. Release*, 2010, **142**, 108–121.
35. X. Zhang, M. Oulad-Abdelghani, A. N. Zelkin, Y. Wang, Y. Haîkel, D. Mainard, J. C. Voegel, F. Caruso and N. Benkirane-Jessel, *Biomaterials*, 2010, **31**, 1699–1706.
36. M. E. Davis, *Mol. Pharm.*, 2009, **6**, 659–668.
37. X. Duan, C. Chan and W. Lin, *Angew. Chem., Int. Ed.*, 2019, **58**, 670–680.
38. C. Wu, X. Guan, J. Xu, Y. Zhang, Q. Liu, Y. Tian, S. Li, X. Qin, H. Yang and Y. Liu, *Biomaterials*, 2019, **205**, 106–119.
39. S. K. Rajendrakumar, S. Uthaman, C.-S. Cho and I.-K. Park, *Biomacromolecules*, 2018, **19**, 1869–1887.
40. J. Chen, H. Fang, Y. Hu, J. Wu, S. Zhang, Y. Feng, L. Lin, H. Tian and X. Chen, *Bioact. Mater.*, 2022, **7**, 167–180.
41. S. Liu, H. Liu, X. Song, A. Jiang, Y. Deng, C. Yang, D. Sun, K. Jiang, F. Yang and Y. Zheng, *Nanoscale*, 2021, **13**, 15789–15803.
42. S. K. Sahoo and V. Labhasetwar, *Drug Discov. Today*, 2003, **8**, 1112–1120.
43. J. Zhou, T. R. Patel, M. Fu, J. P. Bertram and W. M. Saltzman, *Biomaterials*, 2012, **33**, 583–591.
44. M. A. Postow, M. K. Callahan and J. D. Wolchok, *J. Clin. Oncol.*, 2015, **33**, 1974–1982.
45. P. Sharma and J. P. Allison, *Science*, 2015, **348**, 56–61.
46. A. J. Claxton, J. Cramer and C. Pierce, *Clin. Ther.*, 2001, **23**, 1296–1310.
47. M. Puts, B. Santos, J. Hardt, J. Monette, V. Girre, E. Atenafu, E. Springall and S. Alibhai, *Ann. Oncol.*, 2014, **25**, 307–315.
48. N. Kamaly, B. Yameen, J. Wu and O. C. Farokhzad, *Chem. Rev.*, 2016, **116**, 2602–2663.
49. J. Li and D. J. Mooney, *Nat. Rev. Mater.*, 2016, **1**, 16071.

50. H. Wang and D. J. Mooney, *Nature Mater.*, 2018, **17**, 761–772.
51. X. Lu, L. Miao, W. Gao, Z. Chen, K. J. McHugh, Y. Sun, Z. Tochka, S. Tomasic, K. Sadtler and A. Hyacinthe, *Sci. Transl. Med.*, 2020, **12**, eaaz6606.
52. S. A. Rosenberg, M. T. Lotze, L. M. Muul, A. E. Chang, F. P. Avis, S. Leitman, W. M. Linehan, C. N. Robertson, R. E. Lee and J. T. Rubin, *N. Engl. J. Med.*, 1987, **316**, 889–897.
53. S. Nakamura, K. Nakata, S. Kashimoto, H. Yoshida and M. Yamada, *Jpn. J. Cancer Res.*, 1986, **77**, 767–773.
54. L. Chen, R. N. Apte and S. Cohen, *J. Control. Release*, 1997, **43**, 261–272.
55. R. C. Mundargi, V. R. Babu, V. Rangaswamy, P. Patel and T. M. Aminabhavi, *J. Control. Release*, 2008, **125**, 193–209.
56. S. Acharya and S. K. Sahoo, *Adv. Drug Deliv. Rev.*, 2011, **63**, 170–183.
57. Y. Zhang, Z. Zhang, S. Li, L. Zhao, D. Li, Z. Cao, X. Xu and X. Yang, *ACS Nano*, 2021, **15**, 16030–16042.
58. M. B. Heo and Y. T. Lim, *Biomaterials*, 2014, **35**, 590–600.
59. N. Sonawane, F. C. Szoka and A. Verkman, *J. Biol. Chem.*, 2003, **278**, 44826–44831.
60. D. Goula, J.-S. Remy, P. Erbacher, M. Wasowicz, G. Levi, B. Abdallah and B. Demeneix, *Gene Ther.*, 1998, **5**, 712–717.
61. J. Suh, H.-J. Paik and B. K. Hwang, *Bioorg. Chem.*, 1994, **22**, 318–327.
62. W. Godbey, K. K. Wu and A. G. Mikos, *J. Control. Release*, 1999, **60**, 149–160.
63. H. Petersen, T. Merdan, K. Kunath, D. Fischer and T. Kissel, *Bioconjug. Chem.*, 2002, **13**, 812–821.
64. Y. Yin, X. Li, H. Ma, J. Zhang, D. Yu, R. Zhao, S. Yu, G. Nie and H. Wang, *Nano Lett.*, 2021, **21**, 2224–2231.
65. M. E. Martin and K. G. Rice, *AAPS J.*, 2007, **9**, E18–E29.
66. J. Nam, S. Son, K. S. Park and J. J. Moon, *Adv. Sci. (Weinheim, Baden-Württemberg, Germany)*, 2021, **8**, 2002577.
67. S. Prijic, L. Prosen, M. Cemazar, J. Scancar, R. Romih, J. Lavrencak, V. B. Bregar, A. Coer, M. Krzan, A. Znidarsic and G. Sersa, *Biomaterials*, 2012, **33**, 4379–4391.
68. M. H. Kleinman, J. H. Flory, D. A. Tomalia and N. J. Turro, *J. Phys. Chem. B*, 2000, **104**, 11472–11479.

69. J. D. Eichman, A. U. Bielinska, J. F. Kukowska-Latallo and J. R. Baker, Jr., *Pharm. Sci. Technol. Today*, 2000, **3**, 232–245.
70. L. Qin, D. R. Pahud, Y. Ding, A. U. Bielinska, J. F. Kukowska-Latallo, J. R. Baker, Jr. and J. S. Bromberg, *Hum. Gene Ther.*, 1998, **9**, 553–560.
71. S. K. Cho, A. Pedram, E. R. Levin and Y. J. Kwon, *Biomaterials*, 2013, **34**, 10228–10237.
72. J. Song, H. Zhang, D. Wang, J. Wang, J. Zhou, Z. Zhang, J. Wang, Y. Hu, Q. Xu, C. Xie, W. Lu and M. Liu, *J. Control. Release*, 2021, **338**, 583–592.
73. J. Liu, H. J. Li, Y. L. Luo, Y. F. Chen, Y. N. Fan, J. Z. Du and J. Wang, *Nano Lett.*, 2020, **20**, 4882–4889.
74. J. Haensler and F. C. Szoka, Jr., *Bioconjug. Chem.*, 1993, **4**, 372–379.
75. M. L. Patil, M. Zhang, O. Taratula, O. B. Garbuzenko, H. He and T. Minko, *Biomacromolecules*, 2009, **10**, 258–266.
76. C. Z. Chen, N. C. Beck-Tan, P. Dhurjati, T. Dyk and S. L. Cooper, *Biomacromolecules*, 2000, **1**, 473–480.
77. M. V. Maus, A. R. Haas, G. L. Beatty, S. M. Albelda, B. L. Levine, X. Liu, Y. Zhao, M. Kalos and C. H. June, *Cancer Immunol. Res.*, 2013, **1**, 26–31.
78. R. H. Vonderheide, K. T. Flaherty, M. Khalil, M. S. Stumacher, D. L. Bajor, N. A. Hutnick, P. Sullivan, J. J. Mahany, M. Gallagher and A. Kramer, *J. Control. Release*, 2007, **25**, 876–883.
79. X. Liu, Z. Li, J. L. Xian, K. Chen and Y.-L. Wu, *Macromol. Rapid Commun.*, 2018, **40**, e1800117.
80. X. Liu, X. Chen, M. X. Chua, Z. Li, X. J. Loh and Y. L. Wu, *Adv. Healthc. Mater.*, 2017, **6**, 1700159.
81. C. Wang, J. Wang, X. Zhang, S. Yu, D. Wen, Q. Hu, Y. Ye, H. Bomba, X. Hu and Z. Liu, *Sci. Transl. Med.*, 2018, **10**, eaan3682.
82. Y. L. Wu and J. Li, *Angew. Chem.*, 2010, **48**, 3842–3845.
83. F. Huang and H. W. Gibson, *Prog. Polym. Sci.*, 2005, **30**, 982–1018.
84. K. Miyake, S. Yasuda, A. Harada, J. Sumaoka, M. Komiyama and H. Shigekawa, *J. Am. Chem. Soc.*, 2003, **125**, 5080–5085.
85. Q. Lin, Y. Yang, Q. Hu, Z. Guo, T. Liu, J. Xu, J. Wu, T. B. Kirk, D. Ma and W. Xue, *Acta Biomater.*, 2017, **49**, 456–471.
86. C. Nathan and A. Cunningham-Bussel, *Nat. Rev. Immunol.*, 2013, **13**, 349–361.

87. M. Foldvari, D. W. Chen, N. Nafissi, D. Calderon and A. Rafiee, *J. Control. Release*, 2015, **240**, 165–190.
88. G. Chen and M. Jiang, *Chem. Soc. Rev.*, 2011, **40**, 2254–2266.
89. B. Rybtchinski, *ACS Nano*, 2011, **5**, 6791–6818.
90. X. Yan, F. Wang, B. Zheng and F. Huang, *Chem. Soc. Rev.*, 2012, **41**, 6042–6065.

Recent Advances in Thermal Interface Materials*

Jing Cao, Tzee Luai Meng, Xikui Zhang, Na Gong, Rahul Karyappa, Chee Kiang Ivan Tan, Ady Suwardi, Qiang Zhu and Hongfei Liu[†]

Institute of Materials Research and Engineering (IMRE),
A*STAR (Agency for Science, Technology and Research),
2 Fusionopolis Way, Singapore 138634, Singapore
[†]*liuhf@imre.a-star.edu.sg*

Ever-increasing performance and functions of electronic systems are pushing the requirements for heat dissipation of devices at an unprecedented pace. To package individual devices (especially those working in high-power mode), research and development of thermal interface materials (TIMs) have accelerated since the beginning of this century. Thermal conduction, mechanical performance, and electrical insulation are the general specifications of interest besides their durability and reliability in developing TIMs. Mechanical performance is crucial in reducing the thermal interface resistance (TIR) between TIM and its bridged surfaces, i.e., between the device and the heat sink. To fill the gaps formed by the hard surfaces for reducing the TIR, TIMs should be "soft" enough upon external pressures. Nevertheless, depending on practical applications, the selection of TIM might release some specifications to guarantee the others. This review summarizes the latest developments of TIMs, addresses their processing methods and heat dissipation performance, highlights their remaining issues, and provides a perspective on their future development.

[†]Corresponding authors.
*To cite this article, please refer to its earlier version published in the *World Scientific Annual Review of Functional Materials*, Volume 1, 2230005 (2023), DOI: 10.1142/S2810922822300057.

Keywords: Thermal interface materials; thermal conduction; mechanical behavior; electrical insulation; heat dissipation performance; material design and synthesis.

1. Introduction

The active regions of electronic devices are typically made of semi-conducting materials, they unavoidably generate heat during their operations. The temperature in the active regions increases if the dissipation of the undesired heat is not effective and onset of malfunctions occurs when a critical temperature (T_c) of the device is approached. For example, the increased density of hot carriers at high temperature can destroy the switch control of the carrier migration channel in the semiconductors. Likewise, the increased mechanical stress at an elevated temperature can induce dislocations and/or microcracking of the device. The situation is worse for electronic devices with high working current and/or voltage, which give rise to a higher power of heat generations in their active regions. Although wide bandgap semiconductors, such as GaN, SiC, and Ga_2O_3,[1-3] have been employed to fabricate electronic devices that have a higher tolerance of working temperatures, i.e., higher T_c, than those of traditional Si- and GaAs-based electronic devices,[4-8] the issue of heat dissipation is not resolved.

On the other hand, to lower the temperature of the working device (T_w), heat sinks have been developed to accelerate the heat dissipation from the device to its outer environment.[9] Furthermore, a layer of thermal conduction material is usually applied to enhance the heat transport from the working device to the heat sink, which is inserted between the hard surfaces of the device and the heat sink to fill the microscale gaps. This material is called thermal interface material (TIM) by the community of electronic device packaging.[10] The schematic diagram in Fig. 1(a) shows the typical geometry of TIM that bridges the hard surfaces of the device and the heat sink. Both the contacting interface and the TIM play important roles in heat transfer from the device to the heat sink. They are commonly characterized by the thermal interface resistance (TIR) and the thermal conducting resistance (TCR) of TIMs in the literature.

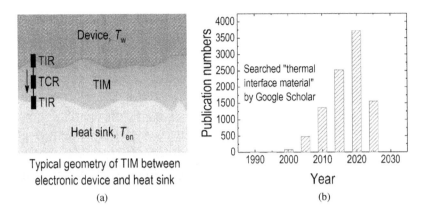

Fig. 1. (a) Typical geometry of TIM between electronic device and heat sink, showing the TIR and the TCR of TIMs and (b) publication numbers obtained by Google Scholar (as of June 30, 2022) with inputting "thermal interface material". The arrow indicates the heat flow direction. The temperature gradient induced by TIM reads $\Delta T = T_w - T_{en}$.

Higher thermal conductivities (lower TCRs) of TIM can dissipate the heat more effectively from the device. Besides thermal conductivity, TIM must be mechanically strong and flexible enough to fill the microscale gap between the hard surfaces of the device and the heat sink to reduce the TIRs. Additionally, the durability and reliability of TIMs influence the reliability of devices and affect the running cost and safety of the electronic system.

Upgradation of semiconductor materials for the fabrication of electronic devices, especially those high-power ones, is quite limited. In this light, the development and upgradation of TIMs are crucial in device packaging for their working at higher environment temperatures (T_{en}). The temperatures T_w, T_c, and T_{en} are linked and shown in the following equation:

$$T_w = T_{en} + \Delta T \leq T_c. \qquad (1)$$

In Eq. (1), ΔT is the temperature difference between the active region of electronic device (T_w) and the working environment (T_{en}, approximately that of the heat sink), i.e., the temperature gradient induced by TIM. T_c is an intrinsic parameter which is constant for a certain electronic device. At a certain working condition, the heat generation and thus the T_w is relatively constant for an individual device. From Eq. (1), a reduced ΔT can effectively increase T_{en}, i.e., the environment temperature of the working device. In this light, TIMs with increased thermal conductions play an important role in reducing ΔT. This is important for increased number of devices assembled in limited space, e.g., on printed circuit boards (PCBs) housed in avionic chassis for enhanced functions and performance. It is shown in Fig. 1(a) that both TIR and TCR must be minimized to reduce ΔT in developing advanced TIMs.

In the past two decades, many kinds of TIMs have been developed, including fluids,[11] pastes,[12] phase change materials (PCMs),[13] ceramic reinforced polymer composites (CRFPCs),[14] three-dimensional (3D) ceramic filler networking composites,[15] flexible thin films with high thermal conductivities,[16] and resilient thermal conductors.[17] Figure 1(b) presents a brief literature survey using Google Scholar with the keyword "thermal interface material". It shows that the investigations on TIMs started in the middle of the 1990s. After that, the research activities on TIMs increased nearly linearly year-by-year. In this review, we summarized the recent development of TIMs, highlighting their advantages and potential issues and providing the readers a perspective on future developments of TIM. Table 1 highlights some common types of commercial TIMs and their typical properties.[18]

Table 1. Commercial TIMs and their properties. TC: Thermal conductivity; BLT: Bond line thickness; TIR: Thermal interface resistance; PCM; Phase change material; and TCA: Thermal conductive adhesive.[18]

Type	TC (W/(m.K))	BLT (nm)	TIR (Km²/K)	Pump-out	Absorbs stress	Reusable	Replaceability
Thermal grease	0.4–4	0.02–0.15	10–200	Yes	Good	No	Medium
Thermal pad	0.8–3	0.2–1.0	100–300	No	Good	Yes	Good
PCM	0.7–1.5	0.02–0.15	30–70	Yes	Good	No	Medium
Thermal gel	2–5	0.08–0.25	40–80	No	Medium	No	Medium
TCA	1–2	0.05–0.20	15–100	No	Medium	No	Poor
Solder	20–80	0.03–0.20	<5	No	Poor	No	Poor

2. Al_2O_3 Reinforced Polymer Composite-Based TIMs

The combination of electrical insulation and the mechanical characteristics of polymers makes them a popular candidate for producing TIMs. High thermal conduction inorganic fillers have been mixed with functionalized polymers to enhance their thermal conductivities. Al_2O_3 is a common ceramic insulator possessing high mechanical and chemical stabilities at elevated temperatures. These characteristics, together with relatively high thermal conductivity, make Al_2O_3 a promising filler for producing CRFPCs toward their TIM applications. Table 2 presents brief comparisons of common fillers that have been developed and used in polymeric matrix to

Table 2. Intrinsic thermal conductivities (κ) of common metallic, carbon, and ceramicfillers that have been employed in polymer matrix for TIM applications.[19]

Category	Fillers	κ (W/m.K)	Electrical insulation
Metal	Al	234	No
	Cu	386	No
	Ag	427	No
	Au	158	No
	Ni	103~200	No
Carbon	CNT	1000~4000	No
	Carbon fiber	300~1000	No
	Gr	Theoretical, 4840~5300	No
	Gr	Experimental, 3080~5300	No
	Graphite	100~400 (in plane)	No
	Carbon black	6~174	No
Ceramic	α-Al_2O_3	30	Yes
	Bulk h-BN	40~250	Yes
	BN nanosheet	~2000 (//), 100 ($\perp$)	Yes
	AlN	100~300	Yes
	β-Si_3N_4	103~200	Yes

(Continued)

Table 2. (*Continued*)

Category	Fillers	κ (W/m.K)	Electrical insulation
	β-SiC	120	Yes
	ZnO	25	Yes
	MgO	48	Yes
	Diamond	1000~2000	Yes
	BeO	270	Yes
	Fused SiO$_2$	1.5~1.6	Yes
	Crystalline silica	3	Yes
	BaTiO$_3$	6.2	Yes

enhance their thermal conductivities. In general, they can be categorized into metals (e.g., Au, Ag, Cu, etc.), carbon-based materials and structures such as carbon nanotube (CNT), graphene (Gr), graphite nanoplatelets (NPs), and ceramics (e.g., BN, AlN, Al$_2$O$_3$, SiC, etc.).[19]

In general, the metal- and carbon-based fillers are electrical conductors, while the ceramic-based fillers are typically electrical insulators. The carbon fillers, except for black carbon, have better thermal conductivities than those of metal and ceramic fillers. The van der Waals crystals (e.g., Gr and BN nanosheets) commonly exhibit anisotropic thermal conductivities along the directions parallel and perpendicular to the basal planes. The incorporation of metal- and carbon-based fillers into the polymer host matrix could deteriorate the electrical insulations as well as the dielectric properties of the polymer composites. This may cause some limitations for their TIM applications. In comparison, the ceramic fillers are typically electrical insulators and some of them have relatively high thermal conductivity, comparable with metals, e.g., BN, AlN, Si$_3$N$_4$, SiC, and BeO (see Table 2). Such high thermal conductivities typically originate from their light atomic mass, which makes their lattice vibrations easily actuated. As a result, high-frequency acoustic phonons are intrinsic in these crystals to mediate heat transfer. The electrical insulation and thermal conduction properties make ceramic-based fillers more unique than metallic- and carbon-based fillers for producing thermally conductive polymer composites targeting TIM applications.

Physically, the thermal conductions in CRFPCs are facing two kinds of resistance. One is the interface thermal resistance between the individual ceramic fillers and their surrounding polymer matrix and the other is the contact thermal resistance between the adjacent fillers contacted with one another.[20] Surface finalizations and geometric modifications of ceramic-fillers have been extensively investigated to enhance the bonding strength between the fillers and their surrounding polymers, which in turn decrease the phonon scattering at the interface and thus reduce the interface thermal resistance. An increase in the filler compositions may increase the overall contact area between the fillers and the polymer matrix, which increase the interface thermal resistance. Meanwhile, the overall filler aggregations increase with the filler compositions, which also increase the phonon scattering and thus the thermal contact resistance. However, the larger thermal conductivity of the fillers than that of the polymers usually requires a high fraction of the fillers to enhance the thermal conductivities of the polymer composites. For this reason, the concentration of Al_2O_3 fillers in Al_2O_3 reinforced polymer composites (AORFPCs) is usually very high. The increased fractions of Al_2O_3 fillers in polymer tend to degrade its mechanical properties such as tensile strength, flexibility, and processability.[21] Nevertheless, it was recently demonstrated that the incorporation of Al_2O_3 fillers into epoxy (EP) with a volume fraction of 36.2% could increase the thermal conductivity of EP by 15 times, while the thermal conduction performance of the Al_2O_3/EP composites was quite stable upon heating–cooling cycles.[22]

As mentioned above, the surface finalizations and geometric modifications of ceramic fillers can have important consequences in enhancing the thermal conductions of the CRFPCs. In this regard, many techniques have been developed for processing Al_2O_3 fillers, including homogeneous precipitations, sol-emulsification-gel, oil-drop, spray, templating, etc.[19] Besides the effect of morphology, size, and dispersion, the surface finalization of thermally conductive fillers has been probably one of the most aggressively investigated topics in producing AORFPCs. Table 3 summarizes the thermal conductivities (κ) of recently reported AORFPCs.[19] The comparisons in Table 3 show that most of the Al_2O_3 filler modification processes can enhance the thermal conductions of the polymer matrix.

However, it is seen in Table 3 that the improvement of the thermal conductivity by surface functionalizing Al_2O_3 fillers is limited and usually less than 2 W/m.K. Hence, more processing strategies are developed to enhance the thermal conductivities of AORFPCs. For example, hybridizing the Al_2O_3 fillers with different sizes and/or with other type of fillers from different categories (see Table 2), including those of low-dimensional materials such as Gr and CNTs. Table 4 presents thermal conductivities (κ) of some recently reported AORFPCs produced by using hybridized Al_2O_3 fillers.

3. Ceramic Composite TIMs Based on Si_3N_4 Matrix and Carbon-Related Fillers

Different from the polymer composite-based TIMs, TIMs based on ceramic composites can work at higher temperatures due to their increased thermal stabilities in air environment. Similar to Al_2O_3, Si_3N_4 is also a ceramic insulator with even higher thermal conductivities than that of Al_2O_3 (see Table 2). Its mechanical properties at high temperatures (e.g., the fractural toughness, hardness, flexural strength, etc.) are comparable with or even better than those of AlN. In general, the thermal conductivity and mechanical properties of ceramics exhibit opposite behaviors as a function of their grain size evolutions, which bring up challenges to the engineering of ceramic materials for TIM applications. Therefore, carbon-related fillers (CRFs) have been proposed to mix with Si_3N_4 for fabricating ceramic composite-based TIMs with high thermal conductivities and excellent mechanical properties.

Like regular ceramic processing procedures, CRFs reinforced Si_3N_4 ceramic composites (CRFSNCCs) can be fabricated through powder synthesis (e.g., ball milling or rolling mixing), compacting

Table 3. Thermal conductivities (κ) of some recently reported AORFPCs.[19]

Al_2O_3	Matrix	κ of matrix (W/m.K)	Modifiers	Loading	κ of ARPCs with modified Al_2O_3 (W/m.K)	κ of ARPCs with unmodified Al_2O_3 (W/m.K)
4 μm	Epoxy resin	0.185	Polydopamine	34.7 vol.%	0.79	0.82
~0.3nm	Nitrile rubber	0.172	Polydopamine	30 phr*	0.211	0.204
0.6~30 μm	Polylactic acid	0.278	Maleic acid	30%	0.66	—
300 nm	Silicone rubber	0.147	Poly(dopamine)	30 vol.%	0.585	<0.585
10 μm	Polyurethane	—	γ-aminopropyltriethoxysilane	—	80.6%*	—
2 μm	Silicone rubber	0.16	PDMS-PHMS-MMA	500 phr	1.40	1.42
2 μm	Silicone rubber	0.16	PDMS-PHMS-GMA	500 phr	1.57	1.42
2 μm	Silicone rubber	0.16	PDMS-PHMS-MPS	500 phr	1.73	1.42
nano Al_2O_3	Silicone rubber	—	Vinyl tri-methoxysilane	70 phr	104.7%*	76.8%*
~30 nm	Epoxy resin	0.16	γ-aminopropyltriethoxysilane	30 wt.%	0.27	0.241
0.2, 1.0, and 40 μm	Silicone rubber	—	Polypyrrole	83 wt.%	1.98	—
200 nm	Bismaleimide-triazine resin	—	KH-560	70 wt.%	0.6	—
200 nm	Bismaleimide-triazine resin	—	KH-550	70 wt.%	1.07	—

(Continued)

Table 3. (Continued)

Al$_2$O$_3$	Matrix	κ of matrix (W/m.K)	Modifiers	Loading	κ of ARPCs with modified Al$_2$O$_3$ (W/m.K)	κ of ARPCs with unmodified Al$_2$O$_3$ (W/m.K)
200 nm	Bismaleimide-triazine resin	—	Graphene oxide	70 wt.%	0.83	—
30 nm	Epoxy resin	0.236	γ-aminopropyl-triethoxysilane	20 wt.%	69%*	21%*
20 nm	Epoxy resin	—	KH570, γ-methacryloxypropyl trimethoxy silane	40 wt.%	0.55	—
20 nm	Epoxy resin	—	KH590, γ-mercaptopropyl triethoxy silane	40 wt.%	0.53	—
20 nm	Epoxy resin	—	A151, vinyl triethoxysilane	40 wt.%	0.47	—
20 nm	Epoxy resin	—	TDI (1,4-phenylene diisocyanate)	40 wt.%	0.41	—
500 nm	Epoxy resin	—	KH570, γ-methacryloxypropyl trimethoxy silane	40 wt.%	0.41	—
500 nm	Epoxy resin	—	KH590, γ-mercaptopropyl triethoxy silane	40 wt.%	0.39	—
500 nm	Epoxy resin	—	A151, vinyl triethoxysilane	40 wt.%	0.42	—
500 nm	Epoxy resin	—	TDI (1,4-phenylene diisocyanate)	40 wt.%	0.39	—
2 μm	Silicone rubber	About 0.16	3-(trimethoxysilyl)propyl methacrylate grafted poly (methylhydrosiloxane-codimethylsiloxane)	500 phr	1.68	—
3 μm	Polyurethane	—	Polydopamine	30 wt.%	138%*	—

Note: *κ enhancement = ($κ_{composite} - κ_{polymer\ matrix}$)/$κ_{polymer\ matrix}$; phr: parts per hundred rubber.

Table 4. Thermal conductivities (κ) of some recently reported Al_2O_3, in combination with other fillers, reinforced polymer composites (AORFPCs).[19]

Al_2O_3	Other fillers	Matrix	Loading	κ (W/m.K)
2 μm	BN nanosheets, average lateral size of 500 nm and average thickness of 2.5 nm	Epoxy resin	65 vol.%, Al_2O_3: BN nanosheets = 7:1	2.43
30 nm	h-BN, 10 μm	Epoxy resin	7.5 wt.% Al_2O_3 and 22.5 wt.% h-BN	1.182
45 μm	BN NPs, (diameter < 500nm, thickness < 30 nm)	Epoxy resin	70 vol.% Al_2O_3 and 12 vol.% BN NPs	3.6
100 nm	BN platelets, mean diameter is 0.5 μm	Epoxy resin	40 wt.%, Al_2O_3: BN = 1:3	0.87
30 nm	BN platelets, average diameter is 3 μm	Epoxy resin	50 wt.%, Al_2O_3: BN = 1:4	0.808
d50: 12.5 ± 5.35 μm	BN, d50: 8.3 ± 5.20 μm	Epoxy resin	52 wt.% Al_2O_3 and 10 wt.% BN	1.65
3 μm	BN platelets, 32.52 μm	Vinyl terminated polydimethylsiloxane	30 wt.% Al_2O_3 and 5 wt.% BN	3.64
14.4 μm	BN platelet, ~45 μm	Epoxy resin	5.32 vol.% BN and 68.82 vol.% Al_2O_3	4.4
—	BN nanosheets	Silicone rubber	30 wt.%, BN nanosheets: Al_2O_3 = 1:1	2.86 (//) and 0.89 ($\perp$)

(Continued)

Table 4. (Continued)

Al_2O_3	Other fillers	Matrix	Loading	κ (W/m.K)
200 nm	BN nanosheets, 1 μm	Epoxy resin	6 vol.% Al_2O_3 and 4 vol.% BN nanosheets	0.47
10 μm	BN platelets, 1 μm	Epoxy resin	30 wt.%, Al_2O_3: BN = 1:1	0.57
—	h-BN, 1 μm	Polybenzoxazine	6.9 vol.% hybrid skeleton	3.24
3 μm	BN, <150 nm	Polyimide	25 wt.% BN@PDA@Al_2O_3	6.41 (//) and 1.01 (⊥)
250 nm	Ag, with an average diameter estimated to about 10 nm	Silicone rubber	30 vol.% Al_2O_3–PCPA–Ag	0.4367
μ-spheres	Ag, 20 nm and 60 nm	Epoxy resins	70 wt.% Al_2O_3 and 1.99 mol.% Ag	1.11
Average thickness of 372 nm and a diameter of 8.2 μm	Ag, 10–20 nm	Epoxy resin	50 wt.% Al_2O_3 and 0.6 wt.% Ag	6.71
17.4 μm	Ag, 50–60 nm	Epoxy resin	70 wt.% fillers with 1.96 wt.% Ag	1.304
5 μm	Graphene NP, several micrometers to 30 μm	Silicone rubber	89 wt.% Al_2O_3 and 1 wt.% graphene NP	3.37

25–30 nm	Reduced graphene oxide, average thickness 2.05 ± 0.44 nm and the lateral size 0.3~1.3 μm	Epoxy resin	0.4 wt.% Al_2O_3 and 0.6 wt.% reduced graphene oxide	23.4%*
—	Carbon fibers, with a length of 150~200 μm and diameter of 10 μm	Silicon rubber	30 vol.% Al_2O_3 and 12 vol.% carbon fibers	7.36
4 μm	Graphene oxide, the lateral dimension is 9.55~40.61 μm	Polyurethane	—	0.502
4.98 μm	Graphene nanosheets, the average lateral size is 4.96 μm	Epoxy resin	42.4 wt.% Al_2O_3 and 12.1 wt.% graphene	33.4 in radial direction and 13.3 in axial direction
2~5 μm	CNTs	Silicone rubber	—	105.5%*
d50: 1.1 μm	reduced graphene oxide	Epoxy resin	50 wt.% Al_2O_3 and 1 wt.% reduced Gr oxide	0.72
0.7 μm and 5 μm	Graphene sheet	Silicone oil	63 vol.% Al_2O_3 and 1 wt.% Gr	3.45
10~50 μm	Graphite nanosheets, the lateral size is 5~10 μm	Silicone rubber	300 phr of Al_2O_3 and 5 phr of graphite nanosheets	2.162
About 20 nm	Graphene oxide	Nanofibrillated cellulose	5.6 wt.% Al_2O_3 and 30 wt.% Gr oxide	8.3 in plane

(*Continued*)

Table 4. (Continued)

Al_2O_3	Other fillers	Matrix	Loading	κ (W/m·K)
About 30 nm	Reduced graphene oxide	Epoxy resin	30 wt.% Al_2O_3 and 0.3 wt.% reduced Gr oxide	0.329
3 μm	Carbon black and CNTs: 20–40 nm in diameter, 10–20 μm in length	Polypropylene	50 wt.% Al_2O_3 and 15 wt.% carbon materials	0.68
1 μm	Carbon fiber, the diameter is 10 μm and mean length is 150 μm	Silicone rubber	5 vol.% Al_2O_3 and 25 vol.% carbon fibers	9.60
5 μm	Carbon fibers	Epoxy resin	74 wt.% Al_2O_3 and 6.4 wt.% carbon fibers	3.84
1–10 μm	CNT	Epoxy resin	8 wt.% Al_2O_3 and 2 wt.% CNT	0.39
0.3 μm	Graphite oxide, lateral size of several micrometers	Polyurethane	Al_2O_3: polyurethane: graphene = 50:50:5 (by weight)	~0.9
—	Graphite, the diameter is 3~5 μm	Phthalonitrile	40 wt.% Al_2O_3@graphite	1.409
~3 μm	Graphene oxide	Carboxyl nitrile butadiene rubber	30 vol.% Al_2O_3@poly(catechol-polyamine)@graphene oxide	0.48
30 nm	Graphene sheet	PVDF	40 wt.% Al_2O_3 coated graphene sheet	0.586
0.5 μm	AlN, 10 μm	Epoxy resin	58.4 vol.% (Al_2O_3:AlN = 3:7)	3.402
10 μm	AlN, 0.1 μm	Epoxy resin	58.4 vol.% (Al_2O_3:AlN = 7:3)	2.842

Recent Advances in Thermal Interface Materials

		Matrix	Composition	κ enhancement*
—	18 μm	Polylactic acid	38 wt.% Al_2O_3 and 2 wt.% AlN	0.715
	AlN, 38 μm	Nylon 6	60 wt.%, Al_2O_3:AlN = 1:2	1.65
	AlN, 14 μm	Nylon 6	60 wt.%, Al_2O_3:AlN = 1:2	2.44
	AlN, 38 μm	Nylon 6	60 wt.%, Al_2O_3:AlN = 1:1	2.29
	MgO, 5 μm	Polycarbonate/acrylonitrile-butadienestyrene polymer alloy	70 wt.%, Al_2O_3:MgO = 7:3	1.95 ± 0.1
60 μm	MgO, 5 μm; graphene NPs, the planar size is in the range of 3–15 μm and the thickness is about 3 nm	Polycarbonate/acrylonitrile-butadienestyrene polymer alloy	70.5 wt.%, Al_2O_3:MgO: graphene NPs = 7:3:0.5	3.1 ± 0.2
5 μm and 40 μm	Gr NPs, mean size of ~40 mm and a thickness of ~100 nm; $Mg(OH)_2$ powders	Epoxy resin	68 wt.% Al_2O_3, 7 wt.% graphene and 5 wt.% $Mg(OH)_2$	2.2
20 μm and 70 μm	Multi-walled CNTs, the length is 10–30 μm and outside diameter is 10–30 nm; SiO_2, 500 nm	Epoxy resin	60 wt.% Al_2O_3, 3 wt.% multi-walled carbon and 8 wt.% SiO_2	1.73
20 nm and 0.5–3 μm	Si_3N_4, 0.3–3 μm	Silicone rubber	30 vol.%, Al_2O_3:Si_3N_4 = 4:26	1.62

Note: *κ enhancement = $(\kappa_{composite} - \kappa_{polymer\ matrix})/\kappa_{polymer\ matrix}$.

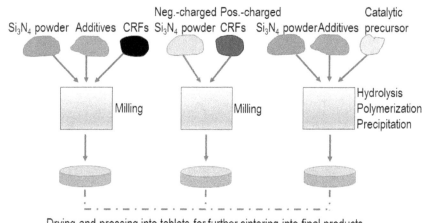

Fig. 2. Schematic diagrams showing the synthesis and sintering processes of CRFSNCCs for TIM applications.[23]

by pressing, sintering, and post-sintering treatment. However, the processing technique plays an important role in properties of the resultant CRFSNCCs, largely deepening on the spatial density of the Si_3N_4 matrix and the process-induced structural damages of the nanometric-scale CRFs. Figure 2 schematically presents a few typical procedures for processing CRFSNCCs, including powder, colloidal, and sol–gel methods. After mixing by ball milling (i.e., the powder and colloidal processes) or growth of CRFs in Si_3N_4 matrix through hydrolysis-polymerization-precipitation (i.e., the sol–gel process), the powder can be pressed into blocks. Finally, the block samples can be densified by hot pressing (HP), hot isostatic pressing (HIP), spark plasma sintering (SPS), or gas pressure sintering (GPS).[23] In the sintering process, the sample temperature is elevated slightly lower than the ceramic's melting point to drive the fusion and bonding of adjacent particles, followed by material densifications via activated atomic diffusions and mass transfer. To mitigate the structural damage of CRFs, both the powder synthesis and the sintering processes must be optimized. In this regard, many sintering techniques such as HP, HIP, and SPS, have been developed for CRFSNCCs. Tables 5–7 present the thermal conductivity

Table 5. Thermal conductivity and mechanical properties of CNT-reinforced Si_3N_4 ceramic composites for TIM applications.[23]

CNT (wt.%)	Sintering parameters	Sintering additives	Bulk density (g/cm³)	Flexural strength (MPa)	Fracture toughness (MPa.m$^{1/2}$)	Hardness (GPa)	Young's Modules (GPa)	Thermal conductivity (W/m.K)
2 MW	1700 °C/HP/1 h	SiC_w, YF_3, MgF_2	3.15	765 ± 58	9.70 ± 0.8	26.9 ± 2.9	244 ± 9.2	99.28
1 MW	1750 °C/HP/1 h	Y_2O_3, Al_2O_3, ZrO_2	3.22	996 ± 25	6.6 ± 0.6	15.0 ± 0.1	—	—
2 MW	1700 °C/HP/1 h	YF_3, MgF_2	3.05	883 ± 46	12.76 ± 1.2	26.4 ± 1.3	260 ± 9	—
1 MW	1500 °C/SPS/5min	Y_2O_3, Al_2O_3	3.17	—	5.3	16.6 ± 0.4	285.73	—
1 MW	1650 °C/SPS/3min	Y_2O_3, Al_2O_3	3.19	—	4.4	19.1 ± 0.6	306.16	—
3 MW	1700 °C/HIP	Y_2O_3, Al_2O_3	2.7	649 ± 50	3.1 ± 0.2	10.04 ± 0.3	—	—
3 MW	1700 °C/HIP/3 h	Y_2O_3, Al_2O_3	3.39	—	—	—	—	19.8
3 MW	1700 °C/HIP/3 h	Y_2O_3, Al_2O_3	2.65	—	5.9 ± 0.3	10.1 ± 0.6	—	—
1 MW	1700 °C/HIP/3 h	Y_2O_3, Al_2O_3	3.00	750 ± 10	—	—	230 ± 5	—
2 SW	1600 °C/SPS/3min	Y_2O_3, Al_2O_3, MgO	—	—	—	—	—	20.16
2 CNFs	1700 °C/HP/1 h	SiC_w, YF_3, MgF_2	3.16	571 ± 39	8.86 ± 0.1	27.2 ± 0.9	250 ± 3.8	110.63

Table 6. Thermal conductivity and mechanical properties of Gr-reinforced Si_3N_4 ceramic composites for TIM applications.[23]

Gr (wt.%)	Sintering parameters	Sintering additives	Bulk density (g/cm³)	Flexural strength (MPa)	Fracture toughness (MPa·m^(1/2))	Hardness (GPa)	Young's modulus (GPa)	Thermal conductivity (W/m·K)
2 GNP	1650 °C/HP/2 h	AlF_3, MgF_2	3.07	617 ± 35	11.26 ± 0.35	18.53 ± 0.2	—	137.47
2 GNP	1700 °C/HP/1 h	SiC_w, YF_3, MgF_2	3.12	583 ± 74	7.58 ± 0.3	24.0 ± 0.7	228 ± 4.0	96.71
5 MLG	1600 °C/SPS/10min	Y_2O_3, Al_2O_3	3.66	506	5.58	16.4 ± 1.5	296	—
5 GrO	1600 °C/SPS/10min	Y_2O_3, Al_2O_3	3.72	258	3.6	8.4 ± 0.1	166	—
4.3 GNP	1625 °C/SPS/5min	Y_2O_3, Al_2O_3	3.18	—	—	17.7 ± 0.4	290 ± 4	—
4.3 rGO	1625 °C/SPS/5min	Y_2O_3, Al_2O_3	3.19	—	—	15.9 ± 0.3	264 ± 8	—
1 GNP	1700 °C/HIP/3 h	Y_2O_3, Al_2O_3	—	—	9.92 ± 0.38	16.38 ± 0.5	—	—
1 GPL	1700 °C/SPS/5min	—	3.175	—	5.8 ± 1.18	20.4 ± 0.37	—	—
1 GNP	1750 °C/HP/1 h	—	—	876 ± 53	8.6 ± 0.4	12.8 ± 0.3	—	—
2.25 rGO	1700 °C/HP/1 h	Y_2O_3, Al_2O_3	3.24	1116 ± 28	10.35 ± 0.4	17.3 ± 0.3	—	—
21 GNP	1625 °C/SPS/5min	Y_2O_3, Al_2O_3	—	—	—	—	—	44.0
4 rGO	1625 °C/SPS/5min	Y_2O_3, Al_2O_3	—	—	—	—	—	14.4
1 GNP	1700 °C/HIP/3 h	Y_2O_3, ZrO_2, Al_2O_3	3.411	—	8.9 ± 0.3	14.6 ± 0.4	—	—

Table 7. Thermal conductivity and mechanical properties of SiC/SiC$_w$-reinforced Si$_3$N$_4$ ceramic composites for TIM applications.[23]

SiC SiC$_w$ wt.%	Sintering parameters	Sintering additives	Bulk density (g/cm^3)	Flexural strength (MPa)	Fracture toughness (MPaxm$^{1/2}$)	Vickers hardness (GPa)	Young's modulus (GPa)	Thermal conductivity (W/m.K)
5 SiC	1700 °C/HP/1 h	AlF$_3$, MgF$_2$	3.083	715 ± 7	8.82 ± 0.08	18.66 ± 0.3	—	105.12
10 SiC	1700 °C/HP/1 h	AlF$_3$, MgF$_2$	3.041	692 ± 11	9.02 ± 0.14	20.53 ± 0.5	—	129.22
15 SiC	1700 °C/HP/1 h	AlF$_3$, MgF$_2$	3.045	686 ± 21	8.87 ± 0.34	19.93 ± 0.4	—	130.12
20 SiC	1700 °C/HP/1 h	AlF$_3$, MgF$_2$	3.05	709 ± 17	10.3 ± 0.2	20.22 ± 0.1	—	113.66
25 SiC	1700 °C/HP/1 h	AlF$_3$, MgF$_2$	3.090	627 ± 15	8.39 ± 0.52	20.84 ± 0.2	—	145.66
10 SiC	1750 °C/PL/2 h	Y$_2$O$_3$, Al$_2$O$_3$	—	570 ± 5	3.45 ± 0.1	—	265 ± 5	—
1 SiC	1750 °C/HP/2 h	Y$_2$O$_3$, C, SiO$_2$	—	786	4.54 ± 0.07	14.46 ± 0.2	299 ± 12	—
5 SiC	1750 °C/HP/2 h	Y$_2$O$_3$, C, SiO$_2$	—	678	3.75 ± 0.53	16.34 ± 0.1	322 ± 10	—
20 SiC$_w$	1700 °C/HP/1 h	YF$_3$, MgF$_2$	3.11	502 ± 3	7.05 ± 0.2	23.1 ± 2.1	241 ± 14	78.4
15 SiC$_w$	1770 °C/HP/1 h	Y$_2$O$_3$, Al$_2$O$_3$	3.23	989 ± 49	9.6 ± 0.64	15.6 ± 0.26	313 ± 2.6	—
10 SiC$_w$	1750 °C/HP/30min	Y$_2$O$_3$, AlN	3.20	680 ± 14	8.33 ± 0.23	16.6 ± 0.3	—	—
5 SiC	1700 °C/HP/1 h	AlF$_3$, MgF$_2$	3.083	715 ± 7	8.82 ± 0.08	18.66 ± 0.3	—	105.12
10 SiC	1700 °C/HP/1 h	AlF$_3$, MgF$_2$	3.041	692 ± 11	9.02 ± 0.14	20.53 ± 0.5	—	129.22

(Continued)

Table 7. (Continued)

SiC SiC$_w$ wt.%	Sintering parameters	Sintering additives	Bulk density (g/cm^3)	Flexural strength (MPa)	Fracture toughness (MPa×m$^{1/2}$)	Vickers hardness (GPa)	Young's modulus (GPa)	Thermal conductivity (W/m.K)
15 SiC	1700 °C/HP/1 h	AlF$_3$, MgF$_2$	3.045	686 ± 21	8.87 ± 0.34	19.93 ± 0.4	—	130.12
20 SiC	1700 °C/HP/1 h	AlF$_3$, MgF$_2$	3.05	709 ± 17	10.3 ± 0.2	20.22 ± 0.1	—	113.66
25 SiC	1700 °C/HP/1 h	AlF$_3$, MgF$_2$	3.090	627 ± 15	8.39 ± 0.52	20.84 ± 0.2	—	145.66
10 SiC	1750 °C/PL/2 h	Y$_2$O$_3$, Al$_2$O$_3$	—	570 ± 5	3.45 ± 0.1	—	265 ± 5	—
1 SiC	1750 °C/HP/2 h	Y$_2$O$_3$, C, SiO$_2$	—	786	4.54 ± 0.07	14.46 ± 0.2	299 ± 12	—
5 SiC	1750 °C/HP/2 h	Y$_2$O$_3$, C, SiO$_2$	—	678	3.75 ± 0.53	16.34 ± 0.1	322 ± 10	—
20 SiC$_w$	1700 °C/HP/1 h	YF$_3$, MgF$_2$	3.11	502 ± 3	7.05 ± 0.2	23.1 ± 2.1	241 ± 14	78.4
15 SiC$_w$	1770 °C/HP/1 h	Y$_2$O$_3$, Al$_2$O$_3$	3.23	989 ± 49	9.6 ± 0.64	15.6 ± 0.26	313 ± 2.6	—
10 SiC$_w$	1750 °C/HP/30 min	Y$_2$O$_3$, AlN	3.20	680 ± 14	8.33 ± 0.23	16.6 ± 0.3	—	—

and mechanical properties of some CNT-, Gr-, and SiC/SiC$_w$-reinforced Si$_3$N$_4$ ceramic composites, respectively.

4. Carbon-Based TIMs

Physically, the heat transfer of condensed-state materials is dominated by free carriers and/or the dynamic lattice vibrations (i.e., phonons). Metallic crystal has plethora of free electrons, and their collisions with the metal cations dominate the intrinsic heat transfer. The higher strength of the metallic bonds gives rise to higher thermal conductivity. In comparison, free carriers in most polymeric materials are usually limited. Therefore, their thermal conduction is typically contributed by phonons. In nonmetallic crystalline materials, the phonon propagations are mediated by the crystal lattice with reduced energy loss and thus effective thermal conductions. However, the phonon transports in polymers are typically scattered at the interface between adjacent functional groups within the polymeric molecules, which results in energy loss and hence reduced thermal conductions. It is the same for amorphous or crystals with reduced grain sizes, where the increased amount of grain boundaries brings up undesired phonon scattering.

Low-dimensional crystals such as two-dimensional (2D) graphene (Gr) and one-dimensional (1D) CNTs may exhibit metallic characters although their free carrier densities are relatively low. As a result, their thermal conductions are also dominated by phonon scattering. Any crystal defects that could scatter the carrier migrations and the phonon transport tend to decrease the heat conductivity of these materials. In practical applications, the intrinsic properties of Gr and CNTs, including structural defects, grain size, layer thickness, geometric configurations, and surface adhesive properties play an important role in their heat dissipation behaviors. Nevertheless, carbon-based crystalline materials, including Gr, CNTs, and their 3D structures such as Gr foams, Gr aerogels, and vertically aligned CNTs arrays have been attracting increased research interest for TIM applications due to their intrinsic high thermal conductivity.

Carbon can be condensed and crystallized in many structures, e.g., carbon black (graphite), graphite NPs, diamond, etc., besides those of Gr and CNTs. These materials possess high intrinsic thermal conductivities and have been extensively investigated and developed in the past decade for TIM applications. Some of these materials, e.g., carbon black, can be coated on TIM to enhance its specific contacting thermal conductance. Similarly, graphite and diamond particles can be reinforced into desired matrix to enhance the thermal conductance. Besides the filler applications in host matrix, Gr- and CNT-based structures can also be directly fabricated into TIMs. Guo *et al.* have recently reviewed the carbon-based TIMs, including their mechanism and thermal conduction properties.[24] It has been revealed that the thermal conductivity of Gr layers significantly decreased after their aggregation from micro to macro scale. The out-of-plane thermal conductivity of the aggregated Gr paper is over two orders of magnitude lower than that of the in-plane ones. This information is helpful in engineering carbon-based TIMs toward enhanced thermal conductions. It has been mentioned above that for TIM application, not only the thermal conductivity but also the mechanical strength and deformation tolerance are the key specifications. In this regard, Gr 3D foam greatly surpasses Gr and CNTs for providing required mechanical properties. However, there are more challenges to fabricate such 3D-foam Gr structures when compared with Gr and CNTs materials.

4.1. *CNT-based materials and structures for TIM applications*

Since the 1990s, considerable efforts have been made to develop CNTs due to their unique physical and mechanical characteristics. Although most of the fundamental research studies on CNT are focusing on their active electronic applications, studies on CNTs for TIM applications have been increasing in the last decade. Theoretical simulations showed that individual CNTs could have thermal conductivity as high as 6.6×10^3 W/m.K at room temperature.[25] However, the 1D configuration and the intrinsic atomic arrangement along the length of CNT give rise to its anisotropic thermal conductivity. The thermal conductivity along the diameter of the CNT is much smaller than that along the length. For this reason, when employing CNTs for TIM applications, their orientation alignments, length-to-diameter ratios (i.e., aspect ratio), and structural defects play a vital role in the overall thermal conductivities. Besides the high thermal conductivity, the low elastic modulus along the diameter of CNTs makes them "soft" under external pressures, e.g., those induced due to the thermal coefficient mismatch between the electronic device and the heat sink materials. Meanwhile, the CNT-based TIMs feasibly deform when compressed and fill the microscale gaps between the device and the heat sink, which can significantly decrease the TIR.

Zhang et al. have grown vertically aligned CNT arrays by microwave plasma-enhanced chemical vapor deposition (PECVD) on Al catalytic substrates. They found that TIMs made from such CNT arrays can effectively reduce the TIR to as low as 7 mm^2K/W between the contact of Al and Si.[26] The reductions are more than those induced by the state-of-the-art commercial TIMs.[26] This advantage of CNT-based TIMs is strongly related to the well alignment, the high thermal conductivity, and the low elastic modulus along the diameter of the CNTs.

Yao et al. have also studied vertically aligned CNT arrays for TIM applications motivated by their high intrinsic thermal conductivity and robust mechanical and chemical properties.[27] The CNT arrays were grown by chemical vapor deposition (CVD) on 10-nm Al_2O_3/0.8-nm Fe/300-nm SiO_2/Si substrate. The 10-nm Al_2O_3/0.8-nm Fe double layer was deposited by electron beam evaporation to catalyze the growth of CNT. The grown CNT arrays were then metalized and transferred onto a metallized substrate, i.e., Si or Cu, leading to the CNT-based TIM, i.e., consisting of Si–Ti/Ni/Au–In–Ti/Ni/Au–VACNT–Ti/Ni/Au–In–Ti/Ni/Au–Cu. The thermal conduction test shows that such fabricated TIM has a low TIR, i.e., 3.4 mm^2K/W,[27] shedding new light on TIMs based on vertical aligned CNTs.

Although the CNT-based TIMs have been demonstrated with high performance comparable to those of commercial TIMs based on thermal greases,[28] the high temperature (i.e., ~1000°C) and catalyst assistant CVD growth may limit their applications. Meanwhile, lift-off and transfer process are typically necessary for fabricating TIMs from the vertical aligned CNT arrays. These processes may introduce chemical contamination and/or topological defects, which, in turn, degrade the thermal conduction properties.

4.2. *CNT-reinforced polymer composites for TIM applications*

Because of the critical high-temperature synthesis method and the necessary lift-off and transfer processes for fabricating the vertical aligned CNT-based TIMs, their large-scale applications are somewhat limited. Instead, thermal conductive polymeric matrix with CNT-based fillers, i.e., CNT-reinforced polymer composites (CNTRPCs), has been developed as an alternative method to fabricate TIMs with enhanced thermal conductivities. Epoxy (EP), polymethylmethacrylate (PMMA), and polyimide (PI) are some of the typical polymers that have been employed as the host matrix for producing CNTRPCs. The thermal conductivity of the CNTRPC-based TIMs can be influenced by many factors, including the intrinsic characters of both the host matrix and the CNT fillers as well as the processing parameters such as the compositions, surface functionalization and dispersions of CNTs, density, orientation, and alignments. Table 8 presents a summary of the thermal conductivities of some typical CNTRPC-based TIMs.[24]

4.3. *Graphene-based materials and structures for TIM applications*

Although the theoretical studies on whether crystals can nucleate and grow in the form of single layer atoms started 100 years ago, experimentally realization of Gr, i.e., a single layer of carbon atoms,

Table 8. Summary of thermal conductivities (κ) of some typical CNTRPC-based TIMs.[24]

Material	Matrix	Fraction	κ(W/m.K)
CNT	EP	16.7 vol.%	4.9
CNT	EP	14.8 vol.%	2.4
CNT	EP	13 wt.%	1.2
CNT*	EP	—	2.0
CNT*	Polyethylene	—	63.7
CNT*	PI	—	10.9
CNT/Gr	PVDF	10 wt.%	1.6
SWCNT	PMMA	1 wt.%	2.4
MWCNT	PMMA	4 wt.%	3.4
CNT	Polyamide	1 wt.%	16.9

Note: The * indicate that there is no fraction data available for the CNT.

was not successful until the beginning of this century. This breakthrough has triggered extensive investigations on Gr and non-Gr 2D materials as well as their applications, including in TIMs. For example, the first thermal conductivity measurement of single-layer Gr was carried out by Balandin's group, which showed a thermal conductivity as high as 5.30×10^3 W/m.K.[29] This thermal conductivity is higher than those of CNTs. Likewise, a more accurate measurement in a vacuum, where the heat loss to the air environment is mitigated, still shows the high thermal conductivity of Gr, i.e., around 2.5×10^3 W/m.K.[30] However, Gr is crystallized in van der Waals layer structures, where the in-plane carbon atoms are bonded by strong covalent sp^2 bonds while the individual layers are adjacently bonded by van der Waals force. For this reason, the in-plane thermal conductivity is much higher than the out-of-plane one in Gr structures. In this regard, to enhance the heat dissipations of Gr-based TIMs, a lot of work has been done to algin Gr structures with their in-plane directions along the heat flow, i.e., across the gap between the working device and the heat sprayer of the heat sink.

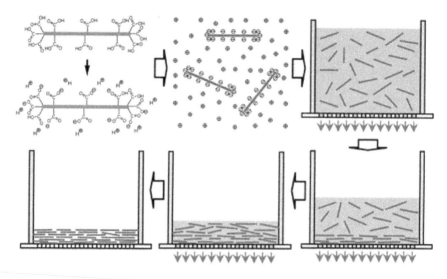

Fig. 3. A schematic diagram showing the process that has been developed to align the fMGs. The A-fMGs, a few tens of millimeters in horizontal directions and a few millimeters in thickness can be cut into smaller pieces and assembled into 3D TIMs with the in-plane direction of the fMGs perpendicular to the surface of the TIMs.[31]

Liang et al. have developed a vacuum filtration method to effectively align the functionalized multilayer Gr nanosheets (fMGs).[31] Figure 3 presents schematics of the typical process for aligned fMGs by dispersing them in water to form an aqueous solution. Following the water filtration method that has been developed to prepare paper-like Gr oxide,[32] the fMGs can be aligned as shown in Fig. 3. The aligned fMGs (A-fMGs) have dimensions to a few tens of millimeters in horizontal dimensions and a few millimeters in thickness. Such A-fMGs coupons can be further cut and assembled into 3D TIMs, in which the in-plane directions of the fMGs can be feasibly adjusted. The experimental results show that TIM based on vertically aligned fMGs has a thermal conductivity as high as 75.5 W/m.K.

Xin et al. have fabricated Gr fibers with high thermal conductivity and enhanced mechanical strength using spin-injection method.[33] They used large-sized Gr oxide (LGGO) sheets to constructure the backbone and fill the gaps between the LGGO sheets with small-sized Gr oxide (SMGO) while keeping the dominant orientation of the LGGO intact. By adjusting the compositional ratios of LGGO and SMGO, the balance between the compactness and the sheet alignment can be optimized in such Gr oxide fibers. By introducing post-spinning treatments such as thermal reduction and annealing, the Gr oxide fibers can be converted to Gr fibers with ordered orientations, high thermal conductivities, and strengthened mechanical properties. Typical intercalated fiber structures produced from LGGO and SMGO feedstocks and their thermally induced structural conversions are schematically shown in Fig. 4. The obtained Gr fibers consisting of crystallite domains with submicron dimensions can have the thermal conductivity as high as 1.29×10^3 W/m.K and tensile strength up to 1080 MPa.[33] Similar to the water filtrated fMGs laminates, the Gr fibers can be cut and assembled with the in-plane directions of the highly ordered Gr sheets along the surface normal direction of TIMs.

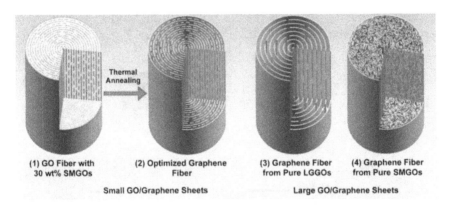

Fig. 4. Schematic diagrams showing the "intercalated" structure of the as-spun Gr oxide fibers and the thermal annealing converted Gr fibers. (1) As-spun Gr oxide fibers produced with an optimized compositional ratio between LGGO and SMGO; (2) Highly ordered Gr fiber with compact structures obtained by thermal annealing of the optimized Gr oxide fibers; (3) Highly ordered Gr fiber with less dense structures obtained by annealing of Gr oxide fibers produced from pure LGGO; and (4) Randomly oriented Gr fiber obtained after annealing of Gr oxide fibers produced from pure SMGO.[33]

Through manipulating the stacked configuration of conventional graphene paper for TIM applications, Dai et al. have recently reported that the thermal conductivity of the Gr-based TIMs can be as high as 143 W/m.K,[34] which outperforms that of many metals. A more important feature is that such Gr-based TIM is soft with a compressive modulus as low as 0.87 MPa (compatible with silicones), which makes them feasibly filling into the microscale gaps of the contact between the electronic device and the heat sink. The general processing procedures of the Gr-based TIM are schematically shown in Fig. 5. First, a crumpled Gr structure is obtained from conventional Gr paper (e.g., those produced by vacuum filtration of aqueous Gr dispersion, see Fig. 3). In this step, the Gr paper was firstly adhered onto a sheet of polyacrylate elastomer that was biaxially in-plane pre-stretched. After bonding of the Gr paper, the stretching force applied to the polyacrylate elastomer was released, and its shrinkage tends to deform the Gr paper into crumpled structures. The polyacrylate elastomer substrate was then removed by soaking in ethanol, and the crumpled Gr paper was further compressed (biaxially in-plane) into a monolith for fabricating the honeycomb panel-like Gr pad (HLGP) as shown in Fig. 5.[34] The trials

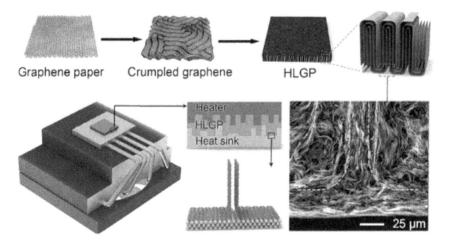

Fig. 5. Schematic diagrams showing the process for fabricating crumpled Gr-based TIMs.[34] The microstructural image is recorded by scanning-electron microscopy (SEM) from the cross-section of the crumpled Gr structures.

with the fabricated Gr-based TIM in the cooling system revealed that the cooling efficiency has been significantly increased, i.e., three times higher than those of the state-of-the-art commercial TIMs.

Gr can also be fabricated into 3D foam structures for TIM applications. A typical process developed by Chen et al. starts from growing Gr layers on Ni foam substrate by CVD at around 1000°C employing CH_4 precursors carried in H_2. After CVD growth, the Gr/Ni foam structure is coated by a protective polymer layer, e.g., poly(methylmethacrylate), i.e., PMMA. Then the Ni substrate is chemically removed in $FeCl_3$/HCl solutions. After removal of the Ni substrate, the PMMA coating is dissolved in acetone. Finally, the free-standing Gr foam is infiltrated with polydimethylsiloxane (PDMS), which can be further cut and assembled into TIMs for thermal conduction applications.[35] The processing procedures are schematically shown in Figs. 6(a)–6(f).[35] One can imagine that the density of so-obtained Gr foam strongly depends on the pore dimensions of the Ni foam substrate, which can be engineered during the Ni foam fabrication. Alternatively, stacking and compressing process can be introduced to densify the Ni foam substrate and thus enhance the density of the final 3D Gr foam structures [see Fig. 6(g)].[36]

The most striking feature of so-obtained 3D Gr form structures for TIM applications is that they can significantly enhance the overall interfacial contact thermal conductivity although their bulk thermal resistance might be rather high. This mechanism originates from the reduced modulus and the increased elongation of the 3D foam structure. These characteristics allow them easily to fill into the microscale gaps on the contacting surface of the device and the heat sink. Another advantage of the 3D Gr foam TIMs is the absorption of thermal stress (i.e., induced by the thermal expansion coefficient mismatch between the semiconductor electronic device and the metallic heat sink), by the foam structure without generating stress cracking. This characteristic enhances the device's service life. In fact, Zhang et al. have reported that by increasing the pressure applied on the Gr foam-based TIMs from 70 to 240 kPa can effectively reduce the TIR by 50% (i.e., from ~0.16 to ~0.08 cm^2K/W).[37] However, there is a trade-off between the thermal interface conductivity and the

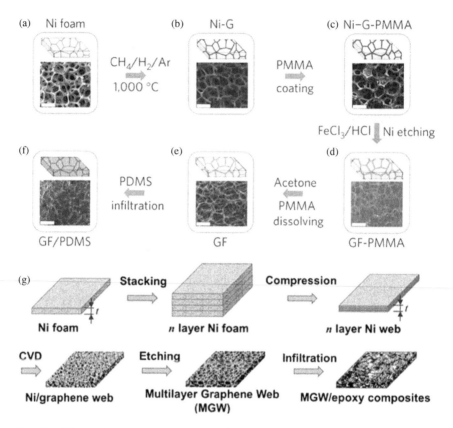

Fig. 6. Schematic diagrams showing the process for fabricating 3D Gr foam structures employing the combination of Ni foam substrate and the CVD for TIM applications.[35,36]

foam density of the 3D Gr foam-based TIMs. An increase in the density will increase the bulk thermal conductivity but also increase the modulus, which tends to decrease the thermal interface conductivity. This should be of concerns when designing the 3D Gr foam structures for TIM applications.

4.4. *Graphene-reinforced polymer composites for TIM applications*

Both the high thermal conductivity and the high mechanical strength of Gr make it recognized as a promising candidate for thermal conductive fillers in the polymeric matrix to form the Gr-reinforced polymer composites (GRPCs) for advanced TIM applications. Such polymer composite-based TIMs have been widely employed in integrated circuits, photovoltaic devices, and batteries. Recent studies indicate that the thermal conductivity of polymeric material can be enhanced by 20~30 times when incorporating Gr fillers, which is much higher than those induced by CNT fillers. The reinforcement process is much easier with Gr fillers than that of CNT fillers. Table 9 presents a summary of the thermal conductivities of some GRPCs.

Table 9. Summary of the thermal conductivities (κ) of same GRPCs toward their TIM applications.[24] EP: Epoxy; PET: Polyethylene terephthalate; PBO: p-Phenylene benzobisoxazole; PCM: Phase change material; TR: Thermal reduced; NPs: Nanoplatelets; and NSs: Nanosheets.

Material	Matrix	Fraction	κ (W/m.K)
Gr*	EP	—	33.54
rLGO	PVDF-HFP	27.2 wt.%	19.5
TR Gr sheets	Poly(PBO)	<5.0 vol.%	50
Gr	EP	10 vol.%	5
Gr	EP	30 vol.%	4.9
Gr	EP	10 vol.%	3.87
Gr	EP	20 wt.%	5.8
Gr	Thermal grease	2 vol.%	14
Gr	EP	25 vol.%	6.44
Gr	PCM	4 wt.%	1
Gr laminate*	PET	—	40–90
Graphite NPs	Polyethylene	21.4 vol.%	4.624
Graphite NPs	EP	30 wt.%	1.698
Gr flakes	EP	10 wt.%	1.53
Carbon NSs	EP	33 vol.%	80
Gr nanoflake	PVDF	25 vol.%	10
Cellulose/Gr aerogel	PCM	5.3 wt.%	1.35

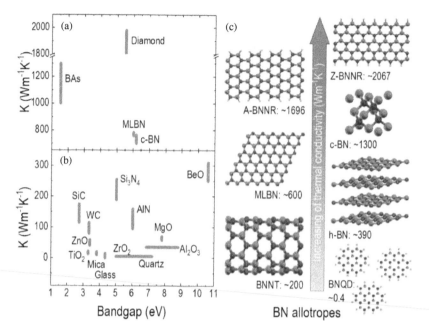

Fig. 7. (a), (b) Thermal conductivity comparisons of some common insulators and semiconductors. (c) Schematic drawing for thermal conductivity of BN allotropes including BN quantum dot (BNQD), BN nanotubes (BNNTs), h-BN, monolayer BN (MLBN), c-BN, A-BNNR, and Z-BNNR.[38,39,44]

5. BN-Based TIMs

In Figs. 7(a) and 7(b) we have summarized the thermal conductivities of some common semiconductors and insulators by plotting them as a function of the bandgap energies.[38,39] BN, as a large bandgap insulator, has a high thermal conductivity, which is only second to diamond. The large thermal conductivity of BN is contributed dominantly by the lightweight of both B and N atoms. Their dynamic lattice vibrations are easily actuated, and in response, high frequency acoustic phonons are intrinsic in BN. Although ultra-high lattice thermal conductivity is usually found in materials built from light weight elements like BN, an exception has been found in BAs consisted of light N and heavy arsenic atoms. This counterintuitive has been attributed to the unique phonon dispersion structures that significantly suppressed the three-phonon process scattering.[40–42] Among bulk BN polymorphs, the highly symmetric c-BN is thermally more conductive than h-BN. In isotope-enriched c-BN, the room temperature thermal conductivity can be as high as 1600 W/m·K, while that of h-BN is about 400 W/m·K. Nonetheless, in reduced dimensions, due to the size and boundary effects, h-BN can have even higher thermal conductivities [see Fig. 7(c)]. For example, theoretical calculations indicated that the thermal conductivity of zigzag-BN nanoribbons (Z-BNNR) is up to 1700~2000 W/m·K, which is about 20% higher than that of armchair-BN nanoribbons (A-BNNR).[43]

h-BN has the same crystal structure as graphite and the low-dimensional structures of h-BN are similar to Gr. In this regard, the process to fabricate the Gr-based TIMs shown in Fig. 6 is also applicable for processing BN-based TIMs. For example, Loeblein et al. have reported high-density 3D BN foam structures synthesized using Ni foam substrate.[45] Ammonia borane complex powders are used as the source materials, which were loaded in a ceramic boat and located away from the heating element of the quartz tube reactor of CVD. Thermal annealing was performed at 1000°C under H_2 flow (50 Pa) for 2 h before starting the BN growth. The growth was started by heating the ammonia borane at 120°C and 60 Pa. After growth the 3D-BN/Ni was dip-coated with PMMA and then soaked in dilute HCl to remove the Ni substrate. Finally, the PMMA protective coating was removed by annealing at 700°C for 1 h in the air ambient. The cross-plane thermal conductivity of the 3D BN foam structures is as high as 62 W/m.K and their surface exhibits excellent conformity. Table 10 presents the cross-plane thermal conductivities of some Gr and h-BN structures.[45]

It is worth mentioning that the CVD method has some limitations in conformally deposition on the networks of foam structures with high aspect ratios. In this regard, atomic layer disposition (ALD) can be introduced to initiate the growth. For example, TiO_2 nanotubes have been grown by ALD using hierarchical nanowires as the substrate.[46] After ALD growth, the nanowire cores can be

Table 10. Cross-plane thermal conductivities of some Gr and h-BN.[45]

Material	κ ($\perp$) (W/m.K)	Remarks
Gr	6	Weak van der Waals forces limit conductivity between layers
h-BN	1.5~2.5	Same as Gr
Gr paper	max. 1~5	Anisotropic behavior and alignment of the Gr particles, which becomes evident in the layered structure under SEM
h-BN paper/ BN-polymer nanocomposites	N.A. (only in-plane)	Like Gr paper, h-BN paper is extremely layered; pure BN has weak strength and is difficult to obtain as freestanding film, thus needs support from another material
Commercial silver epoxy TIM	1.76 (pristine) 9.9 (5 vol.% graphene filler)	Stable in working temperature range of epoxy, which is only 5~75°C
Gr nanocomposite epoxy	5.1 (10 vol.%)	Stable only at 5~75°C
Gr laminate on PET	40~90	Strongly dependent on flake size due to the number of single flakes required not freestanding
Multilayer Gr in epoxy	5	Temperature dependence of values, not freestanding
3D BN nanosheet networks in epoxy	2.4 (9.29 vol.%)	Infiltrated in epoxy anisotropic
h-BN in polyimide	7 (60 vol.%)	Not freestanding and requires high volume fractions
3D-foam	62 ± 10 (BN) 86 ± 10 (Gr)	Freestanding and compressible, stable up to 700~900°C

chemically removed, giving rise to the TiO_2 nanotube structures. Post-growth heat treatments can be further applied to crystallize the TiO_2 amorphous grown by ALD. It is believed that following the same method many other oxides and/or nitrides 3D structures can be grown by ALD employing 3D Ni foam substrates. However, when compared to CVD, ALD usually has a much lower growth rate, typically in the range of 0.1 nm/cycle, i.e., around 0.1 nm/min.[47–51] In this light, CVD is still necessary for growing thick layers on 3D foam substrates for TIM applications. A hybrid process, which can switch the growth mode between the conformal growth of ALD and the high-rate growth of CVD in a single reactor chamber, might greatly promote the research and development of 3D foam-based TIMs.

Although the 3D BN foam structures grown by CVD discussed above have high cross-plane thermal conductivities and excellent mechanical properties for TIM applications, the CVD growth procedures are critical with the high-temperature heat treatment in H_2 environments. Meanwhile, the ammonia borane complex tends to polymerize at elevated temperatures (>70°C), which brings up challenges in controlling the partial pressure of the reaction species. Also, the complete substrate removal of Ni from the 3D BN/Ni foam structures is critical and time-consuming, typically >5 h. To this end, Tian et al. have developed a direct foaming process to fabricate 3D foam structures based on BN and epoxy composites.[52] Figure 8 schematically presents the process procedures for fabricating such 3D BN/epoxy composites for TIM applications.

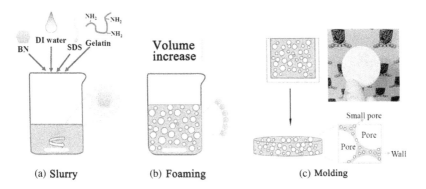

Fig. 8. Schematical diagrams showing the process procedures for synthesizing BN/epoxy composites for TIM applications.[52]

The synthesis starts with mixing 0.5 wt.% sodium dodecyl sulfate (SDS) and 4.0 wt.% gelatin with H_2O. The mixture was then heated at 60°C with mild agitation for the complete dissolution of the solute [see Fig. 8(a)]. The SDS was chosen as the foaming agent as well as a surfactant that prevents the BN fillers from aggregating in the later stage while the gelatin is to protect the BN foam structures from rupturing. Once the solution is ready, micron- and/or submicron BN powders were gradually added to it under stirring at 1500 rpm to make the foams along with the volume increase [see Fig. 8(b)]. As a result, the BN fillers are dispersed in the slurry. The obtained slurry was then cured, solidified, and dried to form the 3D porous BN foam structures. The BN 3D foam and epoxy composites can be obtained by soaking the foam structures in a mixture of epoxy resin, curing agent, and catalytic agent, followed by curing and heat treatment processes.[52] By this method, 3D-foam BN/epoxy composites with the filler fraction of ~24.4 wt.% have been obtained. Their in- and out-of-plane thermal conductivities are about 5.19 W/m.K and 3.48 W/m.K, respectively. This method is feasibly scaled up for mass productions of 3D foam-based (not only BN but also other ceramic fillers) polymer composites.

6. Conclusion and Outlook

Investigations on TIMs have been extensively progressed in the past two decades. Numerous methods have been developed for the processing of metal-, ceramic-, polymer-, and their composite-based TIMs for different applications. Thermal conduction and mechanical properties (e.g., to enhance the thermal interface conductions and absorb the thermal stress) are the key factors that drive the materials design and process while electrical insulations provide an additional advantage to widen the TIM applications. This literature review summarized recent developments in TIMs and addressed their thermal conduction and mechanical properties. They include Al_2O_3-reinforced polymer composites, Si_3N_4-based ceramic composites with CRFs, carbon-based structures with varied configurations (i.e., 1D, 2D, and 3D), BN-based 3D foams, and 3D BN foam/epoxy composites. On one hand, the emerging micro- and nano-engineering techniques lead to dedicated composite design and processes aiming at enhanced thermal conductions and mechanical properties for advanced TIM applications. On the other hand, the growth developments of low-dimensional materials such as graphene, CNTs, 2D h-BN, etc., by CVD with controlled orientation alignment open new routes for designing advanced TIMs. 3D foam structures and their low-cost and scalable process method speed up the design of 3D ceramic foam composite with polymer for TIM applications.

Driven by the increasing functionalities and performance requirements of electronic devices, TIMs with increased tolerance to work in higher temperatures and/or harsh environments are in demand. From this perspective, not only the basic characteristics, i.e., the thermal conductivity, the mechanical properties, and the electrical insulations but also the chemical and thermal stability will be continuously the center of the research focus on TIMs in future. The advantages of recently invented Gr and h-BN 3D foam-based TIMs, i.e., with remarkably increased cross-plane thermal conductivities (62~86 W/m.K), are still limited in laboratory scales. There is a long way to go before bringing them up to practical applications. Compared with the rapid expansion of materials research on TIMs in the past two decades, reliable and universally applicable thermal performance measurement techniques are still lacking. Likewise, efforts are necessary for evaluating the endurance and reliability of TIMs in varying temperatures and environments.

Acknowledgment

This study is supported by the Aerospace Program of A* STAR, Singapore through the project IMRE/14-2P1114 (Grant No. 1421500068).

References

1. H. F. Liu, S. B. Dolmanan, L. Zhang, S. J. Chua, D. Z. Chi, M. Heuken and S. Tripathy, *J. Appl. Phys.*, 2013, **113**, 023510.
2. R. Adappa, K. Suryanarayana, H. S. Hatwar and M. R. Rao, 2019 2nd Int Conf Intelligent Computing, Instrumentation and Control Technologies (ICICICT), Kannur, India, 2019, pp. 1197–1202.
3. R. Singh, T. R. Lenka, D. K. Panda, R. T. Velpula, B. Jain, H. Q. T. Bui and H. P. T. Nguyen, *Mater. Scien. Semicond. Process.*, 2020, **119**, 105216.
4. H. Liu, Y. Jin, M. Lin, S. Guo, A. M. Yong, S. B. Dolmanan, S. Tripathy and X. Wang, *J. Mater. Chem. C*, 2018, **6**, 13059–13068.
5. H. Chen, Y. K. Li, C. S. Peng, H. F. Liu, Y. L. Liu, Q. Huang, J. M. Zhou and Q.-K. Xue, *Phys. Rev. B*, 2002, **65**, 233303.
6. M. Pessa, C. S. Peng, T. Jouhti, E.-M. Pavelescu, W. Li, S. Karirinne, H. Liu and O. Okhotnikov, IEE Proc — Optoelectronics, 2003, pp. 12–21.
7. H. F. Liu, N. Xiang and S. J. Chua, *Appl. Phys. Lett.*, 2006, **89**, 071905.
8. S. Yin, K. J. Tseng, R. Simanjorang and P. Tu, *IET Power Electron.*, 2017, **10**, 979–986.
9. S. Nawaz, H. Babar, H. M. Ali, M. U. Sajid, M. M. Janjua, Z. Said, A. K. Tiwari, L. Syam Sundar and C. Li, *Appl. Therm. Eng.*, 2022, **206**, 118085.
10. D. D. L. Chung, *J. Mater. Eng. Perform.*, 2001, **10**, 56–59.
11. X. Wang, C. Lu and W. Rao, *Appl. Therm. Eng.*, 2021, **192**, 116937.
12. C.-K. Leong, Y. Aoyagi and D. D. L. Chung, *J. Electron. Mater.*, 2005, **34**, 1336–1341.
13. C. Liu, W. Yu, C. Chen, H. Xie and B. Cao, *Int. J. Heat Mass Transf.*, 2020, **163**, 120393.
14. P. Satheesh Kumar, S. Subramania Pillai, V. Subha, M. Sunil Kumar, R. Selvaraj, P. B. Gaikwad and S. Rajkumar, *Mater. Today: Proc.*, 2022, **59**, 1442–1446.
15. H. Yoon, P. Matteini and B. Hwang, *J. Non-Cryst. Solids*, 2022, **576**, 121272.
16. S. Lin, S. Ju, J. Zhang, G. Shi, Y. He and D. Jiang, *RSC Adv.*, 2019, **9**, 1419–1427.
17. S.-W. Wu, T.-C. Chang, Y.-H. Lin, H.-F. Chen and Y.-K. Fuh, *Int. J. Adv. Manuf. Technol.*, 2022, **121**, 3453–3462.

18. M. Rosshirt, D. Fabris, C. Cardenas, P. Wilhite, T. Tu and C.Y. Yang, *MRS Proc.*, 2009, **1158**, 1158-F03-03.
19. Y. Ouyang, L. Bai, H. Tian, X. Li and F. Yuan, *Compos. Part A: Appl. Sci. Manuf.*, 2022, **152**, 106685.
20. X. Xu, J. Chen, J. Zhou and B. Li, *Adv. Mater.*, 2018, **30**, 1705544.
21. J. Song, Z. Peng and Y. Zhang, *Chem. Eng. J.*, 2020, **391**, 123476.
22. Y. Wu, K. Ye, Z. Liu, B. Wang, C. Yan, Z. Wang, C.-T. Lin, N. Jiang and J. Yu, *ACS Appl. Mater. Interfaces*, 2019, **11**, 44700–44707.
23. A. Saleem, R. Iqbal, A. Hussain, M. S. Javed, M. Z. Ashfaq, M. Imran, M. M. Hussain, A. R. Akbar, S. Jun and M. K. Majeed, *Ceram.Int.*, 2022, **48**, 13401–13419.
24. X. Guo, S. Cheng, W. Cai, Y. Zhang and X.-a. Zhang, *Mater. Des.*, 2021, **209**, 109936.
25. S. Berber, Y.-K. Kwon and D. Tománek, *Phys. Rev. Lett.*, 2000, **84**, 4613–4616.
26. K. Zhang, Y. Chai, M. M. F. Yuen, D. G. W. Xiao and P. C. H. Chan, *Nanotechnology*, 2008, **19**, 215706.
27. Y. Yao, J. N. Tey, Z. Li, J. Wei, K. Bennett, A. McNamara, Y. Joshi, R. L. S. Tan, S. N. M. Ling and C. P. Wong, *IEEE Trans. Comp. Packag. Manuf. Technol.*, 2014, **4**, 232–239.
28. S. Ganguli, A. K. Roy, R. Wheeler, V. Varshney, F. Du and L. Dai, *J. Mater. Res.*, 2013, **28**, 933–939.
29. A. A. Balandin, S. Ghosh, W. Bao, I. Calizo, D. Teweldebrhan, F. Miao and C. N. Lau, *Nano Lett.*, 2008, **8**, 902–907.
30. W. Cai, A. L. Moore, Y. Zhu, X. Li, S. Chen, L. Shi and R. S. Ruoff, *Nano Lett.*, 2010, **10**, 1645–1651.
31. Q. Liang, X. Yao, W. Wang, Y. Liu and C. P. Wong, *ACS Nano*, 2011, **5**, 2392–2401.
32. D. A. Dikin, S. Stankovich, E. J. Zimney, R. D. Piner, G. H. B. Dommett, G. Evmenenko, S. T. Nguyen and R. S. Ruoff, *Nature*, 2007, **448**, 457–460.
33. G. Xin, T. Yao, H. Sun, S. M. Scott, D. Shao, G. Wang and J. Lian, *Science*, 2015, **349**, 1083–1087.
34. W. Dai, T. Ma, Q. Yan, J. Gao, X. Tan, L. Lv, H. Hou, Q. Wei, J. Yu, J. Wu, Y. Yao, S. Du, R. Sun, N. Jiang, Y. Wang, J. Kong, C. Wong, S. Maruyama and C.-T. Lin, *ACS Nano*, 2019, **13**, 11561–11571.
35. Z. Chen, W. Ren, L. Gao, B. Liu, S. Pei and H.-M. Cheng, *Nat. Mater.*, 2011, **10**, 424–428.

36. X. Shen, Z. Wang, Y. Wu, X. Liu, Y.-B. He, Q. Zheng, Q.-H. Yang, F. Kang and J.-K. Kim, *Mater. Horizons*, 2018, **5**, 275–284.
37. X. Zhang, K. K. Yeung, Z. Gao, J. Li, H. Sun, H. Xu, K. Zhang, M. Zhang, Z. Chen, M. M. F. Yuen and S. Yang, *Carbon*, 2014, **66**, 201–209.
38. Q. Cai, D. Scullion, W. Gan, A. Falin, S. Zhang, K. Watanabe, T. Taniguchi, Y. Chen, E. J. G. Santos and L. H. Li, *Sci. Adv.*, 2019, **5**, eaav0129.
39. V. Sharma, H. L. Kagdada, P. K. Jha, P. Śpiewak and K. J. Kurzydłowski, *Renew. Sustain. Energy Rev.*, 2020, **120**, 109622.
40. H. F. Liu, N. Xiang, S. Tripathy and S. J. Chua, *J. Appl. Phys.*, 2006, **99**, 103503.
41. H. F. Liu, S. Tripathy, G. X. Hu and H. Gong, *J. Appl. Phys.*, 2009, **105**, 053507.
42. H. F. Liu, A. Huang, S. Tripathy and S. J. Chua, *J. Raman Spectrosc.*, 2011, **42**, 2179–2182.
43. T. Ouyang, Y. Chen, Y. Xie, K. Yang, Z. Bao and J. Zhong, *Nanotechnology*, 2010, **21**, 245701.
44. J. Cao, T. L. Meng, X. Zhang, C. K. I. Tan, A. Suwardi and H. Liu, *Mater. Today Electron.*, 2022, **2**, 100005.
45. M. Loeblein, S. H. Tsang, M. Pawlik, E. J. R. Phua, H. Yong, X. W. Zhang, C. L. Gan, E. H. T. Teo, *ACS Nano*, 2017, 11, 2033–2044.
46. H. F. Liu, Y. D. Wang, M. Lin, L. T. Ong, S. Y. Tee and D. Z. Chi, *RSC Adv.*, 2015, **5**, 48647–48653.
47. H. F. Liu, K. K. A. Antwi, Y. D. Wang, L. T. Ong, S. J. Chua and D. Z. Chi, *RSC Adv.*, 2014, **4**, 58724–58731.
48. H. Liu, R. B. Yang, W. Yang, Y. Jin and C. J. J. Lee, *Appl. Surf. Sci.*, 2018, **439**, 583–588.
49. H. Liu, R. B. Yang, S. Guo, C. J. J. Lee and N. L. Yakovlev, *J. Alloys Compd.* 2017, **703**, 225–231.
50. H. Liu, S. Guo, R. B. Yang, C. J. J. Lee and L. Zhang, *ACS Appl. Mater. Interfaces*, 2017, **9**, 26201–26209.
51. H. Liu, *J. Mol. Eng. Mater.*, 2016, **4**, 1640010.
52. Z. Tian, J. Sun, S. Wang, X. Zeng, S. Zhou, S. Bai, N. Zhao and C.-P. Wong, *J. Mater. Chem. A*, 2018, **6**, 17540.